T0195314

Dead Tree Media

HAGLEY LIBRARY STUDIES IN BUSINESS, TECHNOLOGY,
AND POLITICS

Richard R. John, *Series Editor*

DEAD TREE MEDIA

Manufacturing the Newspaper in Twentieth-Century
North America

MICHAEL STAMM

Johns Hopkins University Press
Baltimore

© 2018 Johns Hopkins University Press
All rights reserved. Published 2018
Printed in the United States of America on acid-free paper
9 8 7 6 5 4 3 2 1

Johns Hopkins University Press
2715 North Charles Street
Baltimore, Maryland 21218-4363
www.press.jhu.edu

Library of Congress Cataloging-in-Publication Data

Names: Stamm, Michael, author.
Title: Dead tree media : manufacturing the newspaper in twentieth-century
    North America / Michael Stamm.
Description: Baltimore : Johns Hopkins University Press, 2018. | Includes
    bibliographical references and index.
Identifiers: LCCN 2017051884 | ISBN 9781421426051 (hardcover : alk. paper) |
    ISBN 9781421426068 (electronic) | ISBN 1421426056 (hardcover : alk. paper) |
    ISBN 1421426064 (electronic)
Subjects: LCSH: Newsprint industry—North America—History. | Newspaper
    publishing—North America—History.
Classification: LCC HD9839.N43 N778 2018 | DDC 338.4/7676286—dc23
LC record available at https://lccn.loc.gov/2017051884

A catalog record for this book is available from the British Library.

Special discounts are available for bulk purchases of this book. For more information,
please contact Special Sales at 410-516-6936 or specialsales@press.jhu.edu.

Johns Hopkins University Press uses environmentally friendly book materials,
including recycled text paper that is composed of at least 30 percent post-
consumer waste, whenever possible.

For Ani

CONTENTS

This book started in 2010 as a simple inquiry into what I thought was a little loose end that I had come across while finishing my first book: How was it that the *Chicago Tribune* came to build a company town in Quebec? This turned out to be a much more complicated question than I initially thought, and tracking down the answers took me in all sorts of unimagined directions over the past few years and changed how I think about North American history. In pursuing this work, I have accumulated many debts and owe many favors to many people. These acknowledgments do not come close to expressing my gratitude toward or reciprocating the generosity of all these colleagues and friends, but I start here by offering my heartfelt thanks to everyone who offered the support, guidance, assistance, and friendship that made writing this book possible.

This book simply would not have happened without the work of M. Stephen Salmon, a now-retired archivist at Library and Archives Canada. Along with his colleague Jennifer Mueller, Stephen acquired and processed the records of the Quebec and Ontario Paper Company, and this became the most important archival material that I would utilize in writing this book. It is an extraordinary collection of corporate documents relating to a wide range of aspects of North American business and media history. I was one of the first researchers to use this collection, which was not open to the public when I started working with it in 2011 and which remains in this state as of this writing. Stephen was instrumental in helping me gain access to the collection and in orienting me to what was in it, and he was also great company on my research trips to Ottawa. Years later, he generously read and commented on the entire draft manuscript and offered a number of suggestions that have improved this final work.

My editor for the Johns Hopkins University Press series, Richard R. John, has been an enthusiastic supporter of this project back to its earliest stages and of my work in general for even longer than that. Over the past few years, he has read what by now seems like countless drafts of this manuscript, and his close readings, criticisms, and constant encouragement have helped immeasurably in making this a better book. I could not have asked for a better critic and reader. Thanks also to Erik Rau. The referees that JHUP solicited offered many helpful suggestions and criticisms, and I thank them for their feedback on and support of the manuscript. At JHUP, Elizabeth Demers, William Krause, Matt McAdam, Juliana McCarthy, Kim Johnson, Morgan Shahan, Lauren Straley, and Gene Taft have been exceedingly helpful in getting the book into print. Jeremy Horsefield did a fantastic job copyediting the text.

At Michigan State University, my department chair Walter Hawthorne has been unfailingly supportive of this project and extraordinarily generous in helping me as I researched and wrote this book.

While serving as the public affairs officer at the Consulate General of Canada in Detroit, Dennis Moore was instrumental in helping me receive a Canadian Embassy Faculty Research Grant that funded the initial research in Ottawa which got this project moving. Thanks also to Roy Norton, who served as the consul general of Canada to Indiana, Kentucky, Michigan, and Ohio.

I made significant progress on an early draft of this project while holding a Fulbright Visiting Research Chair in Public Policy at the McGill Institute for the Study of Canada (MISC) in Montreal in 2012–13. Will Straw was an incredible host at MISC and a guide and companion to the most fun year I have had in the academy. Thanks also to the MISC staff, especially Linda Huddy and Johanne Bilodeau, for helping me have such an enjoyable and productive year. At Fulbright Canada, I am grateful to Michael Hawes and Brad Hector for their support.

I had the great fortune of sharing time at MISC with Karen Fricker, who was an Eakin Visiting Fellow and who is now a professor at Brock University. In subsequent years, Karen generously hosted me on research trips to St. Catharines and arranged what was both an enjoyable and fortuitous talk at Brock, at which I was fortunate enough to meet a few former Ontario Paper Company employees and arrange interviews with them.

In Montreal, my work was aided immensely by the company and ideas of Darin Barney, Sandra Gabriele, Elsbeth Heaman, Andrew Katz, Kari Levitt, Paulina Mickiewicz, Desmond Morton, Stephanie Nutting, Elena Razlogova, Carrie Rentschler, Jarrett Rudy, Rafico Ruiz, Charmaine Sinclair, Johanne Sloan, and Jonathan Sterne. Anne Whitelaw graciously arranged for me to interview her mother, Patricia, who had grown up in Baie Comeau and provided many valuable insights about the town. Bill Buxton has been an invaluable guide to Harold Innis's work, as well as a source of intellectual inspiration throughout my work on this project.

I spent several months in Ottawa, Ontario, on several research trips for this book. It was always great to see Chris Russill, a friend back to our shared time in Minneapolis, and I have benefited a great deal from our many conversations over the past decade about communications and the environment, among other things. I also owe him and Miranda Brady thanks for hosting me in Ottawa. Along with Chris, Andrew Johnston generously arranged for me to give a talk at Carleton University, where I was able to test out some of this book's early arguments in front of a great critical audience. Andrew's wife, Alison Scott, also arranged for me to correspond with her mother, Doreen Scott, who lived for several years in Baie Comeau and gave me some wonderful insights into the city's daily life in the post–World War II period. Doreen passed away in early 2014, and I am grateful to have spoken with her. I also have benefited a great deal from spending time with and learning from Dwayne Winseck, whom I met on an early trip to Ottawa and who has been a friend since. Dwayne read an entire late draft of this manuscript and has had a significant influence on a number of the book's arguments. My time in Ottawa was also better because of Monica Gattinger, Peter Gunther, Michel Hogue, Michael Orsini, and Ira Wagman.

In addition to those whom I've mentioned above, I want to thank all of the people who took the time to sit for interviews with me for this book. I was not able to incorporate all of these interviews into the final version, but I learned something from every one of the people who talked to me. Adrian Barnet came to my talk at Brock University and became not only a great source of material but also a tremendous help in setting up other interviews in and around St. Catharines, Ontario. I owe him and his wife, Jane, special thanks for a great dinner during one of my trips to the area. Thanks also to Bernard Baril, John Houghton, Ross MacDonald and Brigitte Bouchard, Don and Christine McMillan, Bryant Prosser, Ian and Tatin Sewell, Heather Winterburn, and Tony Yau. I spent a wonderful afternoon in 2015 in New York City with Bob Schmon, the grandson of Arthur Schmon, one of the major figures in this book. I learned a great deal about Arthur Schmon in our talk then and in subsequent email exchanges, and I am grateful to Bob for his enthusiasm about my book.

I had the great fortune of meeting Jim McNiven in 2010 when he was a Fulbright Visiting Scholar to Michigan State University and I was just getting going on this project. Jim has been a great source of information in the years since, both via email and over meals in East Lansing, Montreal, and Halifax.

I have had countless conversations with my friend and colleague Steve Lacy while working on this book, and many of them informed its arguments even if we were talking about something else. Whatever the conversations were about, they were always fun.

Mark Kuhlberg generously read and commented on the entire manuscript, and his insights about forestry, newsprint, and Canadian business history have helped improve the book immensely.

I had the wonderful experience of spending February 2015 in Milford, Pennsylvania, as a Scholar in Residence at Grey Towers, formerly the home of Gifford Pinchot and now a National Historic Site, working on two of the book's chapters that are most engaged with forestry. Thanks to Aaron Shapiro for letting me know about this program and to USDA Forest Service chief historian Lincoln Bramwell for the award. Thanks also to Bill Dauer, Lori McKean, Melody Remillard, and Ken Sandri for making my time at Grey Towers enjoyable and productive.

Thanks to Kathy Roberts Forde and Barbara Friedman for the invitation to speak about some of the issues that eventually made their way into chapter 7 of this book at the 2016 annual meeting of the American Journalism Historians Association in St. Paul. A revised version of that talk appeared as "The Flavor of News," *American Journalism* 32, no. 2 (Spring 2015): 208–20.

A number of archivists in the United States and Canada have been immensely helpful in my research. At Library and Archives Canada, Sophie Tellier has answered innumerable questions for me, and I also have benefited from the assistance of Jean Matheson and George de Zwaan. Thanks also to Ginette Robert at Bibliothèque et Archives nationales du Québec, Eric Gillespie at the Colonel Robert R. McCormick Research Center, and the staffs at the Lake Forest College Archives and Special Collections, Library of Congress, New York Public Library Manuscripts and Archives Division, United States National Archives, Nova Scotia Archives, and Société Historique de la Côte-Nord. Closer to home, thanks to the library staff at Michigan State University, especially those in the interlibrary loan department.

For help with obtaining permission to use the images in this book, I offer thanks to Anthony Foster at Our Katahdin and to Gabrielle Bélanger and Alice Minville at Resolute Forest Products. Alice Minville was also instrumental in helping me gain access to the Quebec and Ontario Paper Company records in Ottawa.

I was fortunate to be afforded opportunities to work out the arguments of this book in front of a number of supportive and critical audiences over the past few years. Thanks to those who commented on and challenged my analyses at the Joint Journalism and Communications Historians Conference at the New York University Arthur L. Carter Journalism Institute; the History of American Democracy Conference at the Tobin Project in Cambridge, Massachusetts; the Northeastern & Atlantic Canada Environmental History Forum at the Gorsebrook Institute, Saint Mary's University, Halifax, Nova Scotia; the Fishbein Workshop in the History, Philosophy, and Sociology of Science at the University of Chicago; the Cultural Crossings Conference at the University of Nottingham; the Canadian Business History Workshop at the University of Toronto Rotman School of Management; the Canadian Historical Association Annual Meeting in Waterloo, Ontario; the American Journalism Historians Association Annual Meeting in Kansas City, Missouri; the Business History Conference in St. Louis, Missouri; meetings of the International Communication Association in Seattle and London; and following invited lectures at Brock University, Carleton University, the University of Ottawa School of Political Studies, and the University of Oregon School of Journalism and Communication.

For all manner of encouragement, help, insight, invitations, and kindness along the way, I would like to say thanks to Charles Acland, Gene Allen, (the late) Jim Baughman, Stephen Blank, Claire Campbell, Mike Conway, David Fancy, Neil Harris, Nick Hirshon, Adrian Johns, Rich Kaplan, Juraj Kittler, (the late) Chris Kobrak, Brooke Kroeger, Michael Martin, Matt McKenzie, Paul Moore, John Nerone, Jeff Nichols, Dave Nord, Jeff Pasley, Brian Payne, Robert Picard, Victor Pickard, Jeff Pooley, Rick Popp, Gillian Roberts, Amber Roessner, Mark Rose, Andrew Ross, Phil Scranton, Jonathan Silberstein-Loeb, Will Slauter, Carole Stabile, Leslie Steeves, Richard Sutherland, Dominique Trudel, and Heidi Tworek.

I received generous financial support from Michigan State University while researching and writing this book. A grant from the Provost Undergraduate Research Initiative funded the research assistance done by Ziev Beresh in 2011–12. A Humanities and Arts Research Program (HARP) Development Grant provided a semester leave from teaching just when I needed it, and a HARP Production Grant aided with publication.

I owe thanks to many friends and colleagues associated with Michigan State University, but particularly to Nwando Achebe, (the late) David Bailey, Peter Berg, Howard Bossen, Liam Brockey, Sue Carter, Glenn Chambers, Emily Conroy-Krutz, Pero Dagbovie, Lucinda Davenport, Denise Demetriou, Laura Fair, Kirsten Fermaglich, Lisa Fine, Sean Forner, Eric Freedman, Deb Greer, Matt Grossman, Elyse Hansen, Karrin Hanshew, LaShawn Harris, Eric Juenke, Emily Katz, Steve Kautz, Charles Keith, Christina Kelly, Peter Knupfer, Leslie Lacy, Leslie Moch, Ed Murphy, Jeanna Norris, Folu Ogundimu, Sarah Reckhow, Chris Root, AnnMarie Schneider, Ethan Segal, Lewis Siegelbaum, Mónica Silva, Susan Sleeper-Smith, Mindy Smith, Ronen Steinberg, Gordon Stewart, Steve Stowe, Tom Summerhill, Ramya Swayamprakash, Mike Unsworth, Helen Veit, Naoke Wake, and John Waller.

My family has been a great source of love and support, and I owe tremendous thanks to my aunts, Diane Stamm and Carole Stamm, and to Lusik and Teni Sarkissian. My mother, Michell Smith, has been an inspiration for literally as long as I can remember.

Ani Sarkissian, thank you for all that you have done with and for me.

This book's institutional narrative is organized around the Chicago Tribune Company, often referred to in the text as the Tribune Company. Over their life spans, the Tribune Company's Canadian newsprint manufacturing subsidiaries took on different corporate names and were referred to by a series of acronyms, including Ontario Paper Company, OPC, Quebec North Shore Paper Company, QNS, QNSP, Quebec and Ontario Paper Company, Q&O, and QUNO. Until the Tribune Company took these Canadian newsprint operations public in 1993, they were always wholly owned subsidiaries of the US-based newspaper firm. In most cases in the text, I refer to Canadian subsidiary firms as "Tribune Company" because their operations were ultimately accountable to and were often directed by Chicago management. This was a vertically integrated manufacturing firm owned by and controlled from Chicago by the Tribune Company. In some instances in quotes referring to corporate actors, I have retained the relevant Canadian subsidiary's name in order to preserve original language.

Unless otherwise noted, I have done all translations from French sources. I take sole responsibility for any and all errors of any kind in the text.

Dead Tree Media

What Was a Newspaper?

Until very recently, our interactions with newspapers involved physical objects made of paper. Whether one's printed newspaper was found fresh and unread on a newsstand or in the front yard in a plastic bag, or whether it was picked up secondhand on a bus or in a cafe, the newspaper reading experience was oriented around a material paper object. In the twenty-first century, our relationships with newspapers are increasingly maintained through digital devices, and the screen has replaced paper as the surface on which many people read their newspapers. As audiences navigate among competing news sources across multiple delivery platforms, all news producers struggle to attract their interest. As business firms with a history defined by the production and sale of material goods, newspapers also face the additional challenges of needing to elicit consumer spending on paper objects and advertiser spending on space on that paper. Both sorts of spending have declined dramatically in recent years, and this has placed many newspapers in precarious financial positions.

News in the digital age can seem dematerialized and available freely in what the futurist John Perry Barlow in 1996 described as an environment "blanketed in bit-bearing media." To Barlow, the internet offered a space of freedom from the "increasingly obsolete information industries" controlling the distribution channels of public information by promising that "whatever the human mind may create can be reproduced and distributed infinitely at no cost." The emerging digital age, Barlow forecast, would mean that the "global conveyance of thought" would no longer need "factories to accomplish."[1] Starting in the mid-1990s and accelerating rapidly in subsequent years, the spread of this kind of thinking in the United States has created a host of new challenges for newspapers, and it has marked a fundamental turn in their history. In the twentieth-century media, the newspaper business was built and organized around the operation of factories that produced material objects made of paper. The successful newspaper in the twentieth century was an industrial newspaper. In the twenty-first century, that moment is passing.[2]

Though this book is written at a historical moment of tremendous change in the news business, it is not directly concerned with explaining the present media environment or with forecasting its future. This is not an analysis of what it means that news is today more often read on a screen than on paper. Rather, this book offers an exegesis of what a newspaper *was*. It is a reconstruction of the obscured and forgotten environmental, industrial, and labor processes that together were involved in manufacturing the physical pages on which news and advertising were printed and disseminated as newspapers. It is the history of a business model that was once enormously successful and that is now not.

Twentieth-century newspapers were not just outlets for news but also material products of industrial capitalism manufactured by business firms organized like those in other mass production industries. Producing a printed daily newspaper in the twentieth century was a process of both creating public information through the work of journalists and mass-producing a consumer good through the labor of hundreds and even thousands of skilled and unskilled workers, forming networks stretching from the forest to the delivery person. As one author noted in 1948, "Few people realize that the omnipresent sheet upon which they read the daily news of the world is as truly a forest product as the chair in which they sit, the table from which they eat, or the baseball bat with which home runs are scored."[3]

To make these forest products into printed newspapers required industrial supply chains of the sort celebrated in the *Chicago Tribune*, the newspaper that will feature prominently in the history narrated in this book. In 1929 and 1930, the *Tribune* ran a series of articles under the common title "Trees to Tribunes" showing the natural resources and industrial processes connecting the "northern forest to your doorstop." In the series, the *Tribune* celebrated its development of an elaborate and extensive system that produced newsprint from trees cut in the forests to which it had logging rights in the Canadian province of Quebec. Starting with these trees, the *Tribune*'s supply chain then flowed southwest over roughly 1,700 miles and across an international border along inland waterways from land above the Gulf of St. Lawrence to the docks on the north side of the Chicago River adjacent to the iconic Tribune Tower on Michigan Avenue. "In the depths of the last great wilderness left on the North American continent, your Tribune begins," one story claimed, and the *Tribune* recounted how the work of its lumberjacks began the "struggle with Nature for pulpwood" that formed the basis of its supply chain.[4]

Later articles in the "Trees to Tribunes" series celebrated both the harvesting of trees and the vertically integrated manufacturing process that the company used to transform these raw materials into printed newspapers.[5] From Quebec, the *Tribune* reported, it brought these felled trees on a fleet of company-owned boats to its own newsprint mill on the Welland Canal in Thorold, Ontario. After the trees were manufactured into rolls of newsprint, the finished paper was brought by boat on the Great Lakes to Chicago. At a dock on the Chicago River, "where the Pottawattomies in early days beached their canoes," the newsprint arrived in what was now a "modernist setting," as one reporter described it in 1930. The products of "acres of spruce land on the Gulf of St. Lawrence . . . were fed out of the boat into the presses for this morning's *Tribune* readers." This supply chain, the *Chicago Tribune* told its readers in 1931, required on a daily basis the trees cut from some 50 acres of Canadian forest, and larger Sunday editions could require as many as 225 acres worth of timber.[6]

In its "Trees to Tribunes" series, the *Chicago Tribune* aimed to frame aspects of its production in a way that would make the process seem marvelous. That the reader of the morning *Tribune* was holding something that was recently a living tree in a remote forest on the other side of an international boundary was the result of the company's own initiative and industrial organization. However, in highlighting certain aspects of its production and in framing the narrative as it did, the *Tribune* obscured some important facts about the material and political conditions of its production. Hidden behind this gauzy "Trees to Tribunes" narrative was an international history involving policy, labor, capital investment, and forest management.[7] Considered more critically and expansively, this history offers new insights about the ways in which foreign trade, natural resource exploitation, and industrial capitalism were central to the production of the mass-circulation printed newspaper in the twentieth century.[8]

The kind of newspaper to which the *Chicago Tribune* articles were referring represented the newspaper at a particular moment in its history. This was an industrial newspaper, and it was the highly successful type of newspaper that emerged in the nineteenth century as part of a broader history of industrial capitalism. Like companies in a variety of sectors, newspaper publishers used steam and electrical power and a factory setting in the interest of mass-producing a consumer good. One can see this in a 1922 cross-section diagram of the *Chicago Tribune*'s plant. On the upper two floors, one can see the editorial staff and those involved in creating the newspaper's content. Going down the building, we see the linotype machines involved in preparing that content for production, and on the bottom two floors we see the massive presses and rolls of paper utilized in printing the newspaper.[9]

The global origin of the twentieth-century industrial newspaper is difficult to locate with exact precision, as the mass-produced material object carrying news is a far more recent invention than is the commodification of news printed on paper. But, broadly speaking, the modern industrial newspaper emerged out of two developments in the commercial market for news: the manuscript newsletters

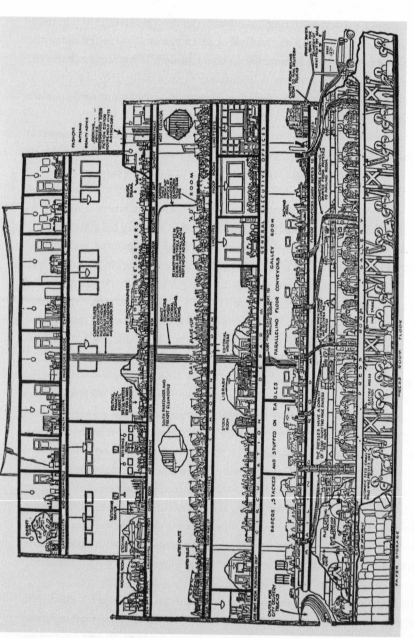

In 1922, the Chicago Tribune Company published a diagram of its newspaper production plant. The wire service operators, cartoonists, and editors are shown throughout the top floors, while the printing presses and rolls of newsprint are at the foundation. Chicago Tribune Company, *From Trees to Tribunes* (Chicago: Chicago Tribune Company, 1922): 86.

appearing across Europe starting in the twelfth century and the mechanically printed news pamphlets that began appearing in the sixteenth century. In the United States since the colonial period, the form in which news has been presented and sold has changed significantly. Early American newspapers were the artisanal products of one person performing almost all the labor from news gathering to printing. These publications included *Publick Occurrences*, the first and only issue of which appeared in 1690, and the *Boston News-Letter*, which began publishing in 1704 and became the first colonial newspaper to sustain regular publication. As the technologies of printing became better developed and more elaborate in the early nineteenth century, a division of labor began to characterize newspaper production, and editorial and production work became increasingly separate realms. In their pages, newspapers developed new ways of organizing and presenting information, for example, by beginning to separate news into topical sections and clearly delineating that news from editorial content.[10]

As objects, newspapers not only changed in appearance but also grew larger and heavier over time. In the early twentieth century, an eight-page paper, the standard size of many local dailies, weighed 1.75 ounces. The more informationally and materially robust sixteen-page urban paper weighed 3.5 ounces, and a thirty-two-page paper some 7 ounces. A metropolitan Sunday paper could weigh 1.5 pounds. As the heft of editions increased, especially among the mass-circulation urban newspapers of the early twentieth century, newspapers communicated abundance in both the range of information and sheer physical bulk. These new urban papers were meant to establish themselves as physical presences, with bigness taken by their publishers to be an indicator of merit.[11]

The increasing size of US newspapers in the twentieth century was the result of them printing not only the work of more robust news-gathering staffs but also an increasing amount of advertising. The history of the mass-circulation daily newspaper is both a history of journalism and a history of advertising. Through display ads offering an increasing abundance of branded consumer products and through classified ads offering a variety of local goods and services, newspapers became catalogs of daily life and guides to the twentieth century's emerging consumer culture.[12] In 1922 in New York City, for example, one survey found that, overall, morning newspapers devoted some 40 percent of their pages to advertising; evening papers, 44 percent; and Sunday papers, 43 percent. In some newspapers, the advertising actually made up more of the overall printed product than did the news. The morning *New York Times*, for example, was composed of 58 percent advertising, and the competing *New York World* was 52 percent advertising. In later decades, these proportions would come to define the US newspaper more generally and on a national level. According to estimates by the federal government, in 1945 some 51 percent of the more than 3 million tons of newsprint consumed in the United States was used to print advertising, and this proportion rose to 59 percent in 1950 and 61 percent in 1960. In 1970, US newspapers used an estimated 62 percent of the more than 9 million tons of newsprint that they consumed to print advertising. Taken as an object, the printed newspaper in the twentieth century increasingly was laden with advertising.[13]

This practice of circulating news surrounded by advertising on printed pages remained remarkably durable in the twentieth century. Indeed, one of contemporary America's most commercially successful printed newspapers is *USA Today*, which began publishing in 1982. Even in the face of decades of challenges from radio and television, each of which stimulated some apprehension about the possibilities of the newspaper's continued survival, the printed newspaper persisted and thrived. Recent events have destabilized the newspaper's dominance of the news business, and it remains unclear whether the printed newspaper as a delivery vehicle for news will remain commercially viable on a mass scale or whether news organizations oriented primarily around selling them will survive. In the digital age, the newspaper, as many readers understand it, is losing its printed materiality, if it has not done so already.

One manifestation of popular concern about the newspaper's mortality is that variations of the phrase "dead tree media" have become part of our vernacular in the twenty-first century to refer to printed newspapers. As one *Washington Post* journalist remarked in 2006, what had become in many ways the "typically depressing newspaper story" about the newspaper business was one that chronicled "plummeting circulation and shrinking revenues as the Internet and other news sources have swiped readers from the 'dead tree' media." In 2007, another *Post* journalist noted that the "noble product that we manufacture and distribute throughout the metropolis—the physical thing so carefully designed, folded and bagged—is now generally referred to in our business as the 'dead-tree edition.' It gets little respect." In many cases, "dead tree media" is used to portray printed newspapers as material anachronisms clinging in vain to relevancy in an era of vibrant and participatory digital media. It is often meant to be cleverly if not sarcastically dismissive of printed newspapers and to conjure thoughts about the medium's survival or demise.[14]

And yet, there is a generally unacknowledged truth contained in the phrase "dead tree media": printed newspapers really are the products of felled trees, as they have been since the late nineteenth century when wood became the dominant raw material used to manufacture newsprint. A newspaper in the twenty-first century is understood to be the product of an organization generating information accessed by the public not only as text and images printed on paper but also

In 1921, the Chicago Tribune Company drew attention to its newspapers' origins in the forests in a trade journal advertisement boasting about its industrial supply chain. Advertisement, Chicago Tribune Company, *Editor & Publisher* 53, no. 38 (Feb. 19, 1921): cover.

as multimedia content perused on digital devices. The displacement of the printed newspaper in the news business has created significant challenges for US publishers, and many are struggling to adapt to the digital-age realities of declining print circulations. Throughout the twentieth century, however, newspaper publishers who wanted to grow their businesses needed factories and adequate supplies of newsprint, and in turn they required a massive labor force and lots of "dead trees." As the Chicago Tribune Company claimed in a February 1921 advertisement announcing that it "owns forests of pulp wood," the firm's lumberjacks were hard at work at the start of its newsprint supply chain. "Cut off from the world by snow and ice," as the company described its winter workforce, "several hundred men in distant Canadian forests are chopping down trees from which

Growing: 140,000,000,000 Newspapers

By rough count . . . Great Northern owns 2,031,000 acres of woodlands in the Great Northwoods of Maine. It draws on five million more. Together this represents 1,711,440,000 trees . . . enough newsprint "on the stump" to print 140 *billion* newspapers. And still supply continues to grow. With the aid of scientific forest management, Great Northern timberland is treasured like the valuable legacy which, indeed, it is. Cut is limited to natural reproduction rate, assuring a perpetual wood supply — good assurance of Great Northern's position of *dependable* leadership in U.S. newsprint production.

Great Northern
P A P E R   C O M P A N Y

Mills in Maine • Sales Offices: 522 Fifth Avenue, New York 36 • Boston • Chicago • Cincinnati • Washington

In 1962, the Great Northern Paper Company claimed that its forest reserves encompassed some 1,711,440,000 trees, which it estimated was enough to make 140,000,000,000 newspapers. Advertisement, Great Northern Paper Company, *Editor & Publisher* 95, no. 9 (Mar. 3, 1962): 25. Used with permission.

Chicago Tribunes will be made late this year." In 1962, the Great Northern Paper Company bragged that the vast scope of its forest holdings meant that it was in effect "Growing . . . 140,000,000,000 Newspapers," a statement that was both similarly boastful to the *Chicago Tribune*'s and equally revealing of the printed newspaper's rootedness in the forest and origins in trees.[15]

In the abstract, the freedom to print a newspaper in the United States is a legal right. In practice, the actual capacity to print mass quantities of newspapers is an economic and organizational challenge facing business enterprises. "Newspapers are not eleemosynary institutions," as the trade journal *Editor & Publisher* noted in 1907; rather, "they are conducted for the purpose of making money. Hence, those who would conduct them successfully must be governed by the same principles as are followed by men engaged in the management of other kinds of businesses." And, as many publishers came to realize, streamlined industrial production was essential to commercial success. As *New York Globe* publisher Jason Rogers noted in 1919, the "first consideration in up-to-the-minute newspaper production is the application of the factory idea to it. A newspaper is a manufactured article, and the sooner more of us recognize this fact and apply it the happier and more profitable will be our business." Though rarely considered as such, the printed newspaper was, like the automobile, a product of industrial capitalism. Just as the automaker was challenged in securing adequate supplies of steel, glass, and rubber, the newspaper publisher needed to manage supplies of newsprint. This newsprint had its natural origins in forests, and this fundamental fact about the news business has been mostly overlooked and forgotten in the digital age.[16]

Toward a New Natural History of the Newspaper

In 1923, the sociologist Robert Park wrote, "The newspaper has a history; but it has, likewise, a natural history." Park's article, "The Natural History of the Newspaper," remains one of the canonical works in the field of journalism history. For Park, describing the newspaper's "natural history" meant describing it in evolutionary terms and showing how different news organizations responded to changes in their local and commercial contexts. As in the natural world, the newspapers that endured did so because they adapted successfully to changes in the areas and circumstances around them. "The type of newspaper that exists," Park asserted, "is the type that has survived under the conditions of modern life." But there is another way of looking at the newspaper's "natural history," and that is to take the phrase literally and think about its origins as trees. "The newspaper of today was a tree yesterday," as a trade journal pointed out in 1923, the same year as Park's article.[17]

In making newspapers into the powerful and profitable institutions that they became in the nineteenth and twentieth centuries, publishers adapted many of the strategies and technologies of other mechanized industries in printing their newspapers and exploited parallel advances made in paper manufacturing. Until the middle of the nineteenth century, newsprint was made from cloth rags. Rag-based paper was and is durable, but the supplies of the raw materials needed to manufacture it proved too limited to meet the demands of publishing for mass circulation. By the 1880s, a series of technological innovations had allowed newsprint manufacturers to switch to a paper made from wood, and this led to a remarkable expansion in the amount of paper available to print newspapers. In 1859, when many newsprint manufacturers were still using rags as their primary raw material, US papermakers produced an estimated 66,000 tons of newsprint. In 1879, when wood-based paper was becoming more common, US production had grown to 149,000 tons, and this increased further to 569,000 tons in 1899. By 1914, through increasing domestic production and a burgeoning import market, some 1.6 million tons of newsprint were available in the United States, and this supply increased to 2.2 million tons by 1920. In the roughly four decades after wood replaced rags as the most commonly used material for making newsprint, US publishers received a massive increase in the amount of paper available to them. This allowed for a huge expansion of the amount of space available to print information for public consumption, and the transition to wood as the basis for making newsprint helped enable the development of the American mass-circulation daily newspaper. Indeed, as the conservationist Gifford Pinchot remarked in 1923, "We are prone to call this the age of steel and of electricity, and to look upon wood as the foundation of an earlier and more primitive civilization. Yet, when one considers the part which wood plays in modern life—housing, transportation, manufactures, and particularly the dissemination of knowledge by means of books and newspapers, this is truly an Age of Wood."[18]

Starting in the late nineteenth century, strategic publishers took advantage of these newly plentiful paper supplies to extend the penny press business model that they had developed in the 1830s, which was based on using cheaply priced newspapers to generate both high circulations and advertising revenues. By the early twentieth century, publishers in the United States such as Robert McCormick, William Randolph Hearst, and Joseph Pulitzer had created metropolitan institutions with circulations well into the hundreds of thousands, and daily sales could approach a million following particularly important events. To take these newspapers to unprecedented levels of circulation and profitability required not only ambitious publishers and talented journalists but also significant capital investments in printing plants that were organized as factories. At a very basic level, it also required increasingly large quantities of newsprint. To produce the *New York Times* in 1919, for example, required some 43,000 tons of newsprint, and the papers in William Randolph Hearst's chain used a combined 200,000 tons of paper that year. In 1921, the *Chicago Tribune* alone used 64,254 tons of newsprint, a figure that had nearly doubled to 126,000 tons in 1928. Other metropolitan papers that year used massive supplies of paper as well, for example, the *Kansas City Star* with 45,000 tons, the *Detroit News* with 54,000 tons, and the *Philadelphia Inquirer* with 57,000 tons.[19]

Considering the importance of these tonnages and the commodity chains leading to them is a way to understand that the newspaper is not just a source of information or the foundation of a daily ritual of reading. It is also an extraordinary feat of mass production, and its history has followed closely those of other industries that have evolved in response to the rise of industrial capitalism. In some ways, mass-circulation newspapers represent a pinnacle of industrial capitalism and mass production. They are on each new day always a new product, and few if any branded consumer goods had this kind of volume and turnover. By the early twentieth century in some American cities, publishers would produce half a million identical copies of a consumer good that would be physically distributed and sold in a single day. On every next day, the process had to be repeated anew, as all of the goods produced yesterday were obsolete. Like a loaf of bread, news had a narrow window within which it had to be sold. As Don Seitz, the business manager of the *New York World,* remarked in 1916, "the newspaper publisher deals entirely in the perishable." And the newspaper is "perishable" not just because it loses its value over time as its news becomes more commonly known, or "stale," but also because it is a natural product. Left to the elements, the newspaper quickly and physically decomposes.[20]

Printed newspapers are products of the natural world, and processes of their manufacture have significant environmental effects. Though they appear every day to readers, the roots of their production were embedded in natural seasonal rhythms. Much of the logging done to cut the pulpwood to manufacture newsprint was carried out in the winter, when logs were easier to move along icy paths to the banks of frozen rivers. When the spring thaw came, these newly risen rivers offered ways of readily conveying the logs to the mills. These mills in turn required massive amounts of water both in the papermaking process and in generating the hydroelectric power necessary to convert the logs into pulp and then newsprint. In essence, nature provided the raw materials for newsprint, the means to get those materials to the mills, and then the power to run the mills. Capital provided the industrial plants and wages for the human labor that organized and exploited nature to begin the process of producing printed newspapers.[21]

Over the course of the twentieth century, as newspaper circulation increased and became more concentrated in urban areas, the effects of that production became increasingly felt on landscapes that were far removed from most readers' immediate environs. In the twentieth century, it was easy for an American reader to pick up a newspaper, peruse its news and advertising, and feel informed about the day's events, all without giving any thought to that newspaper's natural origins. However, in distant forests and along distant rivers, most of which would come to be in Canada, lumberjacks and paper mill workers saw the other end of the process as they turned felled trees into rolls of newsprint to be shipped away to publishers. As industrial supply chains connected trees to factories to readers, as Hearst Newspapers boasted in 1935, the power of the forest was "unchained" as "logs are turned into newsprint and given the living, productive power of publishing genius." What I aim to offer in this book is a new natural history of the printed newspaper which shows both these environmental origins and the industrial processes involved in manufacturing trees into newspapers.[22]

Pulp, Paper, and North American Trade Policy

Regardless of the size of a particular newspaper's circulation, its publisher needed consistent supplies of paper, and the bigger the newspaper and the higher the circulation, the greater the demand for newsprint. Because of publishers' incessant need for newsprint, US newspapers came to have significant effects not only on society and politics through their content but also on the environment and on North American trade through their industrial production. Many of the plentiful trees in the United States proved initially ill-suited to newsprint manufacturing, as the wood from the Douglas fir trees of the West Coast and the pine trees of the South did not readily adapt to newsprint production, and this problem was compounded by the high transportation costs associated with getting the heavy and bulky raw materials and manufactured paper to the Midwest and Northeast, where the major American metropolitan newspapers of the early twentieth century were located. Spruce proved to be an ideal wood for use in newsprint manufacturing, and the densest spruce forests in the world were in the northern climates of such countries as Sweden, Finland, and Canada. Though US publishers could and did get some pulp and paper from across the Atlantic, the proximity of the great spruce forests along and north of the US-Canada border, particularly near the Great Lakes and the navigable waterways connecting them, made Canadian spruce particularly attractive to them. "Canada is essentially the land of the spruce," a trade association noted in 1920, and the exploitation of and trade in these trees would reorient Canada's relationship to the rest of the world.[23]

The twentieth-century development of a Canadian newsprint industry was part of a broad and fundamental shift in Canada's international relations. Since the sixteenth century, trade in staple goods had dominated the economy of the region that would become Canada. The fishing waters off Newfoundland provided the cod that became the first major staple in this history, and that was followed by a fur trade driven by the French. Starting in the eighteenth century, Canadian economic relations with Europe tilted toward the British, who coveted Canadian forest products especially for their use in shipbuilding. By the nineteenth century, Canada found itself in a triangular trade relationship with Britain and the United States, with wheat, meat, and dairy products taking on additional importance. In the twentieth century, the rise of the United States as a global economic power motivated an increasing focus on North American trade, and the United States and Canada eventually would come to have the largest bilateral trade relationship in the world.[24]

The evolution of this trade relationship and of the increasing reliance of US newspapers on Canadian newsprint took place within the broader context of the development of what was called the "pulp and paper industry." A brief word about terminology is necessary here. The "pulp" part of the industry refers to the extraction of cellulose fibers from felled trees, and this can be done through mechanical grinding or chemical processing, depending on the specific type of pulp being manufactured. The mechanical process creates what is called groundwood pulp, and the vast majority of this kind of pulp goes into newsprint production. Chemical processes of extracting cellulose fibers yield several different kinds of pulps, the most important of which for newsprint is sulfite pulp, which is added to groundwood pulp in newsprint manufacturing to increase the paper's durability. Whether the wood is reduced to pulp through mechanical grinding or through the use of chemical agents, the goal is the same: to extract the useful cellulose fibers from the remainder of the organic material not essential for papermaking. On the "paper" side of the industry, manufacturers utilized the various pulps, some of which they made themselves and some of which they purchased from dedicated pulp producers, in manufacturing a range of products, for example, newsprint, writing papers, and cigarette papers. In Canada in the twentieth century, newsprint became the most important and valuable part of the pulp and paper industry. In 1950, it made up roughly 80 percent of the tonnage of all Canadian paper and paperboard production and over half of all combined Canadian pulp and paper production.[25]

With newsprint as its major component, the pulp and paper industry became one of Canada's most important industries in the twentieth century, and by 1950 there were some 130 mills located across the country. As the Canadian Pulp and Paper Association described it in 1950, it was

> not only the major, but the fundamental industrial force which has shaped the social and economic development of Canada. No other industry has, and has had such far-reaching effects upon the economy. . . . The level of Canadian well-being has been raised by this great industry; the nation has attained a first rank position in the trading world owing largely to its activities; and its economic potency has helped bring Canada to nationhood. . . . All together these 130 plants across the nation constitute its largest single industry, Canada's greatest bread-winner: first in employment, first in capital invested, first in wages paid, first in value of production, and first in export wages.

Or, as the somewhat less hyperbolic economic historian Hugh Aitken remarked in 1959, the "manufacture of pulp and paper is today Canada's leading industry, no matter what criterion is used."[26]

As was the case with all industries, the Canadian pulp and paper industry developed as it did because of both the actions of particular business firms and specific policies enacted by the state. Given its massive raw material base, high production capacity, and relatively small domestic market, Canada's pulp and paper producers, including those making newsprint, were strongly drawn to seeking foreign markets for their products. Over time, trade policies enacted on both sides of the US-Canada border allowed US publishers duty-free access to the newsprint produced from the Canadian forests that they coveted. Through tariff reductions and removals, North American policymakers between 1911 and 1913 created the conditions for a continentally integrated economy in the most important and costly input to the newspaper business, and publishers used this new political economy to exploit Canadian raw materials. In 1909, Canadian newsprint imports accounted for only a tiny fraction (1.7%) of overall US consumption, with the overwhelming majority of the 1.16 million tons of newsprint used by US publishers having been manufactured domestically. By 1950, Canada had become by far the world's leading producer of newsprint, with its 5.28 million tons accounting for some 54 percent of total global production. In comparison, the next two highest-producing nations, the United States and Great Britain, accounted for merely 10.4 and 6.2 percent of global newsprint production, respectively. In 1950, after nearly four decades of duty-free newsprint trade with the United States, Canada exported some 90 percent of its newsprint production to the United States, where it accounted for 80 percent of total newsprint consumption.[27]

For American publishers, finding trees to make paper was a problem of what we now call "supply chain management," and they were aided significantly by American and Canadian policymakers in their quest to solve this problem. As one observer celebrated the outcomes of this policy-driven process in 1939, the "forest resources of the continent are more and more approaching an economic unit with common and interdependent markets. Nature has provided the logical framework within which friendly and related peoples operate to their mutual benefit." The actions taken by publishers and policymakers to create a duty-free trade in newsprint would, as this book will show, help structure the evolution of both the US newspaper and the trade relationship between the United States and Canada.[28]

As postal policy had subsidized the US newspaper business starting in the eighteenth century, tariff policy would in the twentieth century. Postal policy aided newspapers by promoting their cheap and wide circulation, while tariff policy allowed newspapers to maintain ready access to the massive supplies of Canadian newsprint that they relied on to produce their newspapers. In the twentieth century, the American mass-circulation newspaper depended on Canadian forests, and access to those forests was structured by the state. Over time, tariff policy would not only aid the US newspaper business but also shape some of the most important developments in the economic and political history of twentieth-century Canada.[29]

Harold Innis and the Internationalization of American History

These intersections between communications media, political economy, and Canadian staple products were at the core of the work of the Canadian scholar Harold Innis. Indeed, my own book's connections and extensions of these themes continue a project that Innis himself was pursuing at the time of his death in 1952. In January of that year, Harold Innis was elected to the presidency of the American Economic Association (AEA), becoming the first person to hold the position while not being a US resident, and toward the end of the year he set about preparing the presidential address that he was to deliver at the annual AEA meeting in December. Innis's health deteriorated as he worked on the essay, and he died on November 8, 1952, leaving behind a fragmentary manuscript that was published in the *American Economic Review* in 1953, in which Innis discussed a subject for which he was well known: the history of the exploitation of "staple raw materials" in Canada. Innis explored the history of a variety of those staples in his substantial body of scholarship, and throughout his career he moved back and forth between considerations of the deep historic trade in fish and fur and analyses of the more current twentieth-century trade in mineral and forest products. As a political economist and Canadian citizen, Innis was convinced that his own era was shaped by trade in pulp and paper, and he also understood that this was an industry with a different kind of political significance than the staples that had dominated previous eras. Beaver pelts sent to France did not change the substance of Canadian culture and politics, but newsprint sent to the United States did, as it was used within an American media system that had created what he called the "possibilities for control over public opinion" across all of North America. As US newspapers dominated North American trade in newsprint, Innis argued, they used Canadian staples to reinforce the United States' domination of North American politics and culture.[30]

Perhaps owing to its author's declining health, Innis's presidential address was ultimately vague and suggestive, even by the standards of his own notoriously oblique prose. But he did gesture toward a theme that had been developing throughout his career, as he argued that the increase in the exploitation of Canadian trees in the interest of supplying paper to US newspapers had significant cultural and political consequences across the continent. As a result of the "enormous expansion in the production of raw material following improvements in technology in the use of wood" to manufacture newsprint, Innis wrote in his AEA presidential address, there was a tremendous "expansion in the production of information by newspapers." Where some might see this as a democratization of knowledge, Innis instead saw it as having the deleterious effect of making industrial newspapers increasingly dependent on selling more advertising and thus more beholden to the idea of profit than public service. "An increase in the circulation of newspapers," Innis argued, and a commitment to "low fixed prices designed to facilitate increased circulation were paralleled by the increasing importance of advertising as a source of revenue and the necessity of developing policies to attract advertising."[31]

For Innis, the AEA address was a version of an argument about the mass press which he had made with much greater force, clarity, and Canadian nationalism in other essays. Throughout his career, Innis developed an argument connecting the exploitation of the most important Canadian staple (trees) with the evolution of the most important medium of mass communication (the daily newspaper) of his lifetime. Indeed, the questions animating many of the inquiries that Innis undertook from the early 1920s to the early 1950s led him to analyze the relationship between staples, communication, and empire by connecting trees and newspapers across the US-Canada border.

Understanding the importance of these linkages requires a brief consideration of Innis's work in the interest of correcting a widely held misunderstanding that it was bifurcated into two incommensurable stages, one defined by a focus on political economy and staples exploitation and the other defined by a late-career shift toward communication.[32] At a surface level, there are tempting reasons to see Innis's career as one defined by discontinuous concerns. Two of his later monographs understood to be the work of a scholar of communication, *Empire and Communications* (1950) and *The Bias of Communication* (1951), have the actual word "communication" in the title. In contrast, someone consulting Innis's early bibliography would find books with titles suggesting far different sorts of inquiries: *The Fur Trade in Canada* (1930), *The Cod Fisheries: The History of an International Economy* (1940), and *The Dairy Industry in Canada* (1937), a volume that Innis edited and for which he wrote an introduction. In addition to the fact that Innis wrote much about a wide range of subjects, he also did so in a style that is elliptical and often plodding. Simply and charitably put, Innis's writing, for all its substantive insights, lacks aesthetic appeal, and the writing is marked by significant impediments to attracting wide readership. Partly owing to the volume of his scholarship and partly owing to the style in which it was presented, Innis has claimed a place in the pantheon of writers who are much easier and more convenient to read about than to actually read. So, when Marshall McLuhan, a flashy popularizer of many ideas that Innis had previously developed, publicly proclaimed that there was an "early Innis" and a "later Innis" in an essay published almost immediately after Innis's death, he helped create a misunderstanding that Innis had a two-stage career, with each phase defined by distinct foci and concerns.[33] More recent scholarship on Innis has demonstrated that there are in fact deep intellectual connections that run throughout Innis's total body of work, and one can see these links by considering Innis's long engagement with the intertwined international histories of the Canadian pulp and paper industry and US newspaper publishing.[34]

As this book will show in chapters 2 and 3, the Canadian pulp and paper industry exploded after 1913 as US and Canadian policymakers used tariff policy to create an integrated continental market for newsprint. During the early years of this process, after serving in World War I, Harold Innis began graduate work in economics at the University of Chicago. He received his PhD in 1920 and returned to Canada to teach at the University of Toronto, where he would remain for his entire career. All around him, many Canadians grew increasingly aware and critical of the nationalist politics expressed by many US newspapers, and many also experienced the growing importance of the pulp and paper industry, both directly and indirectly. As economic historians (and Innis's contemporaries) W. T. Easterbrook and Hugh Aitken noted in 1956, pulp and paper manufacturing "had by the middle of the twentieth century securely established itself as Canada's leading industry, whether the criterion used was value of output, capital invested, wages paid, or contribution to export earnings." Newsprint was the pulp and paper industry's "most valuable product," and its continued success was intimately tied to exports to the United States. "In so far as the Canadian industry is concerned," Easterbrook and Aitken noted, "stability depends largely on the maintenance of effective demand in the United States." In essence, over the course of Harold Innis's adult life and scholarly career, Canadian newsprint was both the most important part of an export- and staples-driven economy and the commodity that was central to the production of US newspapers, which Innis believed had come to have a corrosive influence on the culture and politics of all of North America. Throughout his work, Innis's inquiries into the political economy of pulp and paper were simultaneously also inquiries into both empire and communications, though he would not put those two words together in the title of a book until 1950.[35]

One can see Innis engaging with the pulp and paper industry very early in his career, for example, in the conclusion to *The Fur Trade in Canada* (1930), in which

Innis extended the book's argument forward from its historical starting point in the 1500s and up to the time of his own writing. After nearly four hundred pages of painstaking detail about the history of the fur trade back to the sixteenth century, Innis succinctly applied that framework to Canada's recent economic history. "With the disappearance of beaver in more accessible territory," Innis argued, "lumber became the product which brought the largest returns. . . . The lumber industry has been supplemented by the development of the pulp and paper industry with its chief reliance on spruce." The lessons of the history of the fur trade were for Innis also the lessons of the present trade in pulp and paper. "The economic history of Canada," he concluded in 1930, "has been dominated by the discrepancy between the centre and the margin of western civilization. Energy has been directed toward the exploitation of staple products and the tendency has been cumulative."[36]

Innis's concern with pulp and paper continued throughout the 1930s, and his work during this period shows clear concerns with media and communication. For example, Innis wrote a lengthy entry on the pulp and paper industry for *The Encyclopedia of Canada* in 1937 in which he cited a number of significant examples of American involvement in the Canadian newsprint industry.[37] Innis was actively reading on the subject of newsprint production, and he published numerous reviews of books from what he called a "strangely neglected field" and routinely discussed it in his scholarly articles about press history.[38] By the late 1930s, Innis became quite critical of the Canada-US trade in pulp and paper, and he began to remark on not only its economic but also its cultural and political significance. As he wrote in 1938, the "pressure of overhead costs incidental to large scale capital equipment in newspaper plants and newsprint mills has increased the importance of large scale circulation." In other words, having made significant investments in the industrial plants necessary to produce mass-circulation newspapers, newsprint manufacturers and newspaper publishers needed to make sure that these plants operated as close to full capacity as possible, and Innis believed that both groups thus found themselves imbricated in a newspaper market calibrated more to producing mass-appeal newspapers than producing quality newspapers. Beyond modifying their journalism standards to satisfy the broadest possible audience, Innis argued, mass-circulation newspaper publishers also began to abandon coverage critical of industrial and consumer capitalism so as not to antagonize the advertisers that they relied on to stay profitable. "Dependence on advertising," Innis argued, "particularly from department stores, has become a vital issue in policy as to news, editorials, and features. The cheap newspaper is subordinated to the demands of modern industrialism and modern merchandising."[39]

As Innis's writings about pulp and paper became more polemical over the course of his career, they also by the 1940s took on an increasingly strident and explicit nationalism. In 1946 in *Political Economy in the Modern State*, for example, Innis remarked that the increasing production of Canadian newsprint had enabled the US newspaper to have a much "more intense development of advertising," and he added that the growing tendency among US publishers to produce widely appealing but intellectually vapid newspapers created across the continent a "more unstable public opinion precluding a clear appreciation of our problems and in turn sustained consideration of them." The continued and profitable expansion of the American newspaper was the result of publishers having access to Canadian newsprint, Innis claimed, and he believed that the effects on North American politics and culture were disastrous. By the end of the decade, Innis's intertwined Canadian nationalism and anti-Americanism had become much more overt and much less polite. For example, in 1949 Innis claimed that the growth of the industrial mass press gave those in control of newspapers not only the power to shape public opinion but also the incentive to create a kind of journalism ill-suited for democracy. "Hearst has advised that the important thing for a newspaper to do in making circulation is to get excited when the public is excited," Innis remarked, and this meant that media barons had "exploited" their positions for commercial gain rather than promoting the public good. This had widespread and negative social consequences, Innis concluded, as the "appeal to lower levels of intelligence by unstable people accentuates instability if not insanity."[40]

Ultimately, if we follow Innis's analyses of newsprint and the pulp and paper industry throughout his career, we can see his later work that is obviously and explicitly about communication not as a departure or rupture in his thinking but rather as a crystallization of it. As he had done in *The Fur Trade in Canada* (1930), Innis in *The Bias of Communication* (1951) both operated over a long span of historical time and extended the analysis up to the contemporary moment by linking together his related arguments about staples, communications, and empire. "Large-scale production of newsprint made from wood in the second half of the nineteenth century supported large-scale development of newspaper plants and a demand for effective devices for widening markets for newspapers," Innis argued. "Increased newspaper circulation supported a demand for advertising and for new methods of marketing, notably the department store. The type of news essential to an increase in circulation, to an increase in advertising, and to an increase in the sale of news was necessarily that which catered to excitement. A prevailing interest in orgies and excitement was harnessed in the interests of trade." And, as he had long argued, US newspapers were important both because of the content they produced and because of the ways that they influenced trade policies that gave them favorable access to Canadian newsprint. "In actively supporting a policy of holding down the price of newsprint and of increasing production," Innis argued, "newspapers favoured a marked extension of advertising. The economy became biased toward the mass production of goods which had a rapid turnover and an efficient distributing system. . . . Newspaper civilization had entered its concluding phase of its intensive development in the speculative activity of the twenties. Its bias culminated in an obsession with the immediate." The end result of this obsession, Innis claimed, was that the "conditions of freedom of thought are in danger of being destroyed by science, technology, and the mechanization of knowledge, and with them, Western civilization."[41]

In many ways, these were the themes that Innis was highlighting in the draft of his presidential address to the AEA. As he had been throughout his career, Innis was trying to draw attention to the connections between the pulp and paper industry and North American politics and culture, and he aimed to do so in this particularly prominent public forum. When the draft was published posthumously in the *American Economic Review*, it was appended with a brief essay from his son Donald, in which he argued that there were a number of broader and blunter criticisms about the United States which his father was making in private while working on the essay. Donald related that his father "often remarked that newspapers and superficiality are synonymous. Newspapers must always produce something new and interesting and they must never seem to be interested in anything that is old." Harold Innis's belief that newspapers aimed to permanently create novelty led him to make private remarks far more biting than most of his public statements. "My father used to say," Donald Innis claimed, "that newspapers have to keep stirring up the animals."[42]

As Harold Innis saw it, what US publishers did was in keeping with the patterns of natural resource exploitation and foreign trade which had defined Canadian history. To Innis, the exploitation of "staple products" had shaped Canada's development, from the fisheries and fur trade of the sixteenth and seventeenth centuries to the wheat, mining, and forestry industries of the nineteenth and twentieth centuries. The exploitation of Canadian forests in the interest of pulp and paper manufacturing, Innis argued, created one of Canada's leading manufacturing industries in the twentieth century, but it had also enabled US newspaper publishers to extend their hegemonic reach over the culture and politics of the entire continent.[43] Over time, Innis claimed, the increasing importance of the trade in pulp and paper to the Canadian economy worked to push along the process in which "American imperialism has replaced and exploited British imperialism."[44]

Ultimately, one of Harold Innis's most important lessons is that, in order to understand US history in the twentieth century, one has to understand Canadian history in the twentieth century. Many American readers will be reluctant to do this, despite the fact that calls for globalizing American history have been ringing throughout the profession for some time now. The historians responding to these calls have contributed to a growing body of work expanding our understandings of the connections between the United States and the rest of the world in the twentieth century. And yet, very few scholars have been motivated to globalize American history by looking directly north to Canada.[45] This is a widespread, curious, and stubborn cultural blindness on the part of many US historians and ordinary Americans, and it has contributed to a broader ignorance about some significant aspects of the contemporary global economy. Though few Americans pay much serious attention to Canada's role in North American trade, as of this writing Canada and the United States in fact have the largest bilateral trade relationship of any pair of countries in the world, and the United States imports more oil from Canada than it does from any other country.[46]

These are basic and vital facts necessary to understanding the political, economic, and environmental history of twentieth-century North America. Though contemporary Canadian oil imports may be presenting new and dire environmental challenges, the trade in that commodity is in many ways best understood as the contemporary extension of the long and vigorous resource-based trade relationship that defined US-Canada relations in the twentieth century. Before oil, one of the most important of these staple commodities was newsprint. In some ways, if the Canadian economy in the twenty-first century is inordinately dependent on American drivers, we might say that the Canadian economy in the twentieth century was inordinately dependent on American readers.[47]

The Canadian political economist Mel Watkins remarked in 1963 that the "staple approach to the study of economic history is primarily a Canadian innovation; indeed, it is Canada's most distinctive contribution to political economy."[48] And yet, as other scholars have shown, this is an approach that can take us far beyond Canada, if we follow the commodity chains. Indeed, world-systems theorist Immanuel Wallerstein would remark a decade after Watkins that "in the long run, staples account for more of men's economic thrusts than luxuries."[49] In recent years, historians have been producing new accounts of how staples such as rubber, cotton, aluminum, and guano have been central to the creation of the modern global economy.[50] To consider the intertwined histories of Canadian newsprint production and US newspaper publishing as part of a broader history of North American staples exploitation is to follow some of the most important themes in the emerging new histories of global capitalism.

Forgotten Origins of the Twentieth-Century Newspaper

This book attempts to weave together two histories of the twentieth century. The first of these traces the history of the newspaper in the United States as a product of industrial capitalism. In the twentieth century, the successful American newspaper was organized around the daily manufacturing and sale of factory-produced consumer goods made of paper. The second and related history addressed in this book describes the processes through which publishers gathered the materials that they needed to produce these physical newspapers. The ability of a newspaper to succeed commercially in the twentieth century depended significantly on its publisher's ability to get paper. This is not to suggest that paper in and of itself enabled the development of the American newspaper. A publisher in possession of a printing plant and a large supply of newsprint who had nothing to say that appealed to an audience would soon have a printing plant and a large stock of unsold newspapers, regardless of how cheaply they were priced. However, the converse is also true: absent supplies of newsprint, a newspaper publisher was the owner of a printing plant without the necessary means of using it to reach the public. To use a contemporary analogy, paper might be thought of as akin to bandwidth. A publisher without newsprint was like a computer user without an internet connection. Journalism historians often comment on the increasing costs of starting a newspaper in the twentieth century, but they are less likely to note the increased costs of *maintaining* a newspaper. Industry surveys done from the 1940s to the 1960s, for example, showed that paper and ink remained by far the largest expenses involved in running a newspaper. For mass-circulation newspapers, more than one-third of operating expenses went into printing materials. To understand the history of the twentieth-century newspaper, one must consider the supply and price of newsprint, both of which could fluctuate dramatically for a variety of reasons, among them state policy, war, and collusion among manufacturers.[51]

This book attempts to integrate these two histories—one tracing the newspaper's development as a product of industrial capitalism, and the other showing the international production networks providing publishers with newsprint—through considering both their general and specific aspects. This is an international history involving publishers from around the United States and papermakers and policymakers on both sides of the US-Canada border. When it is relevant, I also situate this continental relationship within an international market for newsprint. This global context is particularly important in the periods surrounding World War I and World War II, when Scandinavian pulp and paper supplies were mostly taken out of the global market.

This international context is balanced by a specific analysis of the Chicago Tribune Company, and this choice of narrative focus is not arbitrary. The Tribune Company was one of the most successful and influential media companies of the twentieth century, and a central element of its success had to do with its owner embracing the manufacturing side of the business. As its "Trees to Tribunes" series of articles had shown, by the 1930s the company had developed a vertically integrated supply chain that was capable of converting the trees from 50 acres of Canadian forest into printed copies of the *Chicago Tribune* every weekday. In 1930 at a speech at Yale University, *Tribune* publisher Robert McCormick remarked, "Newspaper development has been the constant development of mechanical means adapted to the old principles of literature and art." For much of the twentieth century, McCormick's company was one of the most successful in the world at putting this into practice, and an integral component of this was that it vertically integrated across the US-Canada border starting in 1912 in order to control its production process.[52]

The Tribune Company's Canadian operations were massive, in terms of both the size of the financial investment and the amount of raw materials used. In 1940, company executive Arthur Schmon estimated that the company had spent some $153,736,000 for "materials, pay-rolls, and taxes" in nearly three decades of operating in Canada. At a speech in Chicago in 1946, Schmon claimed that, on a daily basis, the Tribune Company's Canadian mills produced enough newsprint to "cover a two-lane highway from San Francisco to New York." To make that newsprint, "we had to cut and transport to our mills last year, over 360,000 cords of wood. This mountain of wood, 36,000,000 sticks four feet long, is enough to fill the stadium at Soldiers' Field here in Chicago to overflowing. To be sure of getting this wood every year we have acquired control of over 7,000 square miles of excellent timber limits—an area approximately ten times the size of Cook County." In 1947, the Tribune Company provided a list (shown here in table 1) of the "major items used in producing the *Chicago Tribune* in a single year," and one can see that 168,331 tons of newsprint was needed for the massive industrial operation that printed 21.6 billion pages of news and advertising over the course of that year. The extensive and costly industrial process of connecting "Trees to Tribunes" left a significant environmental footprint in Canada.[53]

Considering this process illustrates a basic and largely unacknowledged fact about the newspaper business: the metropolitan daily, while in some ways the most local of institutions, has been for over a hundred years an industrially produced consumer good manufactured by corporations employing international supply chains. The *Chicago Tribune* liked to call itself an "American Paper for

TABLE 1
Selected items used to print the *Chicago Tribune* in 1947

Item	Quantity	Item	Quantity
Employees		Equipment	
Mechanical department	1,381	Press units	145
Business offices	388	Typesetting machines	102
Advertising department	307	Stereotype department	36
Editorial department	430	machines	
Circulation department	639	Trucks and automobiles	117
Building department	186	Typewriters and business	700
Materials		machines	
Tons of newsprint	168,331	Telephones	1,135
Pounds of ink	8,500,000	Paper-carrying lake ships	6
Kilowatt hours of power	13,200,000	Output	
Gallons of gasoline	360,000	Printed pages	21,600,000,000
Tons of coal	8,500	Advertising linage	32,851,892
Pounds of zinc	50,000	Lines of type composed	73,994,000
Pounds of brass	10,000	Square inches of	6,151,000
Pounds of copper	17,000	engravings	
Pounds of replacement	380,000	Matrices for press plates	256,000
type metal		Stereotype plates	1,379,000
Pounds of stereotype	96,000,000	News stories	700,000
metal		Photographs	61,000

Source: "What It Takes to Print Trib," *Editor & Publisher* 80, no. 25 (June 14, 1947): 74.

Americans," a slogan that was accurate in reference to its place of publication but disregarded the fact that its material production was dependent on Canadian paper. The *Tribune*'s other immodest slogan—"World's Greatest Newspaper"— was unintentionally a more accurate description of it as an object of international political economy.[54]

In many ways, Robert McCormick's publishing philosophies and practices at the *Chicago Tribune* were at odds with many conceptions of twentieth-century journalism history. The general thrust of this received history is that newspapers in the early twentieth century were defined by blatant and, in many cities, competitive partisanship. After World War I, the sequestering of explicit opinion onto editorial pages happened alongside the abandonment of overtly biased news reporting by journalists. The move toward a dispassionate style of news presentation (often called "objectivity") was a key part of the refashioning of journalism into a form of information production aiming to enlighten readers rather than to lead or incite them.[55] Many newspapers certainly followed this line of development, and quite a few were financially successful in doing so. In thinking about the history of journalism, however, Robert McCormick represents a strong exception to these received and dominant Whiggish narratives about the advancement of the journalism and newspaper publishing businesses toward being defined by the presentation of "objective" news.[56]

To many contemporaries, McCormick ran one of the most constantly and aggressively slanted newspapers in the country, and the biases embedded in *Tribune* coverage were readily apparent throughout McCormick's career as its publisher.[57] Visiting the paper in 1941 for a lengthy profile of McCormick which ran in the *Saturday Evening Post*, journalist Jack Alexander was struck by the degree to which the paper was a product of McCormick's vision and personality. "Because of the publisher's closeness to his paper," Alexander argued, the *Tribune* "fundamentally is Colonel McCormick transmuted into paper and ink. It is, therefore, a confusing compound of paradoxes, whims, prejudices, arrogance and the other components of a strong-willed and perversely individualistic human being. Hence it differs, and sometimes shockingly so, from the average large newspaper, which may reflect the views of the publisher, but rarely his personality." On an everyday and perpetual basis, Alexander found, McCormick exerted strong and direct control over the *Tribune*'s content.[58]

Alexander's findings were echoed widely in other contemporary public portraits of McCormick and his newspaper. One of the most lyrical accounts came in a 1950 *New Yorker* column by A. J. Liebling. In a piece ranging in tone from acerbic to bemused, Liebling described McCormick as "a man with a sense of destiny," and he portrayed the *Tribune* as a singular journalistic achievement. To Liebling, the "most invigorating feature of life in Chicago, otherwise a fairly placid city, is one's daily encounter with the World's self-admittedly Greatest Newspaper, the *Tribune*. The visitor to Chicago, awaking unalarmed in his hotel room and receiving the *Tribune* with his breakfast tray, takes a look at the headlines and finds himself at once transported into a land of sombre [sic] horror, rather like that depicted by the science-mystery magazines." The bias in the news stories was so palpable and particular, Liebling noted, that "the stranger is likely to get the feeling that some of the people and events he is reading about superficially resemble people and events he remembers having read about in the world outside, but he can never be sure." The style and tone of the *Chicago Tribune* struck Liebling as so warped and unique in comparison to its main local competitors, the *Sun-Times* and the *Daily News*, that the "only points on which the *Tribune* agrees with its two rivals are those reported scores in basketball and hockey games." Ultimately, Liebling found the experience of reading the *Chicago Tribune* to be bracing and unsettling. "The effect on the adrenal glands of the morning dip into the *Tribune*'s cosmos is amazing," Liebling concluded. "The *Tribune*'s reader issues from his door walking on the balls of his feet, muscles tense, expecting attack by sex-mad footpads at the next street corner, forewarned against the smooth talk of strangers with a British accent, and prepared to dive behind the first convenient barrier at the sound of a guided missile approaching—any minute now—from the direction of northern Siberia."[59]

Though McCormick and his editorial policies could be mocked easily by critics, he was not running a socially or commercially marginal newspaper. At mid-century, the *Tribune* was the highest-circulating standard-size newspaper in the United States, and it was an overtly and outrageously partisan outlet that was widely and clearly understood to be a platform for its publisher's conservative views rather than fair reporting. The only paper with a higher circulation was the tabloid *New York Daily News*, which McCormick's Tribune Company had founded as a wholly owned subsidiary in 1919, and which became a newspaper profiting from pictures and sensationalism, not objective journalism. Together, these Tribune Company newspapers and the corporation owning them lie almost wholly outside of received understandings of the history of American journalism. To date, there is no scholarly monograph on the history of the *New York Daily News*, despite the fact that it was one of the most successful newspapers of the twentieth century, and the overall amount of public and scholarly attention given to the Tribune Company is dwarfed by the amount given to William Randolph Hearst. Indeed, for many outside the field of journalism history, Hearst may be the most common point of reference as an important twentieth-century publisher, and we may have reached a point where Hearst's considerable significance has actually become overstated simply because of how often and individually he is mentioned in discussions of the political influence of newspapers. Hearst certainly was a powerful figure operating a hugely influential corporation, but his towering presence in the mainstream of twentieth-century US history comes with an assist from Orson Welles, and it also comes at the expense of understanding the rest of the US newspaper business, including newspapers such as the *Chicago Tribune* and *New York Daily News*, both of which had higher circulations than did any of the Hearst papers.

There is thus a major gap in our understanding of the twentieth-century newspaper business, and this book aims to fill that. There is also a huge contradiction in our understanding of journalism history if the animating ethics of the two highest-circulating newspapers are not readily commensurable with the model that scholars have developed. Thus, it is important to reexamine this history in light of the success of the Chicago Tribune Company. Rethinking this history from the perspective of the Tribune Company is a major goal of this book, especially when it comes to understanding the organization as a multinational and diversified manufacturing firm. In many cases, historians of American journalism have written a history of the newspapers that they admire, not the newspapers that many people actually bought. In doing so, they have forgotten that success in the business of selling printed newspapers was not just the result of publishing good journalism but also, and perhaps more importantly, the result of developing an industrial and organizational apparatus to manufacture and distribute journalism and advertising printed on sheets of paper made from trees.

Scope and Organization of the Book

In this book, I aim to integrate one line of analysis focusing specifically on the Chicago Tribune Company with a second line focusing on the international political economy of newsprint production and newspaper manufacturing. The body of the book is divided into three parts, each of which has some temporal overlap with the others. Part I, "The North American Newspaper," narrates the development of the industrial print culture that took shape starting in the mid-nineteenth century with the application of the technologies of the Industrial Revolution to newsprint manufacturing and newspaper publishing. These two industries developed in tandem in North America and were driven by the related growth of the US mass-circulation newspaper and the Canadian newsprint industry that supplied it. By the interwar period, Canada had become the world's largest producer of newsprint, and US publishers the primary consumers of that paper. In tracing Canada's rise to becoming the leading global newsprint producer, one can follow important trends in the geopolitics of journalism and newspaper publishing in the twentieth century.

Part II, "Extending Chicagoland," focuses specifically on how the Chicago Tribune Company evolved during this period from a local newspaper firm into a multinational and vertically integrated newspaper corporation as it built and operated its own Canadian newsprint mills. As we will see, that production allowed the Tribune Company to print both the *Chicago Tribune* and the *New York Daily News*, which it founded in 1919, and soon these would be the two highest-circulating newspapers in the United States. Over time, the move into industrial manufacturing in Canada led the company to become a city and regional planner, as it built the company town of Baie Comeau in Quebec in the late 1930s. Finally, this book's middle section highlights the varieties of newspaper work by drawing attention to the largely hidden industrial labor force in Canada essential to the production of two of the most important newspapers in the United States.

Part III, "The Newspaper beyond the Printed Page," combines the general and specific histories of the first two sections and follows them through to the end of the twentieth century. One of the key lines of analysis in part III involves tracing how the Tribune Company evolved as a media corporation as it adapted its print-based business model to new media, including facsimile broadcasting and the emerging forms of electronic text-based news distribution which predated the internet. On the newsprint manufacturing side in Canada, the Tribune Company's vertical integration and industrial adaptations also took the company into sectors that seem to have little obvious connection to the newspaper business, including commercial food flavoring and aluminum smelting. Over the course of the twentieth century, the Tribune Company extended a business rooted in midwestern newspaper publishing into a diversified and multinational manufacturing corporation, and considering this history gives us a new way to think about what it meant to be in the newspaper business in the twentieth century.

Overall, these three parts aim to trace the history of the industrial newspaper in North America. At turns emphasizing media history, business history, environmental history, and political history, this book is the story of the rise and fall of both the mass-circulation printed newspaper and the particular kind of newspaper publishing corporation that shaped many aspects of the cultural, political, and even physical landscape of North America.

PART ONE

THE NORTH AMERICAN NEWSPAPER

CHAPTER ONE

The Making of Industrial Print Culture

One of the most often cited and enduring representations of the American newspaper comes from the Frenchman Alexis de Tocqueville, who visited the United States in 1831 and published the two volumes of what would be called *Democracy in America* over the following decade. After traveling the country studying Americans' lives, beliefs, and rituals, Tocqueville was struck by how often newspapers created bonds of fellowship among anonymous groups of readers. Tocqueville remarked,

> It often happens that many men who have the desire or the need to associate cannot do it, because all being very small and lost in the crowd, they do not see each other and do not know where to find each other. Up comes a newspaper that exposes to their view the sentiment or the idea that had been presented to each of them simultaneously but separately. All are immediately directed toward that light, and those wandering spirits who had long sought each other in the shadows finally meet each other and unite. The newspaper has brought them nearer, and it continues to be necessary to them to keep them together.

Newspapers, as Tocqueville understood them, created communities of readers, as "only a newspaper can come to deposit the same thought in a thousand minds at the same moment."[1]

Tocqueville's visit to the United States was at a pivotal moment in the history of the newspaper, coming just as technological innovations had begun to transform the business of newspaper publishing. Over the following few decades, wood replaced cotton rags as the main raw material used in newsprint manufacturing, and machinery replaced human labor in both papermaking and newspaper printing. Starting in the mid-nineteenth century, entrepreneurs in the newspaper business transformed newspapers from products of small artisanal enterprises into mass-produced consumer goods manufactured in factory settings. This shift was not immediate or total, and one could still find in subsequent decades in the United States plenty of newspapers with readerships of fewer than one thousand, as many of the papers of Tocqueville's day had. There are still a handful of these kinds of papers around the country even today. But, over the longer term, the kinds of newspapers that Tocqueville observed were ultimately marginal to the future of journalism and newspapers after the application of industrial technologies to newsprint manufacturing and printing starting in the mid-nineteenth century.

One can trace the emergence of this new industrial newspaper in the career trajectory of *New York Times* publisher Adolph Ochs. In a speech in 1916, Ochs reflected on how much the newspaper business had changed since he first became a publisher with the *Chattanooga Times* in 1878. When he took over the paper in Tennessee, Ochs recalled, his printing press was an aging piece of machinery that had "several years' hard usage" on it already, and the *Chattanooga Times'* overall enterprise was modest. Ochs estimated that the total annual operating expenses for staff and materials came out to about $10,000. The paper's daily production required 750 sheets of 22-by-31-inch paper, which amounted to 25 pounds of newsprint. After reviewing these modest origins, Ochs noted in contrast that the *New York Times* in 1916 "regularly employs 1,200 persons," and he claimed that its average payroll was $32,000 per week. In addition to the increased personnel costs, the scope of the *Times*'s industrial plant and operations had also expanded dramatically. The annual bill just for "telegraph and cable tolls" was over $100,000, Ochs noted, and the *Times* used an average of 100 tons of newsprint every day. These numbers continued to climb in subsequent years. In November 1925 alone, the *Times* used 7,650 tons of newsprint, an average of 255 tons per day, in the process of printing some 1,012,555,184 pages. In the exploding scale and scope of industrial operations, the daily newspaper experienced revolutionary changes in the decades surrounding 1900.[2]

For the *New York Times*, as for all industrial newspapers, these changes owed much to the application of technological advances in two areas: papermaking and printing. In the former area, inventors and entrepreneurs developed new processes using wood rather than cotton rags as the basic raw material of production, and paper manufacturers transformed the marketplace for printed materials by taking advantage of the radically more plentiful supply of forest products. Alongside these advancements in newsprint manufacturing, innovations in printing technologies enhanced the ability of publishers to use that paper to print more copies at lower prices. In the realms of both paper manufacturing and printing, key initial advances took place in the early modern period, but the application of industrial technology in the nineteenth century proved transformative. By 1915, as one study of the industry noted, newspaper publishing was a "manufacturing enterprise, governed by the same laws that regulate the production of shoes, plows, or any other commodity."[3]

Taken together, new developments in papermaking and printing allowed commercial newspaper publishers to create a new industrial print culture. This print culture was an urban invention based on the production, distribution, and sale of an unprecedented volume of paper. Its origins lie in the New York penny press of the 1830s, but it reached its fruition several decades later when publishers like Adolph Ochs exploited emerging innovations in papermaking and printing to transform their businesses from small artisanal enterprises—not far removed from what Tocqueville had observed—into mass production firms capable of printing and selling a billion printed pages in a month, as the *New York Times* had in November 1925. This is not a technological determinist argument. In and of themselves, advances in printing and publishing did not do anything. Rather, it was through the adoption and use of these advances by business firms that they took on their importance. Put to use by strategic publishers, these technologies enabled the creation of industrial print culture in the twentieth century.[4]

The Sheet That Fits the Print: A Brief History of Papermaking

The commercial and aesthetic demands of printing different kinds of materials have led publishers to use different kinds of paper. Many hardcover books, for example, are printed using durable and often bright white sheets of paper. The pages containing image plates in the middle of these books, like the entirety of many art books, are often done with an even thicker and glossier page stock, on which pictures will appear more vibrant. Newsprint paper did not and does not need to be of this quality. The most important characteristics of the kind of paper necessary for a mass-circulation newspaper in the twentieth century were cheapness and availability, not durability. That paper needed to be sturdy enough to handle the rigors of industrial printing and of high enough quality so that the ink smudged as little as possible. These were minimum requirements, and some publishers would try to push beyond them to make the printed pages, especially the covers, look better. But the sort of paper most publishers used and coveted did not need to be made to last. It simply needed to be available in adequate quantities on that day, and for the foreseeable future. In many respects, the prominent publishers who developed the modern mass-circulation newspaper, for example, Adolph Ochs, Joseph Pulitzer, William Randolph Hearst, and Robert McCormick, did so as the beneficiaries of innovations in papermaking. There was nothing new to the twentieth century about those who wanted to reach a mass public in print. What was new was the practical capacity to do so, and this came through the development of far more readily available supplies of paper made from wood.

The importance of paper in print culture is something that historians of the subject sometimes elide. For many, the history of print culture involves the consideration of the production of the words on the page rather than the production of the page itself, and the pivotal moment was Johannes Gutenberg's invention of the printing press in the mid-fifteenth century.[5] And yet, as book historians Lucien Febvre and Henri-Jean Martin argue, this print culture flourished not because of the printing press alone but because of the greater availability of paper made from something other than animal hides, which had been among the primary sources of paper for centuries prior to Gutenberg. "What use would it have been," they ask, "to be able to print with moveable type if the only medium was skin"? Given the cost and scarcity of materials to print on, Febvre and Martin assert, printing would have been perpetually limited as a practice without an alternative to animal hides, and it "would have been impossible to invent printing had it not been for the impetus given by paper." A printing press could be put to use in the interest of reaching readers, but only to the degree that printers had paper to print on. This was the case for fifteenth-century book publishers, and it was certainly the case for twentieth-century newspaper publishers. Successful commercial printing on any scale depended on securing adequate supplies of paper at affordable prices. For industrial newspaper publishers, this would take the form of wood-based newsprint, which was developed in the second half of the nineteenth century—fairly recent in the long history of paper manufacturing.[6]

Papermaking has taken various forms going back over two thousand years, and one of the key long-term shifts involved moving from writing on a natural medium to writing on a manufactured medium. In the ancient world, writing was done on and into natural surfaces, for example, carvings into stone and clay and writing on dried papyrus plants. Treated animal skin would later provide the basis for excellent, if rather scarcely available, writing surfaces, in the form of parchment (sheep skin) and vellum (calf skin).[7] Starting around 100 CE in China, papermakers discovered that cellulose extracted from plant matter could be a useful and plentiful raw material. This method proved readily adaptable to large-scale production, and it has formed the basis of papermaking in the following two millennia: a fibrous material is macerated in water into a pulp, that pulp is then flattened as the liquid is extracted, and finally the remaining flat sheets are dried completely. In early Chinese papermaking, rags, bark, and hemp provided the fibrous material central to the papermaking process. Over time, paper manufacturers experimented with other raw materials, but the basic production goal was the same: to extract the cellulose material from a particular substance and then get rid of the excess liquid from the flattened and dried pulp.[8]

One of the most significant developments in the long history of papermaking came in the transition to the use of cotton rags as a raw material by leading European paper producers. Pioneered in China and brought to Europe via the trade networks of the Arab world in the twelfth century, rag-based paper became increasingly common throughout the West in subsequent centuries. Though rag-based paper production spread throughout Europe and across the Atlantic to North America, the basic elements of the process changed little for several hundred years. Until the nineteenth century, the making of each individual sheet of paper was done by hand. In the production process, rags were put into a tub with water and then beaten and shredded with sharp implements into a pulp. After this point, three different workers handled the papermaking process: the vatman, the coucher, and the layman. First, the vatman put a wood-framed mesh screen into the pulp and manipulated the frame so as to evenly distribute the wet pulp across it while removing excess liquid. He handed off the mold to the coucher, who shook it in order to dry the emerging sheet further and then flipped the newly congealed paper out of the mold and onto a sheet of felt. The coucher would then place another sheet of felt on top of the still damp paper and turn around in time to receive another mold from the vatman to keep the process moving. Once the pair of artisans had made and stacked one to two hundred sheets of paper between the alternating sheets of felt, the stack was put into a screw press to squeeze out even more water. The layman, the third member of this papermaking trio, at last would take the fresh paper from between the felt sheets of the stack and place them in a pile. Once pressed again, the paper would be hung to dry until ready for use.[9]

Rag-based paper made in this manner was clean and durable, but the problem for manufacturers, even in the days of artisanal production, was that rags were simply never as readily available in supply as there was demand for paper, and this scarcity constantly bedeviled preindustrial paper production. Paper mills developed local and regional networks that channeled waste rags from households to mills, and the rag picker became a regular feature of many towns and cities. As rags were collected and brought to mills, they were sorted and processed, often by female workers, prior to being pulped and made into paper. In many respects, the foundation of the print culture that developed after Gutenberg's invention of the printing press was the process of recycling and reusing waste rags to make paper.[10]

Rag scarcity became an even greater problem as mechanization was introduced to papermaking in the nineteenth century following inventor Nicholas Louis Robert's patenting of the Fourdrinier machine in 1799. To make paper with this machine, a pulp composed of 1 part fiber and 199 parts water was poured into one end and then moved over a continuously spinning wire mesh surface that began draining water from the pulp. The Fourdrinier machine then moved the emerging yet still soggy paper onto another continuously spinning loop covered in felt, and it was pressed between more felt-covered loops to remove additional water. The paper then emerged out of the other end of the machine and, after being allowed to dry completely, was ready for printing. The Fourdrinier machine allowed paper to be produced much more quickly by replacing the labor of the vatman, coucher, and layman with machinery, and its effects on the papermaking business were dramatic. Fourdrinier machines were first imported into the United States in 1827, and by the 1830s they were in widespread usage in paper mills. By 1845, mechanization had completely transformed paper manufacturing in the United States, and advances in papermaking technology built on the Fourdrinier created steady and significant increases in rates and volumes of newsprint production over the following century. In 1867, the maximum output for most machines was about 100 feet of paper per minute, and this increased to 200 feet per minute in 1880, 500 feet per minute in 1897, 650 feet per minute in 1916, and 1,200 feet per minute by 1930.[11]

As the capacity of papermaking machines increased in the nineteenth century, manufacturers began facing greater challenges obtaining sufficient amounts of raw materials. For US newsprint manufacturers desperate to meet the growing demand for paper, the rag import market became increasingly important, and by the mid-nineteenth century they were getting raw materials from rag markets in Alexandria, Smyrna, and Trieste and throughout East Asia. US rag imports increased from 2.1 million pounds in 1843 to 10.9 million pounds in 1845, 20.7 million pounds in 1850, 75.7 million pounds in 1869, and 105.7 million pounds in 1871. Many in the papermaking business were in a constant state of agitation about securing adequate and affordable supplies of rags to keep their mills functioning at capacity, and their operations seemed perpetually vulnerable to shortages in raw materials.[12]

Advances in newsprint manufacturing leading to a massive increase in supply began around 1840, not long after Tocqueville's visit to the United States, when German inventor Friedrich Gottlob Keller pioneered a method of using ground-wood instead of rags to make paper. Over the following decades, papermakers developed and improved on this process and took advantage of this much-steadier and more plentiful source of raw material. As had been the case with rag-based paper, the initial step in the process in making paper from wood was to extract the cellulose fibers, which made up roughly half of the wood's overall volume. The rest of the wood was composed of a chemical compound called hemicellulose and a binding substance called lignin. In this new wood-based paper production process, felled trees were cut into short pieces, usually 4 feet long, and the bark was removed through a process called rossing. The rossed wood was then forced against grindstones to reduce it to a pulp suspended in water, and this was then drained and screened to remove any extraneous particulate matter. At this point in the production process, a manufacturer had what is known as mechanical or ground-wood pulp. The other part of the later and more advanced method of making newsprint resulted in what is known as a chemical or sulfite pulp. To make a sulfite pulp, paper manufacturers first ground the wood into small chips of less than an inch in length and then put them into a brick-lined steel tank called a digester, into which the manufacturer fed a "cooking liquor" composed of calcium bisulfite. The wood chips and this "cooking liquor" were steamed under pressure for about eight hours, which dissolved virtually all of the substances in the wood except for the cellulose fibers. The resulting solution was rinsed with clean water, and all the sulfite liquor was drained off, at which point all that remained was the chemical pulp. Mechanical pulping extracted shorter and stiffer fibers from the wood, while chemical pulping extracted longer fibers, creating a pulp that, when added to mechanical pulp, made a paper sufficiently durable to go through the industrial printing process without tearing too easily. Newsprint manufacturing relied on a mixture of both kinds of pulp, with the mechanical pulp making up roughly 75 to 85 percent of the total and the chemical pulp the balance.[13]

As was the case with the diffusion of most innovations, the switch from rags to wood as the primary basis for papermaking was a process of gradual adoption over time. One can see this in the lag between Keller's invention in 1840 and the first "commercially successful" use of the process in the United States in 1867 at a mill in Massachusetts. Moreover, as wood-based paper began to be more readily available to publishers in subsequent years, many still purchased rag-based paper. For example, the *New York Times*, which first used wood-based paper in 1873, had shifted entirely to using it in 1874. The *New York Evening Post* had a longer transition, using wood-based paper first in 1875 but not entirely until 1878. Other newspapers had more than a decade of transition using the two types of newsprint, for example, the *New York World*, which used wood-based paper first in 1870 and exclusively only starting in 1881. Ultimately, the shift from rag-based to wood-based paper in printing newspapers was gradual and varied across organizations rather than immediate and uniform.[14] Papermakers continued to use rags in their manufacturing, but increasingly that paper was used for printing books rather than newspapers. Starting in the 1880s, the shift to using wood-based paper in printing newspapers had unmistakably begun, as evidenced in the nearly hundredfold increase in the quantity of wood used in US paper manufacturing in the quarter century after 1879.[15] In the process, this dramatically expanded the amount of paper available to publishers and the amount of information circulating publicly in their publications. According to one estimate, in the century after 1788, when newsprint production in the United States was 100 tons, annual manufacturing increased to 500 tons in 1810, 2,500 tons in 1828, 66,000 tons in 1859, 100,000 tons in 1869, 149,000 tons in 1879, and 196,000 tons in 1889. Production increased even faster in the ensuing twenty years, rising to 569,000 tons in 1899 and 1.2 million tons in 1909.[16]

As wood became the most commonly used source for making paper and as the quantities of paper production increased, newsprint manufacturers began facing the related challenges of ensuring that they had adequate supplies of trees and the necessary quantities of power to convert those trees into paper. That trees were available in the natural world in much greater quantities than rags were did not automatically translate into them being practically available to every paper mill, as the ideal mill needed to be close to specific kinds of trees in order to make the operation commercially feasible. Softer woods from coniferous trees such as spruce, hemlock, and balsam fir proved to be ideal for making paper, for three primary reasons. First, the wood was made up of longer cellulose fibers, which made for a more durable sheet of paper. Second, these types of trees were easier to grind down mechanically into the smaller-sized chips needed. Third, the wood was low in pitch, which meant not only that it was less likely to "gum up" the machine but also that it required little chemical treatment to make into a white sheet. By the late nineteenth century, when wood had become the dominant raw material in papermaking, the industry in the United States had shifted to northern states with large supplies of coniferous trees, most importantly New York and Maine. Given its location and vast swaths of forest, Canada also became an increasingly important part of the continental production system, as both a manufacturing center and a source of raw materials.[17]

As wood pulp allowed papermakers to produce more newsprint, trade journals in the newspaper business began to be filled with accounts from publishers describing their enterprises in terms of both the substance of the journalism and the sheer quantity and mass of the daily output of printed newspapers that they could produce. In 1915, for example, the *Chicago Daily News* boasted that if all of the copies it printed in one day were put in a single stack, "they would make a pile over 4,000 feet high, or more than ten times the height of the Masonic Temple, Chicago's tallest building." By the late 1930s, the amount of paper being used to print American newspapers had reached such levels that it prompted even more staggering claims about distance and scale. In November 1939, paper industry executive Royal Kellogg bragged that US newspapers used enough newsprint that year to cover some 15 million acres of land with a surface of paper. "Those who are statistically minded," Kellogg continued, "can picture this tremendous quantity of newsprint paper in other striking ways. It would, for instance, make a belt nearly a mile wide around the earth at the equator or extend in a strip 500 feet wide from the earth to the moon."[18]

Kellogg's remarks came almost exactly a century after Friedrich Keller's innovations had made it possible for wood to replace cotton rags as the main raw material used in paper manufacturing. Over this period, advancements in papermaking dramatically increased the amount of newsprint available to newspaper publishers and provided the material basis for them to develop the mass-circulation daily newspaper. The transformation of trees into newsprint, in other words, was crucial to the rise of industrial print culture.

Designing the Newspaper Factory

As paper manufacturers developed new methods of making newsprint, a different group of inventors and entrepreneurs were also developing new industrial technologies to print newspapers. As in the papermaking business, these new technologies transformed artisanal enterprises into industrial firms. For publishers taking advantage of increasing supplies of newsprint, new printing technologies allowed them to operate on larger scales and to produce consistently more copies of bigger newspapers.

As was the case with papermaking between the fifteenth and the twentieth centuries, advances in printing moved the newspaper from the print shop to the factory. The foundational development in mechanical printing was Johannes Gutenberg's invention of the printing press in the mid-fifteenth century, which afforded the means of manufacturing greater quantities of printed materials more quickly by replacing the hand copying of individual pages with the mechanical reproduction of multiple copies of the same page from one tray of movable metal type. This was a major development, but it is easy to overstate the suddenness of the change from handwritten to mechanically reproduced texts and the degree to which machinery transformed printing. Handwritten manuscripts remained in wide production and circulation long after the printing press was introduced, and printing remained very much an artisanal craft for centuries afterward. Even into the early nineteenth century, all printers were still setting type letter by letter in a case, inking these letters by hand, placing a sheet on top, and then pressing that sheet onto them with their own power.[19]

Entrepreneurs and inventors began improving printing technologies in the nineteenth century, and they eventually transformed printing shops into industrial enterprises as hand-powered presses were replaced by machines driven by steam and later electricity. Between 1811 and 1813, German inventor Frederick Koenig developed a process of using a rotating cylinder to feed paper sheets through the machine to be pressed onto the type. The first cylinder press was imported into the United States in 1824, and this increased output for publishers from about five hundred to one thousand copies per hour. By the 1830s, R. Hoe and Company had developed a similar machine that could produce two thousand newspapers in an hour. As printing capacities increased, higher newspaper circulations followed. In New York City, for example, the estimated maximum daily circulation of 900 in 1810 rose to 2,000 in 1816, 4,000 in 1830, and 21,000 in 1840 with the application of industrial manufacturing. Around 1840, another set of manufacturing innovations allowed the type itself to be mounted onto the cylinder, which allowed the paper to be fed through the printing press at even higher speeds. Newspaper printing became a continuously flowing process, as rolls of newsprint replaced individual sheets as the paper being fed into the printing presses, which in turn became increasingly able to turn those rolls into greater quantities of printed and folded individual newspapers. Evolving industrial advancements allowed leading publishers to produce up to 77,000 newspapers per day in 1860, 85,000 in 1870, and 147,000 in 1880. By 1890, maximum daily production for the most industrially advanced newspapers was 300,000, and in 1900 it reached one million.[20]

The telegraph may have uncoupled communication from transportation, as James Carey famously noted, but it did not yet uncouple mass communication from industrial mass production. Sent from point to point over the wires, the news still had to be mass-produced on paper in order to reach the public. This application of machinery transformed the newspaper business by allowing publishers to print copies more quickly. It also introduced a new concept of timeliness to news, as it shrank the temporal gap between the receipt of telegraphic bulletins and the mass circulation of them in print. In the era before radio broadcasting was developed in the 1920s, newspapers remained the primary disseminators of breaking news, and this motivated publishers to work to shorten the gap between getting the news over the wires and having it in print and out for sale. For many industry observers, speed of manufacturing became the key to producing a competitive newspaper. The trade journal *Editor & Publisher* noted that, independent of content, the success of a publishing enterprise had to do with the efficiency of its industrial operation. Simply put, the magazine claimed that "a publisher's success or failure can be largely determined by his control of mechanical operations."[21]

In order to respond to this challenge, many metropolitan publishers began organizing their printing plants along the lines of other mass production industries. In 1917, for example, the same year that the Ford Motor Company began building its River Rouge Complex, designed by architect Albert Kahn, the *Detroit News* opened a newspaper factory of its own, described as the "zenith of newspaper planning, architecturally and from the standpoint of operations." The plant's construction cost some $2 million, and one account remarked that it "represents a material fruition of almost half a century of newspaper growth." With the massive structure and its new industrial machinery, the *Detroit News* could produce 432,000 sixteen-page papers in one hour. The industrial design connections between automobile and newspaper production in Detroit would be further strengthened in 1925 when Albert Kahn himself designed a new $4 million plant for the *Detroit Free Press*. The building was, as one account reported, "really built around its remarkable press room," where $1 million worth of printing equipment was capable of turning out 504,000 sixteen-page papers per hour. In 1929, Kahn also designed a new $2.5 million facility for the Hearst-owned *Detroit Times*.[22]

Though the designs of Detroit's newspaper plants harmonized with a local economy based on automobile production, the factory model for newspaper buildings also spread throughout the country. The *New York Tribune*, for example, had been operating since 1845 in a building at Nassau and Spruce Streets which publisher Horace Greeley had commissioned. In 1923, the *Tribune* moved into a new building uptown on W. 40th Street, where the building's designers had applied the "same principles of time saving and money saving that are today worked out in the construction of all factory properties, whether they be for the production of silk, shoes or thumb tacks." The *Tribune*'s new seven-story building was "laid out so that the production line is carried through every department, and floors are laid out so that movement is always toward a central pivotal point, production flowing forward toward output without any backwashes of waste motion." Around the United States, the 1920s witnessed a huge boom in the construction of newspaper factories of varying sizes. In Easton, Pennsylvania, for example, the *Easton Express* in 1923 built a new facility that was "in part an office building, in part an industrial building, and in part a warehouse," reflecting the varied needs of its production. In Long Beach, California, the *Long Beach Press-Telegram* spent almost $1 million on a new plant. In 1924, the *Milwaukee Journal* opened a $2 million facility that could produce 108,000 thirty-two-page papers in one hour, and the *San Francisco Chronicle* spent $1.5 million on a new building and plant that produced 142,000 thirty-two-page papers per hour.[23]

One of the hallmarks of many of these new newspaper factories was that they were situated in their cities so as to be as efficient as possible in turning rolls of newsprint into printed newspapers. As one overview of newspaper plants noted, "With speed of production and speed of distribution as the prime requisites of modern publishing, newspaper publishers are studying plant sites in their cities to find a location which will supply these needs at the smallest capital charge." The most desirable locations were those "convenient to railroad rights-of-way" because newsprint was "bulky and costly to handle and subject to damage if tossed about between freight car and pressroom. A railroad siding in or adjacent to the plant eliminates this source of waste and also cuts handling charges to a minimum." In Philadelphia, the *Inquirer* in 1925 followed this strategy in constructing what was described as a "palatial" new facility built directly on top of and into the city's transportation infrastructure. The twenty-one-story building was topped with a "golden dome" containing a "lantern that flashes its powerful white light in all directions, changing to a brilliant red to flash the signal of each completed quarter hour." But beyond its status as a piece of showpiece architecture, the tower was "built entirely over a network of railroad tracks, most of which are part of one of the Reading's big freight terminals here." The *Inquirer*'s plant was, as one report noted, built with the primary aim of handling incoming supplies of newsprint. The company chose its site not only "with regard to the city's future growth, but also because it offered a practical solution of the problem of prompt and dependable delivery of newsprint paper, one of the necessities of a great newspaper plant. The dual problem of centralized location and facilities for the quick delivery and storage of newsprint paper from convenient freight stations confronts all metropolitan newspapers. Usually the route is from the freight car by truck to a storage warehouse, and thence by truck, through heavily traveled streets, to the newspaper itself." The *Inquirer*'s new structure would "eliminate this expensive and tedious process" because the "building itself is the freight delivery station. Cars loaded with paper from the mills arrive on the tracks under the building and alongside a freight platform. The paper is rolled directly from the cars to elevators, which carry it to the paper storage room on a floor above." The *Inquirer* used an estimated 1,000 tons of newsprint per week, and the new building was designed around a "scheme dictating the movement of the paper from the storage rooms through the presses to the waiting delivery trucks and carriers on the one hand, and from advertising, news and editorial offices, whence comes the material for the printed page."[24]

The year 1925 was a high point for the construction of new and lavish newspaper buildings, as the trade press reported. "New newspaper plants opened this year or to be in operation within the next few months represent values for buildings, real estate, and machinery well in excess of $100,000,000. That figure is conservative." The construction of these sizable and technologically advanced newspaper factories continued even into the Depression. In 1935, the *Los Angeles Times* opened a new building that was meant both to form part of the city's downtown skyline and to take advantage of its transportation networks. The $4 million structure was reported to be the first entirely air-conditioned newspaper plant in the United States, and a battery of ninety-two floodlights kept it illuminated at night for the city's residents to see. Inside, the presses could produce 200,000 thirty-two-page newspapers per hour. "While sentiment and tradition played their parts in determining the location of the new building," one report on the building noted, "efficiency also entered the picture. It is but a stone's throw to the city, county, state and Federal buildings, facilitating the gathering of the volume of news emanating from these offices. From the district highways radiate to all nearby communities, eliminating for the fleet of trucks much of the metropolitan traffic of downtown streets. Close by the new union terminal is under construction, making it but a short haul to all available rail and air transportation."[25]

Over time, the developments in the industrial technologies needed to produce a mass-circulation daily created a sharp divide between generations of publishers. Not only was the cost of equipment a high barrier to entry, but the size of the labor force needed to operate the machinery meant that anyone aspiring to start a paper could not simply do it themselves, as was possible in the nineteenth century. This had been Adolph Ochs's experience as he charted his career in the newspaper business from the *Chattanooga Times* in the 1870s to the *New York Times* in the 1910s, and his story was a representative one.

In 1939, *Emporia Gazette* publisher William Allen White discussed his own similar experience with these changes in an article in *Collier's*, remarking that he "could have bought the materials, presses, type and equipment for the *Emporia Gazette* when I took it over if I'd had seven or eight hundred dollars." The fact that he had $3,000 in 1895 was more than enough capital to continue operations once he completed the purchase. One of the things that gave the young White the "self-reliance" to run the paper once he acquired it was that, because of ten years of previous newspaper experience, "I knew I had done reasonably well everything that I could possibly ask anyone else to do in the publication of that little daily paper. . . . I had swept the office, as a printers devil. I had made up the forms and had put the paper to press. I had set the type. I had washed rollers and helped to make them. I had walked the street for news. I had rung doorbells on front porches and had solicited subscriptions. I had edited other reporters' copy. I had written editorials and had sold advertising." Having spent fifty years in the newspaper business, White had something of an epiphany in June 1935 about how much things had changed. The occasion for his realization was not a bolt of wisdom about journalistic philosophy but rather the more direct acknowledgment that he was detached from newspaper production to the point of being practically unable to imagine doing it without heavy and expensive machinery that he had no clue how to operate. Walking into the *Gazette* offices one day in 1935, White found himself saddened rather than proud on my fiftieth year in the newspaper business to realize that the equipment of a typical country paper in an American town of ten or fifteen thousand would cost $50,000 to $75,000. I remember that when I bought the Emporia Gazette I could do every physical thing that as boss I ordered another man to do. That day in 1935 I could not perform one single, simple mechanical process. What I had done by hand in the passing of half a century men now do by machinery. A linotype sets the type. A stereotype stereotypes it before it goes to the press. As a boy I fed, by hand, my press sheet by sheet. Today it rolls off, thirty thousand an hour.

And his experience, he realized, "was not unique."[26]

In the late nineteenth and early twentieth centuries, newspaper production around the country followed the trajectory that Adolph Ochs and William Allen White had seen. Many newspapers in the late nineteenth century were, like Ochs's *Chattanooga Times* and White's *Emporia Gazette*, firmly based on mechanical production but still relatively modest in scale. Tasks involved in manufacturing a newspaper, while no longer artisanal, were still such that they could be mastered by a single person and carried out on a daily basis by a small staff. By the post–World War I period, that model of the newspaper had vanished in most American cities, and metropolitan newspaper publishing had become a form of industrial mass production. The dramatic increases in plant and equipment expenses made the cost of starting a newspaper prohibitively high for all but a select few wealthy Americans, and those working within newspaper organizations found themselves in increasingly specialized roles where the daily tasks involved the operation of industrial machinery in a factory setting.

Developing the Distribution Network

Like all firms engaged in mass production, newspapers had to find ways to physically distribute their products to consumers. Unlike many firms, however, newspapers had to perform this distribution within an extremely short time frame. As press critic Silas Bent noted in 1927, the newspaper publisher "deals in a perishable commodity, moreover, which loses its bloom when exposed to light. It is the shortest-lived commodity known to the business world." As industrial concerns, newspaper publishers had much in common with the manufacturers of products like apparel and cars, but in many ways they had more operational similarities to those in the milk business than they did to those in the clothing or auto business. Like milk producers, newspaper publishers in the early twentieth century operated mostly in local markets and had an exceptionally limited amount of time to get their product into a consumer's hand before it became unsalable for lack of freshness. For many milk consumers in the twentieth

century, the most important indicator of quality was the date of the milk's production, which let them know whether the milk was safe for consumption. For
newspaper readers, the same was true: the date on the paper let them know
whether the paper was useful. An out-of-date newspaper might not have adverse
health effects on readers, but as a commodity it was equally as unmarketable as
an out-of-date dairy product.[27]

In many cities, the interval in which a daily newspaper had to be sold before
becoming outdated did not even extend to an entire weekday, but rather only
the briefer period of a particular morning or afternoon. Essentially, in the newspaper business anything produced but not sold within a four- to eight-hour
window was useless as a commodity.[28] For much of the nineteenth century, the
distribution system for many metropolitan papers was at best loosely organized.
In many cities, newspapers sold copies at wholesale prices to newsdealers operating newsstands, and these entrepreneurs in turn sold the copies to readers at the
cover price. Interested individual subscribers could take home delivery, but, as
Don Seitz of the *New York World* remarked in 1916, publishers had only begun to
rationalize urban distribution of individual copies in the late nineteenth century.
Prior to around 1885, he claimed, the newspaper "found its way to the reader
largely by chance." Looking back from the vantage point of 1916, Seitz marveled
at the fact that publishers built the sorts of businesses that they did with this type
of system. According to Seitz, the "idea of pushing the paper was usually beneath
the dignity of the ownership. Even such a great seller of news as James Gordon
Bennett, the elder, made the reader hunt for his paper." Seitz found it "incredible" that newspapers had prospered as they did.[29]

In many cities in the late nineteenth century, however, competition would
force several publishers to develop more effective ways to market and sell their
papers. To do so, many began employing a circulation manager for the first time
in the 1890s to rationalize distribution practices. Generally speaking, early urban
circulation managers developed methods of daily local delivery taking two forms,
one involving carriers with specific routes serving subscribers, and the other involving personal sales by newsboys to individual readers. In the early twentieth
century, one of the most important factors structuring the use of a particular
system was the use of mass transit by readers. In cities like Philadelphia and
St. Paul, for example, the carrier system proved more common, as the bulk of
residents did not have long streetcar journeys to and from work. Many wanted the
paper at their doorstep in the morning so that they could read it prior to leaving
the house. In New York, in contrast, the newsboy system better served a population tending to buy papers outside the home in order to read in transit on longer
journeys to and from work. To serve diverse readerships and reach as many

customers as possible, most cities had distribution networks blending the work of carriers and newsboys.[30]

In both systems, publishers sold papers to vendors at wholesale rates, which varied by city and over time but were typically 30 to 40 percent below retail cost. In Milwaukee around 1910, for example, newsboys selling the *Morning Sentinel* paid 62.5 percent of retail cost for their papers, as the *Sentinel* sold them one hundred two-cent papers for the wholesale cost of $1.25. If a newsboy sold all one hundred copies, this would net him $2.00 on the street and give him a profit of 75 cents. What this sort of arrangement meant was that the publisher gave up receiving the full cover price of the newspaper in exchange for ensuring a readily available daily revenue stream by transferring the sales risk to the carrier and newsboy through an incentive-based arrangement. As one survey of the industry noted, "If carriers are paid weekly wages, they handle the paper's money; whereas, if they are on commission, they handle their own money. Hence the advantage of the independent, or commission, system. The carrier must take care of the paper's interests in order to take care of his own." Starting around 1900, the creation by US publishers of this new distribution network led to both improvements in the process of getting newspapers to readers and the establishment of a steady and predictable stream of income that did not require taking on additional payroll costs.[31]

As circulations continued to grow across the country around the turn of the twentieth century, publishers and their circulation managers began employing new measures to rationalize their distribution practices. In Indianapolis, for example, the *News*, an evening paper, had a circulation of 105,000, 65 percent of which was within a 20-mile radius of the city. To distribute all of these papers on a daily basis required a network of sixteen hundred carriers and three hundred newsboys in the city, in addition to eight hundred agents and four thousand carriers for the areas beyond the 20-mile metropolitan limit. In the city, the *News* divided the area into twenty-five districts served by thirty-six substations operated by a mix of *News* employees and dealers working on commission. Carriers, who averaged eleven years of age, reported to their local substation at four o'clock in the afternoon to pick up the papers for their routes, which averaged thirty-seven subscribers each. Many papers used some version of this system. In Chicago, *Daily News* publisher Victor Lawson began rationalizing distribution of his paper in the 1870s and 1880s by selling wholesale to newsdealers who were granted exclusive rights to sell the paper in particular territories within the city. As in Indianapolis, these dealers were responsible for coordinating the network of young carriers, who were estimated to number about three thousand by the early 1890s.[32]

Across the country in the early twentieth century, publishers continued developing ways of organizing and controlling newsboys' work. In Milwaukee, individual newsboys would try to establish a daily presence on specific downtown corners, and one study estimated that more than three-quarters of newspapers were purchased by readers from the same newsboy everyday. In Chicago, some newspapers began issuing cards to newsboys which indicated the corner at which they worked, and the boy had to present the card to get his papers to sell. Particular corners obtained a "quasi-legal position that implied a pecuniary value," as one report put it. These customary rights to sell on particular corners did not, however, give the newsboy any additional benefits or job security, and the system represented another manifestation in which the newspaper rationalized and maintained control over a distribution network while providing no binding job security and few if any benefits to the newsboy.[33]

Many aspects of the newsboy's position in the early twentieth-century newspaper's distribution chain can be seen as exploitative, as these newsboys labored under conditions that few adult workers would abide. As newsboys became an increasingly common urban presence during the Progressive Era, they also became a subject of significant interest among urban reformers who saw them as exploited child laborers. In 1907, researchers at the Federation of Chicago Settlements remarked that the newsboy had become such a common fixture of city life that it was easy to ignore his socially and economically precarious existence. "The newsboy becomes part of our environment," the Federation noted, a "familiar figure, rather under medium size, as we know him best. . . . A veritable merchant of the street scanning every passer-by as a possible customer. . . . The public sees him at his best and neglects him at his worst. He is not considered in the problem of 'child labor,' because he works in the open and seemingly apart from the associations which are so hostile to the health and happiness of the factory child." The Chicago researchers studied a group of one thousand local newsboys and noted that 127 of them were under the age of ten and that many were living a "semi-vagrant" life involving irregular hours and questionable behavior. After visiting the rear of William Randolph Hearst's *Chicago American* building seven times in three weeks, the researchers noted that there were "at least forty and sometimes seventy-five boys, many of them under fourteen years of age. They are smoking cigarettes, eating, sleeping, fighting, or 'shooting craps'; towards morning the most of them will be found sleeping on the floor waiting for the morning editions."[34]

Even those newsboys laboring under less Dickensian conditions found themselves also occupying a gap in labor law, as a study done under the direction of economist John R. Commons noted. Though many states had passed laws barring the use of child labor, these laws did not apply to newsboys' work, as the newsboy was technically "not an employee" of the newspaper but instead classified as a "merchant." Many state child labor laws had explicit exceptions for newsboys, and these policies in effect gave publishers access to a labor pool no longer available to other manufacturers. Many aspects of this "so-called 'little merchant' system" of newsboy labor continued for decades after. In 1934, for example, research conducted by the US Department of Labor reported that the independent contractor system remained in place for newsboys across the country. Though some of the more exploitative conditions of the job of delivering newspapers had been removed by the post–World War II period, the system of using juvenile carriers persisted. Indeed, as late as 1980, historian Anthony Smith noted of American newspapers' distribution infrastructure that the "whole system depends upon child labor. . . . If Congress were ever to cancel or restrict the newspaper's exemption from child-labor legislation, the newspaper would be sunk."[35]

Smith's case was overstated, but it is effective in reminding us of the people and practices that made up the system of producing and distributing industrial newspapers. In the late nineteenth and early twentieth centuries, US newspaper publishers used the parallel advances in papermaking and printing technologies to expand the size and reach of their enterprises. As technology allowed increases in daily production limits, publishers sought new ways of distributing their products with greater speed and efficiency. All of this created an industry that in the early twentieth century looked much less like a Tocquevillian foundation of democracy than a strange hybrid of a mass industrial production enterprise like automobile manufacturing crossed with the problem of perishability facing dairy distributors, with the final product often physically given to a reader by a preteen boy laboring as an independent contractor.

Newsprint and the Economics of the Industrial Newspaper

In the twentieth century, newsprint was the foundation of the modern newspaper enterprise. Publishers needed it in sufficient quantities to be able to print each day, and they faced the daily challenge of physically mass distributing it as quickly as possible. In many ways, publishers had much in common with Henry Ford and James Duke, as all aimed for cheaper production at higher volume. But the newspaper was different from cars and cigarettes in that it was the product of an industry with a peculiar price structure and business model. Mass-circulation newspapers were exceptionally cheap in the United States, and indeed one of the great advances of the American penny press was that it had reduced the cost of the product to the country's lowest possible currency unit. As the *Buffalo Express* noted, the newspaper was "commonly and correctly known as the cheapest thing in the world. . . . If the Mint coined a smaller token than a cent, doubtless that would have been taken as the newspaper unit of price."[36]

Publishers charging a penny were able to do so because of the particular economics of the newspaper business. The newspaper was what media economists call a "joint commodity," as publishers sold not only the physical copies to readers but also advertising space in those copies to businesses. For many newspapers, the revenues generated by the sales of physical copies were lower than the cost of production, and advertising sales generated the profits after operating costs were covered. Unlike manufacturers of cars and cigarettes, in other words, the newspaper publisher earned money in two ways: by selling the consumer a particular commodity and by selling to advertisers access to the purchasers of that commodity. This is essentially the same model that exists today for free internet platforms such as Gmail and Facebook, which have even done away with the penny press's penny: the service costs nothing because it is covered by the revenue generated from advertising. In the twentieth century, however, for both of the revenue-generating parts of the newspaper business—sales of copies and advertisements—a steady supply of newsprint was essential.[37]

The practice of generating revenue jointly by selling newspapers and advertising enabled the commercial development of the US newspaper, and in many important respects this was made possible by the massive increase in the newsprint supply after the introduction of wood-based paper. As wood replaced rags as the main raw material for making newsprint, the increasing quantities of newsprint initially drove paper prices down dramatically, as table 2 shows. In 1860, for example, a ton of newsprint cost $166. Shortages of labor and materials during the Civil War drove the price up dramatically during and immediately after the war, and prices only dipped back to prewar levels in 1876. As wood-based newsprint came onto the market in the 1880s, prices dropped dramatically, going from $129.20 per ton in 1881 to $83 per ton in 1887. From 1897 to 1901, the average cost of a ton of paper stayed at $36. For publishers, this greater availability of cheap newsprint allowed them to price their papers as low as a penny and then generate additional and significant revenues by printing more advertisements. This business model would define the twentieth-century newspaper business, even as prices for newsprint fluctuated. These fluctuations were occasionally dramatic, as chapter 3 will discuss of the period around World War I, and they would create significant operational challenges for publishers who had built their businesses around calculations of paper supplies that they would need to print news and, often more importantly, advertising.[38]

One can see this relationship between circulation, advertising, and newsprint usage in operation in a study of the New York City newspaper market in 1922, as shown in table 3. In the daily morning field, the *Times*, *World*, and *American* were highly competitive in terms of circulation. The *Times* and *World* sold for two cents,

TABLE 2
Newsprint price per ton, 1860–1940 (selected years)

Year	Price per ton ($)	Year	Price per ton ($)	Year	Price per ton ($)
1860	166.00	1890	68.00	1920	112.60
1863	442.40	1893	55.00	1921	111.35
1866	344.00	1896	45.00	1923	81.80
1869	250.00	1899	36.00	1926	71.80
1872	240.00	1902	38.00	1929	62.00
1875	170.00	1905	40.00	1932	48.33
1876	164.00	1908	42.00	1935	40.00
1878	129.20	1911	45.00	1938	49.00
1881	129.20	1914	43.60	1940	49.47
1884	110.00	1917	63.78		
1887	83.00	1919	79.40		

Source: Royal Kellogg, *Newsprint Paper in North America* (New York: Newsprint Service Bureau, 1948), 49–50.

Note: These are estimated prices for newsprint in New York City. Actual prices paid by specific publishers, in New York and elsewhere throughout the country, could vary, and sometimes significantly, based on the quantity of paper purchased and freight charges.

and the *American* for three cents. The *Times*, however, printed substantially more pages than its competitors. For the *Times*, this required using a greater amount of newsprint, which in turn generated higher operating costs. For the *Times*, however, these costs were offset by the higher volume of advertising sales. Whereas the three-cent *American* used 27 percent of its weekday pages for advertisements in 1922, the *Times* used 58 percent of its daily printed pages for advertisements that year. The *World*, in comparison, used 52 percent of its daily pages for advertisements in 1922. Whatever the specific relative balance of news and advertising for each publisher, printing a bigger newspaper did not mean printing more news. Instead, it meant printing more advertisements that had been sold to advertisers at generally higher rates, and this in turn allowed the publishers the option of charging the reader less. In contrast, a five-cent paper like the *Commercial* relied far more on its cover price to generate revenue. The paper's circulation was but a fraction of the mass-circulation dailies, and it used roughly three-quarters of its pages to print news rather than advertisements. Its lower circulation meant that it could command a lower advertising rate. For all of these New York papers, the cost of newsprint forced their publishers to develop a business model that allowed them to cover those paper expenses through a combination of the sales of individual copies and advertising.

Given a street sale price that was inelastic in the short run, publishers could face significant problems when newsprint prices fluctuated too quickly and dramatically, as they occasionally did. As *Editor & Publisher* noted in 1905, the publisher

TABLE 3

Circulation, advertising, and newsprint usage among New York City newspapers, 1922

Newspaper	Total average daily circulation as of Apr. 1	Total pages printed Jan. 1–Aug. 31	Estimated tons of newsprint used Jan. 1–Aug. 31	Percentage of news pages Jan. 1–Aug. 31	Percentage of advertising pages Jan. 1–Aug. 31	Street sale price	Agate line advertising rate (minimum rate for 10,000-line contracts)
Morning newspapers							
American (6 days)	356,127	5,224	14,880	73	27	0.03	0.600
Sunday	1,092,713	3,868	33,813	51	49	0.10	1.250
Total		9,092	48,693				
Commercial (6 days)	11,833	3,148	298	76	24	0.05	0.240
Daily News (6 days)	444,402	5,064	8,913	65	35	0.02	0.670
Sunday	270,479	1,672	1,809	84	16	0.05	0.430
Total		6,736	10,722				
Herald (6 days)	176,082	4,648	6,557	60	40	0.02	0.410
Sunday	196,103	3,450	5,352	61	39	0.06	0.410
Total		8,098	11,889				
Times (6 days)	346,764	6,864	18,802	42	58	0.02	0.553
Sunday	545,276	4,428	19,316	42	58	0.06	0.638
Total		11,292	38,118				
World (6 days)	353,199	5,240	14,806	48	52	0.02	0.440
Sunday	602,359	3,978	19,169	46	54	0.06	0.440
Total		9,218	33,975				
Evening newspapers							
Globe (6 days)	151,626	4,084	4,954	54	46	0.03	0.370
Mail (6 days)	148,663	3,846	4,574	60	40	0.03	0.380
Sunday (6 days)	181,862	5,024	7,309	49	51	0.03	0.410
Telegram (6 days)	105,389	3,818	3,219	57	43	0.03	0.250
Sunday	121,369	760	738	65	35	0.05	0.250
Total		4,578	3,957				

Sources: "English-Language Daily Newspapers of U.S. and Canada with Circulations and Rates," *Editor & Publisher* 55, no. 2 (June 10, 1922): 32; "Advertising Lineage, Circulations, Pages Printed, Newsprint Used, and Its Cost, January 1–August 1, 1922–1920, in New York Papers," *Editor & Publisher* 55, no. 16 (Sept. 16, 1922): 11; "New York Newspaper Circulations, Total Pages, Total Advertising, Dry Goods and Foreign Advertising, and Circulation Rates, Jan. 1–Aug. 31, 1922 and 1921," *Editor & Publisher* 55, no. 18 (Sept. 30, 1922): 12.

selling a mass-circulation daily was "fortunate indeed if he does not actually lose money on his circulation" because of the small and sometimes nonexistent profit margins on print sales. Newspapers were "sold at the smallest margin of profit of any article in general consumption. Any increase in the cost of material or production means the elimination of all profit on circulation." This put a huge burden on publishers to maintain high levels of advertising revenue so as to offset increasing production costs and ensure continued profitability.[39]

Conclusion

For many publishers in the United States, fluctuating newsprint costs and supplies provided an ongoing source of unease about their business. For many, this anxiety would come to project itself past the newsprint suppliers and all the way to the forest. In the twentieth century, US publishers would have to contend not only with paper manufacturers but also with what many around the country saw as an increasingly near and dire shortage of the raw materials necessary to manufacture newsprint. Starting in the 1890s, some publishers foresaw an exhaustion of pulpwood supplies coming in the near future, and many warned of significant effects of a "failure of this supply and a rise in the cost of printing paper upon civilization," as the *Brooklyn Daily Eagle* put it. With newsprint more expensive and in shorter supply, US "newspapers would become compact publications. . . . The penny paper, and probably the two-cent paper would disappear." In addition, all newspapers' contents would have to change. "Copy would be ruthlessly condensed because the limits of space . . . would be small," and "advertising space would be reduced, and the prices for it raised. The effect of such a change upon the habits of thought of a nation would be hardly less radical than the changes of climate from the deforesting of our mountains, which has long seemed probable."[40]

In fretting about newsprint supplies and costs, the *Eagle*'s editorial demonstrates how important newsprint was to the newspaper business in the early twentieth century. The cheap and plentiful newsprint supplies that emerged after wood replaced rags as the standard raw material in manufacturing allowed publishers to print and sell vastly greater numbers of copies each day and to do so at lower prices. It also allowed them to sell vastly greater quantities of advertising to generate the revenues that were in many cases the source of much if not all of their profits. Newspapers were embedded in the foundation of the emerging consumer culture of the early twentieth century as factory-made goods making money from selling advertising for other factory-made goods. For publishers to continue profiting from this business model, they would have to become increasingly adept at using industrial production and maintaining connections to newsprint supply chains stretching back from their printing plants to the forest. In the twentieth century, as the next chapter will show, those supply chains increasingly stretched across an international border and into Canada. For many US publishers, printing their locally circulating newspaper would become firmly embedded in an international context.

CHAPTER TWO

Forests, Trade, and Empire

In the late nineteenth century, Chicago's entrepreneurs and boosters made the city into a center of the industrial capitalist exploitation of nature. As a metropolis and a marketplace, Chicago connected to vast areas of its hinterlands as raw materials were brought in to be processed into commodities and consumer goods that were then sent back out across the country. In 1893 at the World's Columbian Exposition in this staples-driven boomtown, historian Frederick Jackson Turner delivered what would become one of the most enduring statements about American development as he marked the "closing of a great historic movement" and declared that the "frontier has gone," as Americans essentially had completed the settlement of the continent from the Atlantic to the Pacific. As Turner conceptualized US history, the westward-moving settlers formed the vanguard of the industrial developments following in their paths.[1]

Nearly a century later, looking at this same process from the perspective of Chicago, historian William Cronon found it to be the opposite. It was not the yeoman farmer and the rugged individualist who had led westward expansion and American economic development, but instead the urban merchant. Business and industry did not follow agriculture into the hinterland, but rather motivated and organized it. Cronon's history of Chicago remains, nearly three decades after its publication, one of the most incisive understandings of that city's role in American development. But it is a history that, for all its merits, is still representative of a conceptualization of US history which is oriented along an east-west axis.[2]

In reality, within just five years of Turner's speech, a number of important US industries became strongly oriented along a north-south axis as well, and they increasingly penetrated into the economics and affairs of foreign nations. In turning their attentions south, US corporations brought sugar, bananas, and coffee to the United States in great quantities as the state advanced imperial projects during and after the Spanish-American War. When US corporations looked north, they coveted Canada's forests. In Cronon's account, Chicago was the central regional node in the production and sale of meat, wheat, and wood. If we reorient Cronon along a north-south axis and into Canada, however, we can use his framework to see the development of new international commodity chains that would shape the history of the US newspaper in the twentieth century.[3] For understanding the history of the mass-circulation US newspaper, one of Cronon's trinity of commodities—wood—is of particular importance. Chicago had by the 1850s become the "single greatest lumber market in the world," as Cronon remarked. Along the South Branch of the Chicago River, "one entered a world that appeared to consist almost entirely of stacked wood," and there were "seemingly endless heaps of pine lumber ten or more feet high," and the "smell of sap and sawdust hung in the air." By the late nineteenth century, nearby Michigan had been turned into "the Cutover" as a result of this seemingly limitless demand for wood.[4]

In the twentieth century, Chicago's effects on distant forest resources would continue. However, key members of its local business community would increasingly participate in commodity chains turning that wood not just into building materials but also into newsprint. In Chicago, as in other US cities, newspaper publishers extended the processes through which urban capital coordinated distant forest exploitation. In the twentieth century, Canada's plentiful spruce forests offered what seemed like endless supplies of the trees best suited for making newsprint. All that was standing in the way was an international boundary. Publishers would do everything in their power to get ready access to the trees on the other side of that boundary, and their success in doing so would transform the trade relationship between the United States and Canada.

Trees and Tariffs in North America

In order to understand the commodity chains connecting Canadian trees to American newspapers, we need to trace the actions of a cluster of corporate and state actors. The two groups of corporate actors were newspaper publishers and newsprint manufacturers, a pair of manufacturing industries that had a symbiotic and yet almost perpetually antagonistic relationship with one another. Each depended on the other for its financial success, and yet an equilibrium price and interindustry sense of harmony were difficult if not impossible to sustain. The various state actors represented both the two national governments and the eastern Canadian provincial governments, especially in Ontario and Quebec, where the largest spruce stands were located. US newspaper publishers wanted paper made from Canadian trees, and policymakers on both sides of the border had distinct interests in shaping the trade networks in those trees and the products made from them.[5]

The policy mechanisms regulating the commodity chains connecting these trees and publishers were the embargo and the tariff. The embargo is an aggressively protectionist tool used to ban exports and imports. If a country wants to promote domestic manufacturing in a particular industry, for example, it can simply forbid exports of the necessary raw materials involved. Likewise, if a country wants to encourage domestic production of particular commodities, it can enact an embargo on all imports of those goods. Tariffs, on the other hand, have both revenue-generating and protectionist functions. While creating revenues for the state, they can also be used to protect domestic industries from foreign competition by making imported goods prohibitively expensive. As industrial capitalism developed on both sides of the Atlantic in the 1800s and as a global market for agricultural products expanded, the tariff became one of the most important political issues around the world. Within a developing system of international trade, tariffs structured the basic terms for economic development at the national level. If a state enacted high import tariffs on manufactured goods, it created a protective barrier allowing domestic industry to operate with an advantage over international competition. In agricultural markets, countries wanting to protect domestic production of particular commodities established high import duties for things like wheat, corn, cotton, and wood. To remove tariffs is to remove friction from the processes of circulating commodities across international borders, and this creates a system commonly referred to in shorthand as "free trade."[6]

When the two countries participated in trade negotiations in the late nineteenth century, the United States interacted with Canada in a different way than it did with other countries in the Western Hemisphere, where more commonly military engagements were the precursors to economic relations. Starting in 1898, for example, the United States helped create a massive sugar industry in the Caribbean following the Spanish-American War by temporarily occupying Cuba and the Dominican Republic and making Puerto Rico into a part of the US tariff zone as a colony. The war and its aftermath created the conditions for US corporations to develop a Caribbean sugar economy in their interests. The situation was rather different in North America. Unlike the unfolding imperial projects in Latin America, the United States did not engage in military action in or against Canada. Though some individual officials expressed annexationist desires toward Canada, the US government never initiated practical efforts to pursue them. Rather, it aimed to make Canada a trade partner. And Canada, unlike Caribbean nations, had the power to negotiate with the empire in creating the terms of that trade relationship.[7]

The willingness on the part of state actors in the United States and Canada to engage in these negotiations varied over the second half of the nineteenth century. From 1854 to 1866, the two countries operated under the Elgin-Marcy Treaty, which gave the United States permission to fish in some Canadian waters and instituted reciprocal free trade in many commodities. These included, among other things, grain, flour, livestock, eggs, butter, coal, unmanufactured tobacco, and "timber and lumber of all kinds." Trade between the two countries was robust under the treaty. In 1853, before the treaty went into effect, some 6.1 percent of total US exports went to Canada. By 1855, this had jumped to 12.7 percent, and it reached 17.3 percent in 1865. Canadian exports to the United States had been 2.5 percent of total US imports in 1853, and this reached 13.9 percent by 1865. When the United States abrogated the treaty in 1866, trade continued between the two countries, although at a lower volume. US exports to Canada were 7.1 percent of total exports in 1867, while 6.3 percent of total US imports that year came from Canada. Canadian Confederation took place in 1867, after which Prime Minister John A. Macdonald developed the so-called National Policy in 1879. Broadly speaking, the policy was protectionist and nationalist, and it aimed at both stimulating domestic industry and developing the country along an east-west axis through the construction of the Canadian Pacific Railway, which spanned the country from Atlantic to Pacific when completed in 1885.[8]

In the United States, the Republican Party was largely in favor of high tariffs, while Democrats tended to be in favor of lower tariffs, and these stances on the tariff formed key elements of each party's overall national identity in late nineteenth-century political culture. During the Civil War, the federal government had enacted a series of high tariffs to fund the war effort, and these stayed in place after. With Republicans dominating US federal elections for much of the half century following the start of the Civil War, protectionism reigned. As Canadian officials made several attempts to reinstitute reciprocity in the decades following the US abrogation of the Elgin-Marcy Treaty, they found themselves rebuffed by US policymakers who enacted the increasingly protectionist McKinley Tariff of 1890 and Dingley Tariff of 1897, which together, as one historian remarks, "raised the principal of protection to the status of a fetish."[9]

The provisions of the Dingley Tariff created particular problems for US newspaper publishers. Under the Dingley Tariff, US paper manufacturers paid a duty of $1.67 for every ton of mechanically ground wood pulp (one of the key ingredients needed to manufacture newsprint) that they imported, and US publishers paid $6.00 per ton for imported newsprint. Publishers decried both duties on the grounds that they caused them to pay higher prices for newsprint, and they assailed what they believed to be an unfair protection for domestic paper manufacturers that insulated them from foreign competition. For US paper manufacturers, the duties on the finished newsprint allowed them more control of the domestic market. Given also that, at the time, they could import from Canada the pulpwood logs they needed as raw materials free of any duty, the arrangement seemed completely satisfactory to them.[10]

Though they were enthusiastically supportive of US tariff policy, one of the things that troubled US newsprint manufacturers around the turn of the twentieth century was that they began to see a dwindling supply of the pulpwood that they needed as raw material. Having made significant capital investments in their mills and then cut pulpwood indiscriminately around them, newsprint manufacturers after 1900 faced a future in which they would be increasingly reliant on Canadian pulpwood imports. By 1908, rampant pulpwood cutting in the United States had created a situation so dire that official estimates of the pulpwood supply began predicting imminent dates of its exhaustion. In May 1908, for example, Gifford Pinchot, the head of the US Forest Service, told Congress that the existing stands of pulpwood in the primary paper manufacturing states could be gone within three decades at the current rates of consumption. "The essential facts," Pinchot warned, were that "as to New Hampshire we estimate that at the present rate of consumption the supply will last twenty-five years; for Vermont, eleven years; for New York, eight and one half years; for Pennsylvania, nine years; for Minnesota, nine years; and for Maine, twenty-eight and one-half years."[11]

To Canadians, the fact that they had extensive forest reserves put them in a powerful bargaining position, and many began encouraging trade policies that over time would induce US-backed manufacturing to come north across the border. As early as 1898, for example, Canadians understood the strong position that they were in with regard to American corporations wanting their pulpwood. In November, Ontario lumber executive E. H. Bronson encouraged Prime Minister Wilfrid Laurier to enact a high and "practically prohibitive" export duty on pulpwood exports in order to prevent the US mills from buying Canadian wood. An export duty, Bronson argued, would "protect the owners of this wood against what might be termed to a large extent the commercial confiscation of it by United States users." Bronson already sensed that US mill owners were anxious about dwindling sources of domestic raw materials and believed that production would soon shift to Canada. The Americans wanted newsprint, Bronson believed, and there was "no reason why this business should not be done by mills located in Canada and manufacturing our own woods."[12]

One of the important policy details about the North American trade in trees, pulp, and paper had to do with the particular nature of Canadian federalism and resource policy. In the United States, forest policy was a matter of federal jurisdiction, but in Canada it was subject to provincial authority. In effect, any trade agreement about forest products ultimately relied on the cooperation of the provinces, regardless of agreements made by the federal government. This was particularly important when it came to the matter of embargoes and export restrictions on pulpwood. In the early twentieth century, the primary manifestation of this was the ban on the export of pulpwood from so-called Crown lands, or publicly owned forests created by the Crown Timber Act of 1849, the fundamentals of which remain intact to the present day. Under this act, the Canadian provinces leased or licensed forest lands to the private parties that cut timber on them. In 1924, a Royal Commission estimate was that some 83 percent of pulpwood trees in Canada were on Crown lands, meaning that the state controlled the overwhelming majority of the timber necessary for newsprint manufacturing. For the Canadian lands in private hands, owners were allowed to export wood to the United States in whatever form they liked. In contrast, on the vastly more plentiful Crown lands, provincial governments retained land ownership but licensed temporary and provisional control to private parties, which paid what was called a "stumpage" fee on trees cut on the tracts that they leased. For much of the nineteenth century, the main pulpwood provinces allowed the exportation of pulpwood from these Crown lands, but this began to change toward the turn of the twentieth century as pulpwood became a more valuable commodity. In Ontario, influential members of the provincial business community began seeking ways to stimulate industrial growth and transform the province's economy from one based on the export of staples to one based on the export of manufactured goods. One significant manifestation of this project was that the government of Ontario in 1898 began placing a "manufacturing condition" on trade in certain raw materials, beginning with pine timber and then extending to spruce pulpwood in 1900. In practice for the newsprint market, this meant that felled trees could not be exported to the United States from Ontario's Crown lands in their raw state but instead had to undergo some domestic manufacture into pulp or paper. Other major pulpwood provinces would follow Ontario's ban on pulpwood exports from Crown lands, with Quebec instituting an embargo on pulpwood exports in 1910 and New Brunswick in 1911. To many Canadian officials, these policies had the explicit goal of encouraging US investment in Canadian newsprint manufacturing. As Premier of Quebec Lomer Gouin flatly remarked, Quebec had enacted the pulpwood ban "in order to prohibit the exploitation of unmanufactured pulpwood in order to encourage the development of the paper industry in the province."[13]

These embargoes on raw materials did not have the immediate effect that proponents of a Canadian newsprint manufacturing industry had hoped. Pulpwood exports from privately held forest tracts in Ontario were not subject to the embargo, and some lumber companies simply operated in open violation of the law, often with no response from the state. A number of Ontario officials wanted to encourage settlement of the northern reaches of the province, and to do so they enacted policies allowing homesteaders to cut and sell for export the pulpwood on the tracts that they needed to clear in order to farm. In other cases, companies exploited a loophole in laws relating to mining claims that allowed pulpwood to be cut and exported from these lands. Some lumber companies began obtaining mining licenses on particular parcels that they had no intention of actually using to mine, and they would instead cut the pulpwood, which was exportable, and send it to the United States. Even after the enactment of provincial embargoes, lax enforcement and loopholes allowed the amount of pulpwood imported into the United States from Canada to increase from 369,000 cords in 1899 to 1.05 million cords in 1922.[14]

Given the constantly and rapidly increasing American demand for newsprint, many Canadians wanting to promote domestic manufacturing clamored for stronger regulations to promote raw materials being manufactured at home, and thus the tariff emerged as a crucial mechanism in the manufacture of newsprint. In 1902, the Canadian Paper Manufacturers' Association (CPMA) passed a resolution demanding a high duty on pulpwood exports from Canadian private lands to the United States. Though the Ontario law had ostensibly barred pulpwood exports from Crown lands, the CPMA called the prohibition useless and clamored for federal action. The *Toronto Mail and Empire* supported the proposal, arguing that "more and more the United States draw on our forests for the raw material to keep its pulp mills and paper mills going." This meant, the paper claimed, that a "great American industry flourishes by the grace of the Dominion government. There is no reason . . . why all the pulp and paper made from our timber should not be manufactured in Canada. An export duty would work like magic to bring the United States pulp and paper mills to water power centers on our side of the border."[15]

By 1908, many Canadians were concerned that the exploitation of their forests in the interests of serving US newsprint manufacturers had reached such a point that it was dangerous to both the Canadian economy and its environment. As journalist Emerson Bristol Biggar noted, the American disposition toward the forests that had produced what William Cronon would later describe as the midwestern "cutover" was threatening to encroach into Canada. It was "but a few years, for example, since official documents spoke of the timber supplies of the State of Michigan as inexhaustible; but to-day large numbers of the wood-working establishments of that State have to import their raw materials from other parts of the continent, and the saw mills have had to depend for their operations on logs imported from Ontario." New York pulp mills, Biggar claimed, "having eaten into the heart of the Adirondack Mountains have now to turn to Canada for a greater proportion of their supplies," and some pulp mills in Wisconsin, which was "supposed also to have an inexhaustible supply of raw material," were having to bring wood all the way from Quebec to stay in operation. The massive increase in US newspaper circulation was creating a similarly exploding need for newsprint, he claimed, and the "indiscriminate wiping out of the forest for the mere purpose of selling pulpwood, without the creation of a pulp and paper industry, and without regard to the effect on climate, rainfall and water-powers, is national suicide." For their part, many Canadians believed that shifting North American newsprint production from the United States to Canada through trade policy would have both environmental and economic benefits. As Biggar believed, if "Canada's pulpwood were all manufactured at home, industries would arise whose annual value would be millions of dollars, and yet all this could be accomplished while so regulating the cutting of trees as to maintain the present rate of reproduction, and so conserving the value of our forests forever." There would be not only the creation of a "great Canadian pulp and paper industry" but also the promotion of the "greater purpose" of "national self-preservation."[16]

In the first decade of the twentieth century, US papermakers imported from Canada increasing amounts of the pulpwood and pulp that they needed to manufacture newsprint. But the trade in these products remained structured by protectionist policies on both sides of the border, as US import tariffs on newsprint, combined with Canadian provincial embargoes on pulpwood exports from Crown lands, kept some limits in place on the scale and scope of the trade. In the United States, the main beneficiaries of this system were the paper manufacturers, and this was to the increasing ire of their biggest customers, US newspaper publishers. This antagonism would explode into a public and political campaign against monopoly motivated by publishers' naked commercial self-interest.

Publishers against the Paper Trust, 1898–1909

When one thinks about the relationship between journalism and the antitrust movement around the turn of the twentieth century, perhaps the first things that come to mind are the attacks on monopolies by journalists such as Ida Tarbell, Ray Stannard Baker, and Upton Sinclair and cartoonists such as Frank Bellew, Joseph Keppler, and G. Frederick Keller. In words and images aimed at a mass audience, this print media crusade motivated a host of reformist policies aimed at controlling and even breaking up monopolies. Across a range of sectors, the "Trust" became the enemy of the farmer, the worker, and the consumer, and many journalists and illustrators aimed to mobilize public and political action supporting antimonopoly policies.[17]

While scathing reports and cartoons made up one important facet of the relationship between journalism and the antitrust movement, far less historical attention has been given to a contemporaneous and bitter feud between newspaper publishers and what they often referred to derisively and specifically as the "Paper

Trust." Newspapers were not disinterested crusaders in the fight against monopoly. As manufacturing corporations, newspapers were firmly embedded in the
era of rising antitrust sensibilities, and many of their attacks on monopolies became more than moral campaigns against corporate size or economic concentration. In order to get the paper to print articles against societal ills created by the
"Trust," newspaper publishers were forced to rely specifically on the Paper Trust,
and in many cases their attacks on that monopoly stemmed directly from commercial self-interest.

The primary corporate target of newspapers' enmity in the fight against the
Paper Trust was the International Paper Company (IP), a conglomerate formed
on January 31, 1898, through a merger of fourteen paper companies in the northeastern United States. Upon its formation, IP became the biggest paper company
in the world, and the company estimated that its seventeen pulp and paper mills
were responsible for some 60 percent of the newsprint used in the United States.
To their chagrin, many US publishers soon found themselves forced to deal with
a company that they began describing in language resembling that used by popu-
lists and antimonopolists. In 1905, for example, the *Denver Post* portrayed the
"paper trust" as an "octopus," and in 1904 the *New York World* tried to humanize
the cost of monopolistic control of newsprint manufacturing: "Every child on his
way to school with a bag of books has paid tribute to the trust. Every library and
every college is its victim. Every workingman who buys in his weekly or daily newspaper the gathered chronicle of the world's doings, a boon an Emperor's ransom
could not have purchased when the old men of to-day were babes, pays the trust a
tax." As the *World* argued, the Paper Trust was a problem not only for the newspaper business but also for the rest of society. "In a Republic whose very existence
depends upon the general dissemination among its citizens of accurate knowledge
of public affairs, what could be more indefensible than a tax upon intelligence
and upon general information, levied without legal right by a private conspiracy
for gain?"[18]

The power that IP would come to have over the newspaper business in the early
twentieth century drove publishers to seek alternative sources of newsprint to
avoid what they believed to be IP's monopolistic control of the US newsprint market. One of the most important avenues that this quest took was a lobbying campaign to encourage free trade in pulp and paper, the idea being that the removal
of duties would stimulate the construction of newsprint mills north of the US-
Canada border which would be operated by competitive firms offering lower prices
than IP. In order to accomplish this goal, US publishers began pressuring North
American policymakers to lower or eliminate duties on newsprint. In late 1898,
the American Newspaper Publishers Association (ANPA) asked US trade negoti-

ators to advocate for duty-free imports of newsprint from Canada to the United States, arguing that this would give the press the "strong and permanent assurance of protection" from IP and its domination of the US newsprint market. John Norris, the business manager of the *New York World*, began a personal campaign to lobby North American policymakers on this issue, and he told Prime Minister Wilfrid Laurier in January 1899 that he was planning to go to Washington to "call on the President" and other government officials in order to "demand free paper." In November 1901, with US publishers having not yet secured better access to Canadian paper and facing rising newsprint prices, Norris told Laurier that many of them were "anxious to promote the efforts for a reciprocity treaty" that included newsprint among the commodities that would have duties lowered or removed. Laurier expressed little enthusiasm about the idea of pursuing a more general US-Canada reciprocity agreement at that juncture, but Norris and the ANPA remained undeterred. Norris went on to become the business manager of the *New York Times* and one of the most important figures in a publisher campaign for free newsprint imports which would last some fifteen years before achieving success.[19]

Paper manufacturers fought back strenuously with their own public relations and lobbying campaigns. In 1904, one of their trade journals blasted the efforts of Norris and *New York World* publisher Joseph Pulitzer for trying to get the newsprint duty removed. Paper manufacturers adopted a defensively nationalist response, remarking that those running their industry were "gentlemen who are trying to do a legitimate business, and are endeavoring to make a legitimate profit. And this is a type of gentlemen who are building up the country, who are paying big revenue into the Government, who are employing thousands and thousands of people, who are paying hundreds of thousands of dollars of taxes, who are purchasing millions of dollars worth a year of appliances and supplies to enable them to operate their mills." Attacks by Pulitzer in his "penny sheet" were "ridiculous," and Norris's public statements were "idiotic." In February 1907, IP executive Chester Lyman wrote to John Adair, a congressional representative from Indiana, that most US newspapers were "veritable mints for their proprietors" and that their publishers' and representatives' public statements had "misled hundreds of thousands of people" about the issue. These efforts by publishers resulted in some public support from Democratic members of Congress, though as the minority party in both chambers they had no legislative success in getting duties on newsprint removed despite proposing several bills.[20]

Meanwhile, publishers continued to report that they were paying rising prices for newsprint, and they blamed collusion among paper manufacturers for this. For their contract prices between 1907 and 1908, for example, the *Stamford Daily*

Advocate claimed an increase from $42 to $50 per ton, the *Duluth Evening Herald* from $38.50 to $48 per ton, and the *New Orleans Picayune* from $48 to $57 per ton.[21] In response to these rising prices, the ANPA formed the Committee on Paper as a special lobbying group led by John Norris. Newly organized and emboldened, publishers continued to agitate for a federal investigation into newsprint prices and to demand that the duties on imported pulp and newsprint from Canada be "immediately repealed." As one publisher exhorted his colleagues, "We've got to go to Congress and cut down the barriers under which these paper makers are hiding. . . . The paper makers don't need the protection of the tariff tax. Let's carry the matter to Congress and try to get the tariff on wood pulp and white paper removed. There is enough timber in Canada to supply the demand for white paper for years. We must show that the press has influence. Let's get in line and do something." Soon after, the trade journal *Editor & Publisher* made similar threats against elected officials in claiming that "if the tariff is not removed there will be a decided change in the personnel of the next Congress."[22]

This agitation from publishers motivated Congress to take up the matter in 1908, when the House passed a resolution seeking to investigate the ANPA's various allegations about the Paper Trust and to consider whether removing a duty on pulp and newsprint from Canada was advisable. The six-member commission headed by Illinois Republican James Mann conducted a mammoth investigation that began in April and ran for almost a year. Throughout the hearing, publisher after publisher reported the challenges they were facing as a result of newsprint prices that seemed to be rising for no good reason. *Topeka State Journal* publisher Frank MacLennan, for example, remarked that he had been assured after IP's founding in 1898 that "this combination of paper mills would result in great economies" for the producer which would be passed on to newsprint purchasers. MacLennan was then paying $35.50 per ton, based on a contract signed in the fall of 1897. His newsprint costs continued to rise as IP came to dominate the market, and by 1908, MacLennan claimed, he was paying $52 per ton for paper. Though the exact figures would change from paper to paper, numerous publishers offered parallel tales of the previous decade. All of this, John Norris told the Mann Committee, was because paper manufacturers were a "group of lawbreakers, with cunning methods of evasion of the criminal statutes." Their actions, he claimed, had "oppressed an industry which leads all others in the diffusion of intelligence, and which affects most deeply the growth of knowledge and the development of the masses." The solution, Norris repeatedly argued, was to allow pulp and paper to come to the United States from Canada free of the duties leveled on both. Over time, Norris argued, this would increase US domestic supplies of newsprint and in turn lower prices for publishers.[23]

After concluding its investigation, the Mann Committee in February 1909 offered two broad recommendations aimed at satisfying both paper manufacturers and newspaper publishers. The first of these encouraged the development of better conservation practices at the federal and state levels in the United States so as to protect and increase spruce supplies and thus lower domestic manufacturing costs, and the second advocated a revision of the tariff schedule so that forest products from Canada could be imported into the United States with lower duties. Though the committee recommended that the existing wood pulp duty of $1.67 per ton be retained, it also suggested that the duty on newsprint be lowered from $6.00 to $2.00 per ton, the caveat being that the Canadian provinces had to remove their export restrictions on pulpwood. For the committee, the bargain was that Canada would benefit from increased production of newsprint paper with the lower duty, and at the same time US mills would be able to get the pulpwood that they needed to maintain domestic manufacturing at competitive costs.[24]

Reaction to the Mann Report was strong on the part of both publishers and paper manufacturers. For the most part, newspaper publishers were enthusiastically supportive.[25] In contrast, papermakers blasted the report for threatening the ruin of their industry. The American Pulp and Paper Association claimed that Congress showed no concern for the "welfare of the paper industry and of the communities in which it is located in this country" and instead demonstrated its support of publishers as a "privileged class, which is to be encouraged at the sacrifice of any and all other interests." The *Paper Mill and Wood Pulp News* similarly presented the situation as a battle of manufacturers against media barons as it assailed the "millionaire publishers" who were using Congress for their own financial gain and at the expense of hardworking papermakers.[26]

The broader battle over the tariff became one of the central issues in the 1908 presidential campaign. Though it had not yet manifested itself in the nationwide electoral success of low-tariff Democrats, public opinion was increasingly supporting the rejection of protectionist policies on antimonopoly grounds, and more Republicans in Congress were beginning to tilt that way. For many, protectionist policies had come to be seen as aiding and abetting the creation and success of monopolists, and many came to see antitrust policies and tariff reductions as interrelated parts of the fight against monopolies. As the *New Yorker Staats-Zeitung* editorialized in March 1908, the "tariff is the father of the trusts. When a monopoly has been created by a trust, such a monopoly can always be traced back to the tariff. This applies indisputably to the paper trust."[27]

Republican William Howard Taft won the 1908 presidential election, though his commitment to tariff reform proved halting, to the chagrin of many Americans. His administration's first major attempt at tariff revision, the Payne-Aldrich

Tariff of 1909, did little to change the fundamental policy of protectionism or to lower the duties on many specific goods. If Americans in general were not happy with the Payne-Aldrich Tariff, publishers especially found it wanting in the way it treated newsprint. The tariff on imported newsprint did go down from the Dingley rate of $6.00 per ton to $3.75 per ton, which offered publishers a savings on the prevailing rate but still left them paying much more than they wanted.[28]

In the 1910 midterm elections, to some degree because of anger at Republicans who had failed to deliver on the promise of lower tariffs, the Democratic Party made significant electoral gains, taking over the House and leaving Republicans with a narrow majority in the Senate. Given this new political calculus, and perhaps reading public opinion, President Taft began initiating a push for a new trade agreement with Canada which he thought might mollify Americans seeking to end protectionism in North America. Over the following two years, policymakers in Canada and the United States began discussing trade reciprocity, and newsprint remained a major subject in the debates. Through their newspapers, publishers had a larger platform than paper manufacturers to promote their cause, and they did so with gusto. In the process, they would become instrumental in shaping the vigorous debates about North American trade which would take shape in 1911.[29]

Ending Continental Protectionism

The negotiations moving toward a US-Canada reciprocity agreement formally began in early 1910, when Taft sent a small delegation consisting of Henry Emery, the chairman of the US Tariff Board, and Charles Pepper, an official at the State Department, to Ottawa to meet with Prime Minister Wilfrid Laurier and his minister of finance (and former premier of Nova Scotia) William Fielding. Negotiations were productive, and on March 19, 1910, President Taft personally met with Fielding in Albany to continue them. By the end of the year, there was significant momentum behind a reciprocal trade agreement, and in early 1911, representatives from the United States and Canada met again in Washington to negotiate. This time, it was Fielding and Canadian minister of customs William Paterson, along with James Bryce, the British ambassador to the United States. On the US side, it was Secretary of State Philander Knox and a staff of trade experts. On January 21, 1911, the representatives of the two governments reached an agreement on reciprocity. The agreement itself was simply a pair of letters, one from Fielding and Patterson and the other from Knox, outlining the pact. As one US Tariff Board report later noted, these two letters were the "only important official documents relating to the conference." The overall agreement was brief and consisted of four schedules of tariff rates to be enacted by concurrent legislation in each country. . . . The unusual form of this arrangement has caused much misconception in both countries concerning the nature of its obligations. Neither a treaty, nor, indeed, any other form of contractual relation was contemplated, but simply an endeavor on the part of the representatives at the conference to secure the enactment of similar tariff reductions by their respective countries. Each country could adopt the measure in whole, in part, or not at all, imposing no obligations upon the other, except the honorable one of fair dealing. The passage of the proposed laws by one country did not bind the other to accept the arrangement, and even if it were accepted, either country remained at perfect liberty to withdraw at any time.

The form of the trade agreement had advantages for both the United States and Canada. As William Howard Taft noted, this method was much more expedient because the agreement "covered so many different items." Were Congress to take up each one in turn, the "individual objections by senators and representatives would have been so many that we could never have reached an agreement at all." For the Canadians who had actually suggested this format, the form had a similarly attractive flexibility, and it would obviate the need to get British approval of a formal treaty.[30]

In the actual agreement, many items had duties that had been negotiated downward, and almost all of the items on the free list were "natural products." Iron and steel sheets were among the "only manufactured goods of much importance" that were placed on the free list, the others being wood pulp and newsprint. The newsprint provisions, while supportive of US newspaper publishers' interests, were in some specific respects written in such a way as to be only mildly effective, as they stipulated that the duty-free Canadian pulp and paper had to be made from wood that was not subject to export restrictions. This essentially meant that if Canada wanted to export pulp and newsprint to the United States duty-free on a mass scale it had to remove the provincial restrictions on pulpwood exports from Crown lands, as these lands had the overwhelming majority of the wood that would be used to manufacture newsprint. The problem, the Canadian negotiators noted, was that these provincial export restrictions were outside of federal control. As Fielding and Patterson wrote in their official reciprocity agreement letter, "We have neither the right nor the desire to interfere with the Provincial authorities in the free exercise of their constitutional powers in the administration of their public lands." Still, US negotiators hoped that Ontario and Quebec could be induced by the large export market to change their policies in the future and that the reciprocity agreement would open the door to a dramatically increased flow of pulpwood, pulp, and newsprint to the United States.[31]

Taft sent the reciprocity agreement to the US Congress on January 26, 1911, and presented it as a sound economic policy that would benefit both Americans and Canadians. "The guiding motive in seeking adjustment of trade relations between two countries so situated geographically should be to give play to productive forces as far as practicable, regardless of political boundaries," Taft stated. "We have reached a stage in our own development," he went on, "that calls for a statesmanlike and broad view of our future economic status and its requirements. We have drawn upon our natural resources in such a way as to invite attention to their necessary limit." In response to these new circumstances, Taft argued that a "farsighted policy requires that if we can enlarge our supply of natural resources, and especially of food products and the necessities of life, without substantial injury to any of our producing and manufacturing classes, we should take steps to do so now." Opening up the border to trade in natural products would work to the economic and political benefit of both countries, Taft claimed, as it would encourage them to become better "commercial friends" rather than evolve into countries divided by a "perpetual wall" in their trade. "We have on the north of us," Taft concluded, "a country contiguous to ours for three thousand miles, with natural resources of the same character as ours which have not been drawn upon as ours have been." As Taft presented it, reciprocity with Canada would help US consumers by increasing the supplies of many natural products. And, he stated, the entire arrangement was based on a shared "identity of interest of two peoples linked together by race, language, political institutions, and geographical proximity." Given the economic imperatives, the bonds of fellowship across the international border, and the benefits of geography, Taft argued, reciprocity should be seen as a smart policy benefiting a broad swath of Americans.[32]

This kind of internationalism was a departure for US policymakers in the early twentieth century, as being covetous of raw materials in the Western Hemisphere had more commonly led to military engagements than to negotiations over trade agreements. Though it was certainly not the only reason, this common "race" that Taft mentioned was a significant factor both for him and for some in his party, perhaps most prominently Senator Albert Beveridge of Indiana. In 1898, during what would be a successful run for the US Senate, Beveridge gave one of his most famous speeches, "The March of the Flag," in which he described Americans as "God's chosen people" and the "ruling race of the world." Beveridge advocated a continuation of the imperial project that had begun with the Spanish-American War, and he claimed that Americans' "world duties" involved bringing "Liberty and Civilization" to foreign nations. Brushing aside the critics who suggested that the United States should not be a colonial power, Beveridge remarked that the "opposition tells us that we ought not to govern a people without their consent. I answer, the rule of liberty that all just government derives its authority from the consent of the governed, applies only to those who are capable of self-government." Asking the people of Puerto Rico or the Philippines to govern themselves, Beveridge argued, "would be like giving a typewriter to an Eskimo and telling him to publish one of the great dailies of the world."[33]

In February 1911, Beveridge displayed a much different attitude toward Canada, remarking on the Senate floor that the people of the United States and Canada shared "blood, language, institutions, religion, industrial methods, and social customs." As such, Beveridge promoted not occupation or annexation but rather "closer trade relations." The racism that Beveridge and other imperialists used to justify US actions in Latin America was absent entirely, and Beveridge went so far as to claim that there were "wider industrial and social dissimilarities between some localities of our own country then [sic] there are between the Republic and the Dominion, taken as a whole." Canada had the natural resources that the United States needed, Beveridge argued in 1911, but he was confident that the United States could get these through trade negotiations between friendly nations. "We have used up our natural resources so rapidly that the belated policy of conserving them has become one of our greatest national anxieties," he noted. "Perhaps no other single material problem more deeply concerns the great body of our people. But our immediate neighbors and blood kinsmen on our north have enormous natural resources which as yet hardly have been touched. We need those resources. Our Canadian neighbors are willing to give them to us in exchange for our products, which Canada needs."[34]

Congressional opposition to the reciprocity agreement came most prominently from those representing farmers and those criticizing newspaper publishers. James Good, a Republican representative from Iowa, claimed that US farmers would now be "obliged to compete not only with those now engaged in agricultural pursuits in Canada" but also with a host of new farms owned by immigrants who would now go to Canada instead of the United States. Stoking fears of strategic immigrants flocking to Canada in order to farm, Good claimed that, if reciprocity passed, there would soon be "thousands upon thousands of farmers who will in the future buy the cheap farms of Canada and sell their products in our market freely and unhampered. Foreigners desiring farms will not, as they have in the past, come to the United States, where rich farm lands were combined with the best market in the world for agricultural products, but will naturally turn to Canada, where equally rich lands may be had for less than one-tenth the price demanded in the United States, for the same great market will be free to the farm products of both countries." All of this new production, Good and other congressional critics argued, would flood the American market with agricultural products, drive prices down, and ultimately harm American farmers.[35]

Besides the defense of American farmers, congressional criticism of reciprocity also manifested itself in hostility toward newspaper publishers, as many in Congress claimed that publishers tried to intimidate them into supporting the reciprocity agreement in order to get the removal of a duty on newsprint. John Swasey, a Republican representative from Maine, one of the most important paper manufacturing states, was shocked at the "virulence of the attack" against those opposed to removing the duty on newsprint. "In all the history of American politics there was never so wicked and unjust and deceptive misrepresentation as in the last campaign, brought about absolutely by the American press." Joseph Cannon, a Republican from Illinois and former Speaker of the House, remarked that the proposed reciprocity agreement did such a great injustice to American farmers in order to benefit the newspaper press that it "should be labeled 'the publishers' pact,' whereby agricultural products are traded off for the publishers' profits."[36]

Amid the ongoing debates, congressmen were outraged when they obtained a copy of a private memo from ANPA president Herman Ridder to all of the association's members, in which he had written to publishers that it was "of vital importance to the newspapers that their Washington correspondents be instructed to treat favorably the Canadian reciprocity agreement, because print paper and wood pulp are made free of duty by this agreement." Progressive Republican Robert La Follette assailed the press for this, claiming that Ridder's exhortations had resulted in biased coverage of the bill throughout the country's newspapers. "A false impression regarding this bill has gone forth broadcast," La Follette claimed, "because the press of the country has a direct money advantage in it."[37]

The belief that US publishers promoted reciprocity to serve their own interests circulated widely not only among protectionist US policymakers but also in Canadian publishing circles. In January 1911, *Montreal Star* publisher Hugh Graham reported to W. S. Fielding on a meeting he had just emerged from in which the "group of newspaper men in discussing the question of reciprocity this morning so thoroughly endorsed the opinion that I am tempted to write you." Reciprocity was "to the fore now," Graham had learned, "because the newspapers of the United States are being hit. The publishers of Republican and Democratic papers have joined hands. The one aim is cheap paper. Anything else thrown in is to make weight and to save their faces and disarm suspicion and criticism." Graham noted that he was an ANPA member and claimed that his experience of being privy to private debates in these circles meant that "I can speak knowingly when I say the whole reciprocity question turns on paper. If they get paper, nothing else matters, and if they cannot get it without the inclusion of other things to conceal the selfish motive, they will be quite willing to go to great lengths. Our forests are in the balance."[38]

Publishers were certainly not the only group lobbying for or against reciprocity, but they were uniquely positioned to influence the content of the debates, both through their coverage and through their lobbying of members of Congress. Surveys of editorial opinion showed it to be in the aggregate strongly in favor of reciprocity, as a poll done by the *Chicago Tribune* in early June of 1911 found support for reciprocity among newspapers outpacing opposition by roughly a three-to-one margin.[39] As were many US newspapers during the congressional debates, the *Chicago Tribune* remained steadfastly supportive of reciprocity, calling it an act of "farsighted statesmanship" that chose to take the "continental view of two peoples essentially one in race, in law, and broad political and social ideals, in economic conditions and commercial needs" rather than succumb to the nationalist tendency toward the "guerilla warfare of selfish privilege and local interest."[40]

Congressional debates about the reciprocity bill continued throughout the first half of 1911, and their intractability forced Taft to call a special session of Congress to pass the bill, which it did, and which he eventually signed on July 26, 1911. The passage of the Canadian reciprocity bill made it the sole tariff reform for that session, as the Democratic House had proposed a number of tariff revisions that had been thwarted by the Republican-controlled Senate. President Taft was pleased that the reciprocity agreement had passed, noting that he believed that it would open a "great epoch in the history of our relations with Canada," and that he was "very sure it will introduce a great business between Canada and the United States."[41]

The *Chicago Tribune* similarly celebrated the bill as "one of the most important measures passed by congress in recent years" and claimed that it would create newly harmonious relations between the United States and Canada. "A tariff barrier between two neighboring peoples, similar as to character and economic and social conditions cannot be defended by protectionist doctrine, and is as obviously foolish and injurious as a tariff barrier between New York and Massachusetts or Illinois and Indiana." In the long run, the *Tribune* claimed, the bill was a "sane, sound, conservative act of state," and the economic benefits would soon make all critics wonder why there was such animosity about it.

> The pathetic picture of the worthy American agriculturist carried down in a tidal inundation of 'cheap' Canadian wheat will soon be hanging on the walls of memory as a work of imagination. . . . The cost of living will be about where it is now. The Canadians will buy some of our corn. They will sell us some of their barley. A

greater exchange of garden truck along the border and of fruit will be brought about because of the difference of seasons. But a year from now the American farmer will be wondering rather foolishly who got him out of bed, thrust the shotgun into his hands, and unchained the dog.

This rosy forecast was not to be shared by all Canadians weighing reciprocity north of the border.[42]

Rejecting Empire

In the early twentieth century, Prime Minister Wilfrid Laurier was a wary and strategic supporter of a closer trade relationship with the United States. As Laurier told Governor General Albert Grey in 1905, he supported more open trade with the United States because it was obvious that, for example, "Nova Scotia and British Columbia were the natural coal-fields for Massachusetts and California, and that Pennsylvania and Ohio were the natural coal-fields for the centre section of the Dominion." Thus, Laurier claimed, "it would be to the advantage" of the United States and Canada to "make a reciprocal arrangement with regard to coal. The same as to vegetables—for instance strawberries in the United States ripened earlier than those in the Dominion and . . . should be imported free." But Laurier was quick to add that he wanted to do this in the interest of promoting cooperation only as it benefited Canadians. "Canada only had one desire," Laurier stated, "and that was to hit the United States as hard as possible on trade."[43]

With these goals in mind, Laurier responded positively to Taft's 1910 overtures regarding reciprocity, and he was broadly supportive of the agreement that resulted from the negotiations. Once the US and Canadian delegations signed their letters of agreement in January 1911, the matter went to the Canadian Parliament for approval at the same time as it did to the US Congress. Though they took place on the same timetable, the debates took much different courses in each country. In Canada's parliamentary system, the debates proved so acrimonious and intractable that they eventually led to the dissolution of the government and the calling of a federal election. In the US debates about reciprocity, key divides had to do with conflicting assessments of the policy's effects on different sectors of the economy. For Canadians, however, this was not just a debate about, for example, the economic effects of free trade in wheat in a particular legislative district. It was about the country's economic and political relationship with a border country that in recent years had undertaken imperial projects throughout the hemisphere. The election became a referendum on Canada's trade and relationship with the United States and would mark the end of Wilfrid Laurier's fifteen-year term as prime minister. It ultimately put Canada on course for a trade relationship with the United States which would in many respects last until the late 1980s.[44]

In 1911, Laurier's push for reciprocity would fail in Canada for a variety of reasons. To some, the basic economics of the deal did not make sense to Canada. Since the enactment of the National Policy in 1879, many argued, the country had been doing just fine without free trade with the United States. Conservative member of Parliament (MP) Robert Borden, who would eventually succeed Laurier as prime minister after the September election, argued that, under the National Policy, the practices of "building up our own country, developing our own resources and keeping our people at home" had brought prosperity to the country. What Laurier offered was "absolutely inconsistent with that which has been, as I have understood, the purpose and object of the Canadian people for the last thirty years at least."[45]

In addition to opposing reciprocity on the grounds that the economics did not make sense, many Canadians also saw it as weakening the political and economic ties with Britain which had defined their national identity even after Confederation. As of 1911, it was not uncommon in Canada to find people with grandparents who fought against the United States in the War of 1812 or who themselves had been born as Britons prior to Confederation in 1867. For many in Parliament in the early twentieth century, the connection to Britain was both a part of their national identity and a practical economic policy that ensured Canada's prosperity. As Conservative MP J. E. Armstrong noted, since the United States abrogated the Elgin-Marcy Treaty in 1866 "we have gone on increasing in prosperity. Canada should have an ideal, and that ideal should be to become a great nation, while at the same time remaining a daughter in her mother's house. Everything a nation needs is produced in the British Empire. Australia has the meat and the wool, Canada has the grain, the raw material and the lumber, Great Britain herself is the great manufacturing centre, and the clearing house of the world. Canada and Egypt can grow more and better cotton than the United States. Surely this is a world combination that cannot be beaten." The proposed reciprocity agreement, he claimed, was the "first step towards the disintegration of the empire."[46]

For some Canadians, reciprocity raised not only the unwelcome possibility of separation from Britain but also fears that it was a first and fateful step toward an inevitable annexation by the United States. As Conservative MP Thomas Beattie remarked, the reciprocity agreement "if carried out will prove disastrous to Canada. I believe it has been concocted by shrewd United States politicians, thinking and believing in the end it might lead to annexation of this country to the United States." In a publicly circulated pamphlet, the Canadian Reciprocity League added that it was "our belief that the agreement would weaken our British connection and would ultimately lead to annexation with the United States." This fear of annexation was not as overwrought or irrational as it seems today, given what was then the perception around the world of the United States as having become an imperial power. In the United States, members of Congress from both parties spoke about North American relations with what many Canadians found to be an outrageous and distressing level of candor. For example, South Dakota Republican representative Eben Martin remarked, "I have often said that I would like to see Canada annexed to the United States. Her people and our own are of kindred blood and have a common history and common ideals." Perhaps the most prominent of these remarks came when Champ Clark, a Democratic member (and later Speaker) of the House of Representatives, made the intemperate and buffoonish remark in February 1911 that he supported reciprocity because "I hope to see the day when the American flag will float over every square foot of the British–North American possessions clear to the North Pole. They are people of our blood. They speak our language. Their institutions are much like ours. They are trained in the difficult art of self-government."[47]

Clark's remarks would prove endlessly controversial on both sides of the border, despite efforts in the United States to downplay their significance and intent. The *Chicago Tribune* called them merely "untimely jocular chauvinism," and the *New York Herald* dismissed Clark's remarks as "flapdoodle oratory." American attempts to defuse the controversy failed miserably in Canada, where many took Clark literally and seriously and where his name and his quote were repeatedly invoked in Parliament and in public discourse. As the vociferously antireciprocity *Montreal Star* remarked, "Champ Clark spoke his mind freely, he spoke frankly, and he spoke for his fellow-countrymen. What Champ Clark said, millions of the people of the United States—in fact, all the people with but a few exceptions—think and believe, and are acting upon. Champ Clark's reason for supporting reciprocity is the real reason for the support of practically every United States citizen who favors reciprocity." For the *Star*, and for many who followed this line of thinking, rejecting reciprocity meant not only rejecting a trade relationship with the United States but also stopping a slide toward being made a part of the United States. The issue of annexation hovered behind much of the debate about reciprocity, though Clark represented a more outlandish edge to it in the United States. Indeed, James Bryce, the British ambassador to the United States, remarked that it was "an absurdity, an obvious political artifice," and that "no sensible U.S. people are thinking of it." For the most part, Bryce claimed, US policymakers were more concerned with establishing a trade relationship than they were with taking over a country.[48]

But if annexation was never a serious practical threat, the invocation of it offered Canadians a language to talk about the idea of the nation during debates about a trade deal. Some of this was strategic: to encourage resistance to US imperialism offered a more immediately mobilizing political discourse than did one based in hashing out tariff rates for commodities like wheat, wool, or groundwood pulp. Even when the antireciprocity discourse avoided crassly pandering to sentimental nationalism, it still offered a potent way to argue that the policy seriously threatened Canada's national independence. To enter into the reciprocity pact was to give up sovereignty, many believed, as the vastly greater size and economic power of the United States would eventually draw Canada so far into its orbit that it would erode Canada's capacity for national self-determination. One can see this in the remarks of Conservative MP T. S. Sproule, who claimed that the trade pact

> will Americanize our trade, it will Americanize our people in social, commercial and industrial life, to the disadvantage of our people, and it will result in injuring the great towns and cities of Canada and building up the great towns and cities of the United States. . . . We will have 95 millions of people overshadowing, overtowering and dominating the seven or eight million people that we have. It will Americanize them, it will wean them away from their British allegiance, it will wean them away from their loyalty to the mother country, and it will pave the way for unrestricted reciprocity and annexation. That is the goal that many of those who are advocating this treaty are aiming at to-day. That is the goal they desire, that is the goal that many of them would have us reach.[49]

Throughout the summer of 1911, Conservatives in Parliament hammered away at the reciprocity agreement. Laurier and the Liberal Party had a majority, but the customs of Canadian parliamentary procedure prevented Laurier from calling for a vote to end discussion. As W. S. Fielding noted, unlike "most of the important legislative assemblies of the world," Canada had never adopted a rule allowing the majority to call for a closure of discussion on an issue. Instead, Canadian parliamentary procedure allowed for "unrestricted debate, believing that the good sense of Members on both sides of the House would prevent any abuse of this liberty." In the case of the reciprocity debates, however, Fielding argued that Conservative opponents had "not hesitated to grossly misuse the freedom so allowed," as "often as the motion was made to advance the reciprocity resolutions, Opposition Members rose and continued to talk, not, usually, on reciprocity, but more frequently on some matter designed to divert attention from that question." Against the backdrop of intractable parliamentary debate in Canada, on July 26

President Taft signed the reciprocity agreement and on July 29 Prime Minister Laurier dissolved Parliament and called for an election to be held on September 21. It was "generally agreed," the *Chicago Tribune* reported, that "the two month campaign before the country will be vigorous" and that reciprocity "probably will be the sole issue." The stakes were high but clearly understood by all Canadians, the paper reported. "Should the present Liberal government be returned with anything like a working majority, it will mean that a vote can be taken on the reciprocity resolution and that the trade agreement will go into effect. A Conservative victory at the polls means the passing of the Laurier government and the permanent shelving of the reciprocity pact."[50]

When the debate moved to the campaign trail, many of the Canadian statements against reciprocity continued along the lines of those made in Parliament. Perhaps the grandest public statement against the reciprocity agreement came two weeks before the election, when the prominent British imperialist Rudyard Kipling wrote an open letter against it which was published on the front page of the *Montreal Star*. Twelve years after using verse to exhort the United States to "Take up the white man's burden" in colonizing Cuba and the Philippines, Kipling used a public letter to urge Canadians to resist a version of that same American imperialism in their country. "It is her own soul that Canada risks today," Kipling warned of reciprocity. "I do not understand how nine million people can enter into such arrangements as are proposed with ninety million strangers on an open frontier of four thousand miles, and at the same time preserve their national integrity." For Kipling, the matter stemmed from the fact that the Americans, "by their haste and waste, have so dissipated their own resources that even before national middle age they are driven to seek virgin fields for cheaper food and living." Kipling saw no reason for Canada to give away the "enormous gifts of her inheritance and her future into the hands" of the Americans, and he claimed that Canada had nothing to gain from it "except a little ready money, which she does not need, and a very long repentance."[51]

Ultimately, this anti-US discourse was successful, and a majority of Canadian voters recoiled from reciprocity and its Liberal proponents by electing a majority Conservative government in 1911. The popular vote totals were not overwhelming, with the Conservatives receiving 50.9 percent of the overall ballots; however, within Canada's parliamentary system, this resulted in a significant change in the makeup of the government. Prior to the election, Liberals outnumbered Conservatives 135 to 85, while the new Conservative majority after the election was 134 to 87. "Reciprocity is repudiated," the *Montreal Star* announced on its front page. "Canada did not sell her soul. . . . The ghost of Annexation has been laid for a generation, probably for ever."[52]

Canada's unwillingness to ratify the reciprocity agreement certainly did not end trade with the United States. From 1880 to 1911, Canadian imports from Britain declined by more than half as a portion of Canada's total imports, from 47.8 to 24.3 percent, while during the same period imports from the United States increased from 40 to 60.8 percent of the total. (Canadian exports to each country declined slightly.) Over the longer term, these trends would continue, even without a reciprocity agreement. By 1929, only 15 percent of Canada's imports would come from Britain, while 68.8 percent came from the United States that year. In 1911, this emerging shift in the balance of trade between Canada and the United States was enough cause for alarm to many Canadians that they voted to prevent it from expanding more widely or quickly. Ultimately, it would be more than seventy years before representatives of the two countries would again broach such an ambitious trade pact.[53]

The Underwood Tariff, Forest Conservation, and the Creation of a Continental Newsprint Market

In the more immediate term in the United States, the debates about reciprocity with Canada formed part of a broader clash between protectionist Republicans and Democrats favoring lower tariffs. The Republican Taft would lose to the Democrat Woodrow Wilson in the 1912 election, and his defeat at the polls had much to do with his tariff policy. Many Americans remained increasingly attracted by the Democratic Party's promise to remove tariffs in order to lower the prices of consumer goods, which many believed were artificially inflated by protectionism. Many in Taft's own party considered him a traitor to Republican principles, and this included former president Theodore Roosevelt, whose entry into the campaign on a third-party ticket helped Wilson win the election. In Congress in 1912, Democrats expanded their majority in the House and gained a majority in the Senate.[54]

If many North American reciprocity proponents were disappointed by Canada's rebuke of the 1911 trade agreement, that rejection meant something rather different for US newspaper publishers because of what the US Tariff Commission later described as the "special provision" for newsprint which negotiators had put into the pact. The provision "was enacted by the United States independently of the rest of the agreement," and it unilaterally went into effect "in spite of the failure of Canada to accept the remainder of the measure." Thus, as a result of the US Congress approving the reciprocity agreement in July 1911, newsprint paper manufactured in Canada became a duty-free import, at least conditionally, despite the fact that Canadians had rejected the full accord. According to the specifics of the US policy, the duty-free newsprint from Canada had to have been manufactured from wood cut on private rather than Crown lands. The fact that the pulpwood on Crown land was vastly more plentiful than on private land dictated that quantities of imported Canadian newsprint would initially be limited. Despite this condition of the agreement, increasing quantities of newsprint paper began to come to the United States from Canada, and one historian describes this as the result of the major "loophole" that was the "only legislative result of the months of negotiation and agitation" over reciprocity. There is no direct evidence in the papers of the US and Canadian principals in the reciprocity negotiations as to why this newsprint provision was inserted and enacted as it was, and in general the archival records relating to the negotiations are scant. There is, however, as presented in this chapter, strong indication that public and private lobbying by newspaper publishers motivated US policymakers to enact a policy creating a loophole that clearly and strongly benefited the domestic newspaper business. Whatever the motivation, after the United States passed the reciprocity bill, paper began coming across the border in increasing quantities, but it was still not quite the full free trade in newsprint which US publishers had wanted.[55]

Soon after taking office, President Wilson called a special session of Congress to enact a comprehensive tariff revision. The House bill, sponsored by Alabama Democrat Oscar Underwood, offered free trade in numerous commodities and lifted tariffs on all imported newsprint, regardless of any Canadian provincial export restrictions. To offset the loss of revenue from lowering tariffs and to take advantage of the recently ratified Sixteenth Amendment, the tariff measure, the Underwood-Simmons Tariff Act of 1913, also instituted a progressive income tax. Given Canada's proximity and the fact that some duty-free newsprint had already begun conditionally crossing the border after July 1911, the Underwood Tariff would be a boon for North American trade in it. The newspaper industry lauded the bill for its promise to "remove all danger of monopoly" from newsprint manufacturing. The bill passed Congress and became law on October 3, 1913, thus making all imported Canadian newsprint duty-free, regardless of any manufacturing or export conditions placed on pulpwood, pulp, or newsprint in Canada. In effect, policy decisions in the United States and Canada between 1911 and 1913 created an integrated continental economy in the most important and costly input to the newspaper business, and they allowed US newspapers ready access to the vast Canadian spruce forests that they coveted for the paper needed to keep their businesses operating. "The days for a tax upon knowledge are gone," the ANPA claimed.[56]

The lifting of duties on newsprint imported into the United States provided an almost immediate boost to US newspaper publishers and to Canadian newsprint manufacturing. As one columnist noted in the *Toronto Globe* in January 1913, the previous year had "witnessed the greatest development the Canadian pulp and paper industry has ever known." Throughout the 1910s and into the 1920s, as chapter 3 will discuss, as corporations began building newsprint mills in Canada, the result was that US newsprint production essentially stagnated while Canadian production rose dramatically. Within a decade after policymakers made the North American newsprint trade duty-free, the result was a completely integrated continental economy in the most important input to the newspaper business in the United States.[57]

The creation of this cross-border newsprint industry was the hard-won result of lobbying by the US newspaper business. Many publishers were willing to use their newspapers' content and their personal influence with policymakers in the service of their commercial aims, and in doing so they threw themselves into the center of ongoing debates about two of the most important issues of the day: trade and conservation. The period around the turn of the twentieth century saw the rise of the mass-circulation metropolitan newspaper and the creation of a duty-free North American newsprint market, as well as the rise of popular and official conservationist movements on both sides of the US-Canada border. This continental mood manifested itself in, among other things, the creation of a system of national parks in the United States, the development of scientific forestry as an academic discipline in North America, and a range of new agencies and policy mechanisms in both the United States and Canada designed to better manage the natural resources on which a burgeoning industrial capitalist economy depended. In the decades around 1900, as the metropolitan press expanded both the magnitude of its social influence and its newsprint-driven penetration of the forests, environmentalists and policymakers in North America sought mechanisms to better manage those forests. These two processes were linked together in significant ways through ethics of strategic continentalism and conservationism circulating on both sides of the border. In both the United States and Canada, there was significant agitation for conservationist policies from the newspaper press, and the motivation behind this was less silvicultural altruism than economic self-interest.[58]

In the United States, the burgeoning conservation movement led by US Forest Service chief Gifford Pinchot aimed at finding better ways of managing the forests for productive commercial uses. Forests, Pinchot argued, were not just antidotes to the city as natural spaces but also enablers of metropolitan life as natural resources. "The products of the forest are among the things which civilized men can not do without," Pinchot stated in 1905. "Wood is needed for building, for fuel, for paper pulp, and for unnumbered other uses, and trees must be cut down to supply it." The problem, he believed, was that the United States lacked the proper policy tools to effectively manage that cutting. "The question is not of saving the trees," Pinchot remarked, "for every tree must inevitably die, but of saving the forest by conservative ways of cutting the trees." In developing new policies to promote sound forestry practices, Pinchot developed a forestry philosophy that made him for decades one of the most prominent advocates of conservationist policies with a utilitarian bent. This was not just a "moral issue," he believed, but a practical one, and Pinchot found no contradiction in developing a philosophy toward US forests which was attentive to both their natural beauty and their economic potential. The problem, he argued, was that the United States had "attacked our natural resources as no other nation ever has." While, on the one hand, this had given the country "prosperity, vigor, intelligence, and power such as we could not secure under any other circumstances . . . we have at the same time been preparing a condition which if not checked will make vigor, power, and efficiency impossible for the people who are coming after us. It is a serious indictment and a true one."[59]

Across the border, the contemporary status of conservationism in Canada was quite different, as the state already owned the overwhelming majority of the country's forests as Crown lands. Fewer Canadians expressed the "crusading spirit" that Americans did, and many instead concerned themselves with promoting more effective economic utilization of their forests. In 1906, for example, J. F. Mackay, the business manager of the *Toronto Globe*, gave an address to the Canadian Forestry Association entitled "The Publishers' Interest in Forestry," in which he admitted that it was "no sentimental reason that I come before you on this occasion. There is little of the aesthetic about the newspaper publisher's interest in forestry. I suppose it might be said to be purely commercial selfishness." With Gifford Pinchot present in the audience, Mackay encouraged Canadians to continue following policies that made the best productive use of forest resources that the state controlled. In a dig at the Americans, Mackay noted, "Fortunately, we have not far to go in seeking an object lesson in the utter folly of defying nature's laws in these matters. The phrase 'an inexhaustible supply,' which we glibly use in Canada to day in talking of our pulpwood forests, was just as freely used in the United States but a very few years ago. Thus chloroformed, they failed to adopt proper forest regulations." There was not, Mackay claimed, "any class of men who should, for their own sakes, regard this subject with more favor than the publishers of newspapers," and he urged his peers to be mindful of the "direct and vital" link between their businesses and the forests and to continue to support "proper forestry regulations."[60]

In the United States, Mackay's American counterparts in the newspaper business followed this line of thinking and supported conservationist policies because they benefited their own industry. Many publishers exhibited strategic conservationism in the interest of supporting the campaign for reciprocity with Canada. Helping their businesses by giving them cheaper paper imported from Canada, US publishers argued, would also save American forests. In January 1911, ANPA president Herman Ridder exhorted members to support the reciprocity agreement because it promised a way both of "conserving our forests and of removing a tax upon knowledge." The ANPA's campaign harmonized well with Gifford Pinchot's efforts to reverse what he called the "coming timber famine, caused by the reckless waste of the richest heritage of forests ever given to a Nation." To Pinchot, the promotion of better forest management practices in the interest of conservation was not just a task for the government but also "truly a problem for the newspapers, for they must know that without timber for other purposes there cannot be any pulpwood."[61]

US newspapers knew this very well, and they actively promoted the causes of conservation and forest management throughout the early twentieth century. For newspapers, finding solutions to these related problems meant that promoting the conservationist movement also meant promoting their own self-interest. As an editorial in the trade journal *Editor & Publisher* noted, "Whatever the tree may mean to society in general, it counts for something very special to those who live by the printed word. There is no substitute in the world for pulp for paper. . . . For sheer self-preservation, aside from a thousand other public reasons, editors and publishers should arouse public sentiment . . . in behalf of our neglected forests." In the early twentieth-century United States, newspapers used their editorial support of forest conservation efforts as a complement to their widespread activities promoting free trade policies that benefited their businesses.[62]

Conclusion

The strategic support of conservationism and continentalism was not unanimous among US publishers. Some blamed their metropolitan peers for wasteful production methods using unnecessarily large amounts of paper. As Henry Stoddard, the editor of the *New York Evening Mail*, noted in 1907, too much paper was being used by publishers paying no regard to the raw materials that went into producing their newspapers. "Sunday papers so big as to be a burden to the reader, evening papers issued at all hours of the day and night . . . these are the real cause of the trouble," Stoddard noted, and this promiscuous printing meant that "our forests are rapidly disappearing because of this demand." Other newspapers chided the ANPA and its more continentalist members for the shamelessness they exhibited in campaigning for free trade in newsprint. As the *New York Sun* editorialized, "We are, naturally, poignantly concerned to get our paper as cheaply as possible, and as moral scientists we share . . . anxiety about the deforestation of the country." But, the *Sun* noted, "we cannot for the life of us see any reason why there should be any special legislation on the part of Congress in behalf of newspapers. There cannot be and there shall not be any privileged class. There is nothing that the press so utterly abhors and so frantically detests as privilege. . . . We desire no pecuniary benefits or advantages that are not common to the whole people. We consider the proposition immoral."[63]

This type of altruism was uncommon, and far more pervasive were consistent and aggressive editorial campaigns aiming to promote forest and trade policies that served the commercial needs of newspaper publishers. For US newspapers in the early twentieth century, the desire for ever-increasing supplies of newsprint motivated editorial and reporting stances in favor of conservationist and continentalist policies. In both areas, altruism was supplemented by a great degree of commercial self-interest. And this self-interest, indeed, was a common thread among many newspaper publishers in the two decades following the Spanish-American War. The motivations for American imperialism were multiple and various, and newspapers were powerful voices in the public discourse about it. But it is important to note that aspects of this imperialist mood were focused directly north at Canada, and the political project to cultivate closer and deeper trade relations with Canada, whatever form those relations took, had much to do with US newspaper publishers wanting cheaper newsprint and, by extension, more ready access to Canada's plentiful spruce forests.

CHAPTER THREE

The Continental Newsprint Market
and the Perils of Dependency

In the nineteenth century, virtually all the newsprint consumed in the United States was produced domestically, and the overwhelming majority of the trees used as raw materials in producing that newsprint were from domestic forests. As of 1899, some 83 percent of the 569,000 tons of newsprint consumed in the United States had been manufactured with domestic pulpwood, and the balance was produced domestically from imported Canadian pulpwood and wood pulp. The United States in the nineteenth century was, for almost all practical purposes, newsprint independent, and US newspapers were printed on newsprint manufactured at home. After 1900, however, as US newsprint manufacturers engaged in furious pulpwood cutting to meet the increasing demands of newspaper publishers, a significant portion of readily available domestic pulpwood supplies dwindled, and mill owners became increasingly dependent on imported raw material. By 1904, domestic pulpwood supplied the raw material for 72 percent of US newsprint production, and by 1909 this had declined further to 67 percent. Tariff removals starting in 1911 made newsprint imports to the United States from Canada duty-free, and this motivated a wave of newsprint mill construction in Canada. By the mid-1920s, the United States was importing more newsprint from Canada than it was producing at home, and by 1950, US mills were producing just one-fifth of the newsprint used by domestic newspapers (see table 4). In the first half of the twentieth century, US newspaper publishing became dependent on Canadian newsprint.[1]

This process had begun with policy changes on both sides of the US-Canada border. By instituting embargoes on pulpwood exports from Crown lands, Canadian policymakers created a strong incentive for corporations to create a newsprint manufacturing industry in that country. In turn, US policymakers removed the duty on imported newsprint, thus creating the conditions for an integrated continental newsprint market. Motivated by these policy changes, North American newsprint firms after 1911 brought the vast majority of continental production to Canada, in the process creating dramatic increases in both the scope of Canadian newsprint manufacturing and the supply of newsprint available to American newspapers. As one US publisher remarked in 1921, "For newsprint

TABLE 4
The North American newsprint trade, 1869–1956 (selected years)

Year	US newsprint consumption (tons)	US newsprint production (tons)	Canadian newsprint production (tons)	Canadian newsprint exports to United States (tons)	Canadian imports as percentage of US newsprint consumption
1869	N/A	100,000	N/A	0	0.0
1889	N/A	196,000	N/A	0	0.0
1899	569,000	569,000	N/A	0	0.0
1904	883,000	913,000	N/A	1,000	0.1
1909	1,159,000	1,168,000	N/A	20,000	1.7
1913	1,473,000	1,305,000	402,000	218,000	14.8
1914	1,547,000	1,313,000	470,000	275,000	17.8
1917	1,779,000	1,359,000	722,000	491,000	27.6
1920	2,197,000	1,511,968	938,000	679,306	30.9
1923	2,777,315	1,485,000	1,330,138	1,108,466	39.9
1925	3,013,715	1,530,318	1,618,805	1,315,404	43.6
1926	3,409,713	1,684,218	2,068,208	1,750,749	51.3
1929	3,757,806	1,392,276	2,984,327	2,326,502	61.9
1932	2,832,176	1,008,588	2,186,120	1,647,216	58.2
1935	3,344,680	912,392	3,082,994	2,122,099	63.4
1938	3,422,269	820,055	2,892,984	1,940,303	56.7
1941	3,929,773	1,014,912	3,770,665	2,987,235	76.0
1944	3,242,891	719,802	3,264,581	2,529,693	78.0
1947	4,752,904	825,554	4,820,164	3,897,300	82.0
1950	5,936,941	1,014,703	5,278,585	4,748,228	80.0
1953	6,142,896	1,083,982	5,721,296	4,861,372	79.1
1956	6,899,020	1,717,243	6,468,815	5,229,748	75.8

Sources: Earle Clapp and Charles Boyce, *How the United States Can Meet Its Present and Future Pulpwood Requirements,* US Department of Agriculture, bulletin no. 1241 (Washington, DC: US Department of Agriculture, 1924), 82; Royal Kellogg, *Newsprint Paper in North America* (New York: Newsprint Service Bureau, 1948), 29; Newsprint Association of Canada, *Annual Newsprint Supplement, 1957* (Apr. 1958), 2–4, vol. 25358 (acquisition no. 2006-00392-5, box 69), folder Labour Negotiations: General Ontario Paper Co. Thorold 1957–58, QOPC, LAC.

purposes, North America must be considered as one unit. The United States contributes nearly all of the demand," and Canada would contribute increasingly large amounts of the supply. In many respects, Canadian policymakers and US publishers achieved the outcomes that they wanted from their efforts to promote the duty-free importation of newsprint from Canada to the United States.[2]

At the same time, however, the process of negotiating the geographic shift in manufacturing would prove quite fraught for US publishers, and many would have to reckon with the consequences of having obtained what they had desired. The continentally integrated newsprint market would be marked by dramatic spikes in prices and periods of scarce supplies, some caused by the outbreak of World War I, and some caused by the changing patterns of capital investment. What these market fluctuations meant was that publishing a US newspaper would be subject to a new international politics of trade and resource management. This was a defining characteristic of US newspaper publishing after 1911. State policy had made the products of Canadian forests more readily available to US newspaper publishers, and within a decade of newsprint trade being made duty-free, the US publishers' boosterism of the newsprint trade had given way to fears of being dependent on foreign production of newsprint. "The newsprint trail leads straight to the woods," the trade journal *Editor & Publisher* noted in 1920, and those woods were increasingly in Canada. "In ten years," the magazine lamented, "we have driven into Canada two-thirds of an industry that in 1910 was wholly our own. Only a third of the newspaper issues in this country last year were printed on the product of our own forests. The once independent American press, so proud and prodigal, is already reduced to dependence upon a neighbor for its very existence. The great voice of freedom that has been heard around the world has become subjective to and can in an instant be silenced by an alien hand."[3]

As the United States and Canada were developing trade policies integrating the production of newsprint and newspapers in North America, this circuit was also becoming part of a larger global network of newsprint production as major British publishers began developing their own operations on the continent. Starting around 1903, the British press baron Alfred Harmsworth (the future Lord Northcliffe) began planning construction of a newsprint mill in Newfoundland. By 1908, Harmsworth's company had secured leases on roughly 2,300 square miles of timberlands, and it built a mill under the auspices of a newly created subsidiary called the Anglo-Newfoundland Development Company. By 1912, with the mill in operation, Harmsworth was exporting some 3,000 tons of newsprint and 2,700 tons of pulp to the United Kingdom each month, and this helped him sustain one of the most successful media businesses in the world.[4]

The reliance of Harmsworth and other British publishers on North America for raw materials and newsprint was part of an expanding trade relationship in forest products between Canada and Britain. As a 1910 US Forest Service survey noted, the United Kingdom, being largely barren of forest resources, bought much of its wood and wood products on the world market. This was a significant volume of purchasing, and the Forest Service noted that UK purchasers were responsible for "nearly half the total export of all the countries of the globe." As a result, "the wood prices in the English market affect practically the whole world." The heavy reliance of British publishers on foreign forest resources, in Scandinavia and especially in Canada, made them significant competitors of the United States in the global market for newsprint.[5]

As British publishers began carving out a significant presence in Canadian paper manufacturing, newsprint manufacturers across North America also began cautiously exploring the possibility of expanding production in Canada. Even though tariff removals had allowed Canadian newsprint to come to the United States duty-free, it would take several years for production to shift north to fully take advantage of this policy change. Paper mills required significant capital investments and time to build, and manufacturers had to purchase timberlands or arrange leases with provincial governments. US production had nearly doubled between 1899 and 1909 from 569,000 tons to 1.1 million tons annually, but after 1911 domestic mill construction slowed and then halted. Not long after the United States began allowing the duty-free importation of newsprint in 1911, rumors began circulating in industry circles about major US producers moving to Canada, and US paper manufacturers began reducing newsprint production as they planned diversification into other sectors of the paper business. Canadian newsprint exports to the United States, nonexistent in 1899, were estimated at 20,000 tons in 1909. These imports jumped more than tenfold to 218,000 tons in 1913, and then to around 1.1 million tons in 1923 and 2.1 million tons by 1935. By 1950, the United States was importing some 4.7 million tons of newsprint from Canada, which amounted to 80 percent of the newsprint used by American publishers, all while US domestic production essentially held steady. In the first half of the twentieth century, the United States would become utterly dependent on Canadian newsprint production to sustain domestic newspaper publishing. With British publishers making significant demands on Canadian newsprint supplies as well, US publishers found themselves in challenging and occasionally perilous conditions when trying to obtain adequate supplies of newsprint. In the decades after 1911, publishing a local newspaper in the United States became a geopolitical problem.[6]

World War I and the Global Newsprint Market

As the United States was in the early stages of developing this closer trade relationship with Canada in newsprint, the outbreak of World War I in Europe in July 1914 would throw the market into chaos and demonstrate the substantial degree to which the production of the daily newspaper occurred within a global economy. In hindsight, World War I would prove to be the last global conflict in which the printed newspaper was the primary way to inform the mass public about the war's ongoing events. Wireless telegraphy became an increasingly important tool for military communication, but among civilians it remained largely something used by amateurs and experimenters. Radio broadcasting, to some degree helped along by wartime experimentation, would become a mass phenomenon around the world in the 1920s. During World War I, however, the primary medium for providing current information to the mass public was still the printed newspaper.

In the United States, initial reports after the war broke out were cautiously optimistic that there would not be disruptions in newsprint supplies or prices. One International Paper (IP) executive noted that domestic manufacturers were "confident of their ability to cope with the situation as it exists today or may develop in the future." Were the European conflict to drag on, the emerging instability might cause a "hardening of prices," but he was sure that there was a "disposition on the part of the manufacturers in this country to refrain from taking advantage of the situation." Many publishers were also similarly optimistic that the American newsprint market would remain uninterrupted, and in some ways these attitudes reflected US attitudes toward the war more generally: this was a European and not an American problem, and the United States could reasonably sit on the sidelines as a concerned but officially neutral observer. "If the European disturbances should continue for some time the situation might change," *New York World* executive Don Seitz remarked. "France and England are short of paper, or soon will be, but this does not apply to the United States."[7]

By the late summer, this early optimism had begun to fade as observers took stock of the European war, the resolution of which no longer seemed as imminent as originally thought. In the global newsprint market, Great Britain's dependency on supplies of wood pulp from Sweden and Norway to support its domestic newsprint manufacturing placed it in a precarious position. Were these connections to be severed by the war, as was seeming increasingly likely, Britain would have to rely more on Canadian pulp and newsprint. By mid-August, estimates out of London suggested that British papers had only six weeks of paper supplies on hand, and by the end of the month the mood among US publishers had darkened considerably.[8]

Amid this rising anxiety about newsprint supplies, the US newsprint market stayed relatively stable in the early stages of the war. In fact, from 1914 to 1915 the average price per ton for newsprint in New York City actually declined from $43.60 to $41.78. Domestic production held steady while consumption increased, and Canadian imports filled the gap. By 1916, however, newsprint prices rose dramatically, going to $51.78 in 1916 and to $63.78 in 1917. Newsprint manufacturers blamed the price increases on higher production costs, with IP president Philip Dodge claiming in June 1916 that the "fundamental reason for advancing prices is that it costs more to make paper" because of wartime shortages of men and materials.[9]

As the European war continued with no sign of an end, US publishers' concerns about newsprint price increases were important parts of growing concerns about the generally increasing cost of doing business during wartime. After the war broke out in 1914, for example, Allied demand for agricultural products started driving US prices up steadily and dramatically, and the wholesale prices for wheat, cotton, and wool all would more than double between 1914 and 1917. In the newsprint market, the German war effort had specific and immediate effects on global supplies and prices. Prior to the war, Germany, Norway, and Sweden had all been significant pulp exporting countries for the rest of Europe. Once Germany became a belligerent, virtually all of those supplies were cut off from the global market. Sweden and Norway remained officially neutral, though the bulk of Swedish pulp was soon being exported to Germany. This was a major blow to British newsprint manufacturers, which imported an estimated two-thirds of their pulp from Sweden. In Norway, some newsprint manufacturers found themselves cut off from the Russian pulpwood supplies that they relied on, making manufacturing more costly. Though Norwegian producers still exported some newsprint to Britain, prices increased significantly.[10]

All of these European developments had significant effects on the North American newsprint market. Being closer to Canada, US publishers did not face the same kinds of shipping challenges that their European counterparts did, and most could get the same quantities of paper in 1917 as they had in 1914. However, what began to bedevil many was not supply but cost, as many papers could not get sufficient amounts of newsprint at prices that were manageable to sustain their businesses. In mid-1916, for example, Virginia publisher Robert Barrett claimed that he had been buying paper for between $40 and $55 per ton prior to the war, but he claimed that it was now "impossible" to get it for less than $80 per ton, and he claimed to have been quoted prices as high as $120 per ton.[11]

The problem of rising newsprint prices which Barrett faced was common and particularly acute among other smaller publishers who were buying from brokers known as "jobbers." One of the peculiar and defining characteristics of the newsprint market was that there was no central exchange at which paper could be purchased, and all arrangements for pulp and newsprint were done on a direct contract basis between individual buyers and sellers. For larger newspaper firms, this usually meant dealing directly with mills or with major corporations (like IP) that were willing to make annual contracts for significant tonnages of paper. For smaller publishers, obtaining paper usually meant buying from a third-party jobber who had bought large quantities of newsprint from a mill and then sold these to newspaper clients in smaller individual lots. As US newsprint demand came close to matching North American manufacturing capacity, both manufacturers and jobbers took advantage of the tight market and began charging higher prices.

In an attempt to cut production costs amid rising paper prices, publishers around the country began finding ways to limit their newsprint consumption. Some papers began reducing the size of headlines so as to fit more news onto less paper. Others cut down the size of daily issues by simply printing less material on fewer pages. Responses to this challenge varied from publisher to publisher, though in most cases publishers eliminated reading material in favor of printing more advertising so as to generate more revenue per pound of newsprint used. One survey of New York City newspapers is illustrative of these practices. In comparing November 1915 and November 1916, the survey found that the total number of pages printed by a sample of major papers was almost identical, but that the amount of reading material printed on those pages declined 8.5 percent, while the amount of advertising material increased 10.7 percent. For most publishers, when newsprint was more expensive and space was at a premium, that space increasingly tended to be filled by advertising rather than news.[12]

As US publishers developed business strategies to manage wartime paper costs, many also began to believe that newsprint manufacturers were colluding to charge higher prices for newsprint. Their strident public accusations eventually prompted an investigation by the Federal Trade Commission (FTC), to which publisher after publisher claimed to be at the mercy of jobbers who continued to raise prices. For example, R. E. Turner of the *Norfolk Pilot* claimed that in 1916 he had a contract for the 1,350 tons of paper he needed for the year at $44.60 per ton. For 1917, he claimed, he could only get a contract for 1,060 tons at $70.60 per ton. Because of this, he claimed, "We are wiped out, so far as profits are concerned." Newsprint manufacturers defended their higher prices by blaming the federal government for having passed the Underwood Tariff, which IP president Philip Dodge called "grossly unfair legislation" that had created a situation in which it was impossible for domestic newsprint producers to compete with duty-free Canadian imports. US manufacturers, Dodge claimed, were "entitled to a reasonable return on our honest capital," but because of the tariff changes allowing duty-free Canadian newsprint imports, "we can not get that return on that capital in the United States under the existing conditions."[13]

The FTC was unsympathetic to the defenses offered by newsprint manufacturers and brokers. Jobbers, the FTC claimed, had exaggerated newsprint scarcity and created a climate of fear among small publishers, and larger manufacturers had colluded to raise prices. On the heels of these findings, the Department of Justice (DOJ) began considering its own investigation, a significant enough threat to manufacturers that they agreed to set the price of newsprint at $50 per ton, roughly what it had been in 1916 before wartime increases. Most publishers were ecstatic with the arrangement, with *Editor & Publisher* calling it in March 1917 the "advent of a new era in American business life. It is an epochal event. It is in line with the 'new spirit' in commercial life, which aims at the correction of evils through conciliation and concession."[14]

These optimistic pronouncements about the benefits of the voluntary public-private regulation of the newsprint market were short-lived. The United States declared war in April 1917, and later that month the DOJ upset the tentative FTC settlement by issuing criminal antitrust indictments against seven major newsprint manufacturers. The manufacturers, most of which believed that their participation in the FTC arbitration would terminate the DOJ's ongoing investigation, responded by pulling out of the agreement reached with the FTC. That same week, Woodrow Wilson used an executive order to establish the Committee on Public Information, an agency designed to generate propaganda in favor of the war effort. Publishers immediately tried to use their patriotic participation in the war effort as a way to gain greater federal attention in their campaign to get cheaper newsprint. "If newspapers are a public necessity," *Editor & Publisher* claimed, "and are to play a vital part in winning the war, unusual measures are justified in assuring to them a reasonable supply of news print. Legislation to this end will be patriotic legislation." Regulating the price of newsprint and ensuring adequate supplies were, the magazine noted, "proper Governmental functions."[15]

With the DOJ indictment in place and the FTC arbitration stagnant, the FTC released its report in June 1917, in which the agency alleged that major manufacturers had conspired to fix newsprint prices. To correct these distortions in the price of newsprint, the FTC advocated government control of newsprint supplies through an agency set up for the duration of the war. In November, with the DOJ case headed to trial, the accused publishers agreed to a plea bargain in which they paid small fines and agreed to sell newsprint for $60 per ton from January 1 to April 1, 1918. After that, the FTC would maintain authority to set prices until three months after the eventual end of the war. As had been the case after the FTC's March 1917 arbitration settlement, publishers' hopes for stable and more affordable newsprint supplies would be short-lived. As the war continued, the FTC held additional hearings about newsprint in June 1918, after which it ruled that the price should be increased to $62.80 per ton. Few in either the newsprint manufacturing business or the publishing business were satisfied by the price, and legal battles over newsprint prices continued throughout the summer of 1918. Manufacturers appealed in federal court for permission to raise prices, and they succeeded in demonstrating that their increased production costs warranted a price increase. By fall 1918 the regulated price was $75 per ton, a figure that held through the end of the war.[16]

As publishers' complaints about high newsprint costs continued unabated, federal control of newsprint manufacturing and newspaper publishing increased within broader wartime attempts to promote centralized control of industrial production. In July 1917, the federal government had created the new War Industries

Board (WIB) as a means of achieving this goal, and in August 1918, the WIB and the ANPA agreed that publishers of daily newspapers would voluntarily curtail newsprint usage. Despite this agreement, newsprint supplies remained tight throughout the year. In November 1918, the WIB further tightened its control of the newsprint market by instituting control over all newsprint distribution in the United States.[17]

Ultimately, these various attempts at federal management of the newsprint market proved to be at best mildly effective in helping US publishers navigate a tumultuous newsprint market, and the stronger efforts by the WIB came too late. After all of the haggling between government and manufacturers which had gone on over newsprint supplies and prices between 1916 and 1918, by the time that roughly seventy publishers arrived in Chicago to meet with representatives of the WIB's Pulp and Paper Division to formalize an agreement on newsprint prices, the war was ending. As one report noted, the meeting happened "on the very day when the armistice was signed, and the streets of Chicago were filled with rejoicing multitudes while the meeting was in progress." The WIB still wanted the agreement signed to ensure a smooth transition to a peacetime market, but the papermakers refused. The WIB decreed that some degree of federal oversight of the newsprint market would continue anyway, and it set February 1, 1919, as the expiration date of that oversight in order to try to maintain some market stability. In reality, the following two years would be anything but stable in the newsprint market.[18]

Scarcity amid Plenty: Canadian Publishers versus Canadian Papermakers

Though the end of World War I brought a welcome sense of stability to many sectors of the North American economy, the North American newsprint market remained chaotic and unpredictable. The outbreak of war in 1914 had interrupted a transition in the location of capital investment in newsprint mills, and this affected publishers on both sides of the US-Canada border. In both countries, virtually all publishers were faced with the same challenge of getting sufficient quantities of newsprint at affordable prices. On each side of the border, however, that challenge was understood in differing nationalist terms. US publishers came to worry about being dependent on foreign supplies of newsprint, while Canadian publishers came to be worried about the fact that their domestic paper manufacturers were so dependent on supplying US markets that they neglected the home market. For North American newspaper publishers, the shared challenge would be developing relationships with a newsprint manufacturing industry in transition.

As Canada was in the early stages of a process that would by the mid-1920s make it the global leader in newsprint production, many newspaper publishers in

Canada found themselves facing the same challenges as their US counterparts: newsprint prices were too high and supplies were often too low. Many Canadian newspapers claimed that they were being harmed, as Canadian newsprint manufacturers preferred to cater to the much larger and more lucrative US market. Throughout 1916, this simmering dissatisfaction manifested itself in public pleas from Canadian publishers to their federal government to regulate newsprint prices and ensure them adequate supplies. As the FTC was beginning its investigation of the US newsprint market, the Canadian government began a parallel but not formally linked inquiry of its own. In the United States, the FTC brokered a newsprint price of $50 per ton, and the Canadian government responded to demands from domestic publishers by setting the price of newsprint at the same $50 per ton, a figure that Canadian papermakers claimed was too low for them to make a profit. Their subsequent appeal to the government resulted in the appointment of R. A. Pringle, a lawyer and former member of Parliament, to investigate the situation and determine a fair price for newsprint. Pringle determined that $50 per ton was in fact a fair price, and he ordered Canadian newsprint manufacturers to sell it to publishers at that price. Pringle also told publishers to contact him if they found they could not get their regular supplies at that price, and he promised within two days to have a government agent at the mill to compel shipment. This regulated price for the Canadian domestic newsprint market remained in effect until the Pringle Commission raised the mandated price to $57 per ton in the summer of 1918, and then $69 per ton in September 1918, in responses to manufacturers' claims that production costs had increased significantly.[19]

Pringle's regulation of newsprint prices continued into the postwar period, and it became controversial for publishers and policymakers on both sides of the border. One of the flash points was the conduct of the Fort Frances Pulp and Paper Company, a US-owned mill in northwestern Ontario which had become a major source of newsprint for publishers in Manitoba, Saskatchewan, and some northern midwestern American states. Publishers on the Canadian side of the border began complaining to Pringle that Fort Frances Pulp and Paper was curtailing newsprint shipments because it preferred to export to the United States, where it could charge higher prices for newsprint. In November 1918, E. H. Macklin, the president of the Manitoba Free Press Company, wrote to Pringle that he was worried that his paper would be "put out of business by the Fort Frances Mill's refusal to send us paper." A Fort Frances representative, Macklin claimed, "declared he would not ship us paper and he asserted that he would pay no attention to any order that you issued." The whole situation, he claimed, was a "perfect nightmare to us here," and he pleaded with Pringle for help.[20]

The problems facing western Canadian publishers continued throughout the immediate postwar period. In December 1919, Pringle reported to Canadian minister of finance Henry Drayton that the "supply of newsprint to Canadian publishers is becoming very critical owing to the attitude of the manufacturers. All the manufacturers apparently are desirous of availing themselves of the very high prices obtainable in the United States for newsprint at the present time." Some were cooperating with his office, Pringle claimed, but he complained that he now found himself "in the position that some of the manufacturers are absolutely refusing to obey my orders, and unless drastic action is taken many of our Canadian publishers will find themselves without paper." By January 1920, the situation for these publishers had become critical, and one report suggested that "virtually all daily newspapers in Manitoba and Saskatchewan will have to suspend publication" because they had been unable to get adequate supplies of newsprint. In early 1920, supplies fell so short that, on one particular Monday, three daily newspapers in Winnipeg had to pool their supplies in order to print reduced-size editions. If no additional newsprint supplies were forthcoming, the publishers claimed, the newspapers would have to cease publication after Tuesday for lack of paper. They blamed Fort Frances Pulp and Paper, for a number of years the major paper supplier to the city's newspapers, for having recently "shown a determination to cut off the newsprint in order to take advantage of the higher prices in the United States due to the present scarcity of newsprint." The Canadian government responded to the emergency by ordering an embargo on all exports to the United States of newsprint produced by the Fort Frances Pulp and Paper Company in order to address the "severe paper famine," a move that had ramifications for the company's customers south of the border. For some midwestern American newspapers, the Detroit Free Press reported, this action taken by the Canadian government meant that they would be "40 per cent short on their supply of paper." Fort Frances Pulp and Paper refused to comply with the Canadian government's order, and in late January three Winnipeg dailies, the Free Press, Tribune, and Telegram, all suspended publication for lack of paper. The city's papers eventually had to shut down for five days before newsprint deliveries were resumed under government order.[21]

Pringle's ongoing efforts to divert newsprint produced by Fort Frances Pulp and Paper away from the US market and to Canadian customers eventually got the US State Department involved in trying to broker a solution, and the situation took a rather bizarre turn. One day in late January 1920, managers at Fort Frances Pulp and Paper loaded onto railcars all of the paper that they had in stock and prepared to ship it to the United States, and then they ripped out the rail lines connecting the mill to the Canadian National Railway tracks leading to Winnipeg.

When a member of Pringle's staff arrived with the local sheriff, Fort Frances Pulp and Paper managers refused to let them in the building and "threatened forcible resistance if any seizure was attempted." The fiasco forced Pringle to re-sign a few days later, though his successor was able to get the flow of paper sup-plies restarted to Manitoba publishers.[22]

After production resumed, the three Winnipeg publishers collectively wrote to the Canadian government to express their outrage at being mistreated by a do-mestic newsprint manufacturer serving American customers rather than the home market. "The people of Canada," the Winnipeg publishers argued, "have witnessed the spectacle of a Canadian newsprint company, operating its plant with the assistance of wood cut in Canadian forests, driving its machinery with power generated from water flowing over Canadian falls, enjoying the protection of the laws of Canada, forcing the newspapers of its Canadian customers to sus-pend publication day after day while not a single United States newspaper-client of this Canadian Company reported being obliged to miss one issue." To add insult to injury, the publishers claimed, while the Winnipeg papers had been "strangled and out of existence," US papers from just across the border exploited the lack of local circulation by sending copies of their papers north "to flood the city of Winnipeg with thousands of copies of their editions—editions printed on newsprint paper supplied by the Canadian mill to which Winnipeg newspapers were required to look for their supplies; a mill of which the Winnipeg news-papers had been almost life-long customers."[23]

The difficulties that Canadian publishers faced in obtaining adequate news-print supplies from domestic producers were not unique to the western part of the country, nor did they end after Pringle left his position. In June 1920, Gaston Maillet, editor of the Montreal newspaper *Le Matin*, wrote to Prime Minister Rob-ert Borden that he had faced the same problems and "been literally soaked to-day by a paper company." How could it be, Maillet asked, that, in the "greatest pulp-wood producing country of the world, publishers should be the slaves of paper trusts? You are, my dear Sir Robert, the man in the very position to fix those Shy-locks, or at the very least compel them to supply Canadian Publishers before any foreign competitors, many of whom are, as you are aware, great enemies of the Empire." As his American counterparts used political threats in attempts to in-fluence their domestic newsprint market, Maillet warned Borden that, should he not assist Canadian publishers, "you must be assured that you cannot much rely on the support of editors' ink."[24]

Gaston Maillet's outrage at the Canadian newsprint market was an example of an increasingly common perspective among publishers in the 1920s. Many won-dered why their country's trees were being used by manufacturing firms more interested in serving foreign rather than domestic customers. As they observed the process of Canada eclipsing the United States as the continent's leading source of newsprint, many Canadian publishers responded with a nationalist defense of their enterprises. While they ran their publishing businesses in a country producing increasing amounts of newsprint, they wondered how it was that they could find paper being scarce, and many saw the root of the problem in the continental market for newsprint. Supply might be moving to Canada, but the overwhelming majority of demand came from the United States, and many Canadian publishers found it challenging to adapt to these new market conditions.

Relocating Newsprint Production to Canada in the 1920s

On the US side of the border, American publishers were also anxious about their increasing dependence on Canadian newsprint after World War I. Many US publishers feared not only a growing dependence on foreign production in general but also the immediate fact that they would be competing with British publishers to buy Canadian newsprint. In June 1920, Lord Riddell, the publisher of the mass-circulation *News of the World*, estimated that it would be eighteen to twenty-four months before European production could meet UK demand. As *Editor & Publisher* noted in August, these British demands on Canadian production "create not only a more critical situation for the immediate present, but a deadly menace for the near future. It is a menace to more than mere business. It is a deadly menace to the outspoken American independence and to the very existence of our press itself. The press of the United States cannot afford to be commercially dependent upon a foreign country for so vital a necessity as its print supply; far less can the people of the United States afford to have their popular voice subject to being silenced at the dictate of an alien power." In addition to being concerned about the international competition for North American supplies, publishers in the United States were also chagrined that newsprint prices were not coming down after the war ended. Indeed, prices skyrocketed from $64 per ton in 1918 to $79 per ton in 1919 and then to $112 per ton in 1920. The last price was roughly 250 percent higher than it had been in 1914. As had been the case during World War I, smaller newspapers were hit particularly hard by market conditions, and many reported that they could not get contracts for paper at any price.[25]

With prices spiraling upward with no end in sight, the US Senate began considering government action in 1920, to some degree at the behest of publishers who saw that as the only viable option. In May, Frank Munsey, publisher of the *New York Sun*, *New York Evening Telegram*, and *Baltimore News*, testified to a Senate subcommittee that large US publishers had "only started on this drunken orgy, you know, of using paper," and he remarked, "I see no possible remedy for the situation left to individual publishers. I think the only way the thing can be handled at all is for it to be handled by the restriction of the National Government, on the theory of the good for all the people." After taking testimony from numerous other publishers, the Senate declared the newsprint situation to be "deplorable," and its report claimed to have found that manufacturers were again engaging in collusive practices. Despite harsh language directed at newsprint manufacturers, however, the Senate ultimately declined to take any direct and immediate action, the logic being that the paper market would adjust without direct regulation. Like publishers in Canada, US publishers in the years around World War I struggled to stay in business in the face of increasing paper costs and occasionally scarce supplies. For most, paper was available in theory, but many found that it was not so in practice, as prices were simply too high. Voluntary mechanisms to limit consumption proved unsuccessful, and federal efforts proved similarly ineffective in the near term. All the while, the shared condition for newspaper publishers in North America was that manufacturing was increasingly concentrating in Canada.[26]

As Canadian raw materials increasingly became the basis for the North American newsprint industry, US papermakers pursued two strategies: some converted domestic mills to other kinds of paper production, while others built new mills in Canada. These mill conversions resulted in some newly profitable businesses for US firms manufacturing wood pulp into goods other than newsprint. In 1919, for example, the paper company Kimberly-Clark introduced Kotex sanitary napkins, and soon after it began producing Kleenex tissues. Increasingly throughout the 1920s, another common strategy among leading US paper companies was to build branch plants in Canada to manufacture newsprint to be exported duty-free to the United States. Most were not overly enthusiastic about the geographic reorientation of newsprint production, but many came to see it as simply the best practical option, given the newly prevailing policies regulating the trade in timber, pulp, and paper. For most manufacturers, the primary concern was obtaining timber at the lowest price possible so as to produce newsprint as cheaply as possible. Many would have preferred simply importing pulpwood from Canada to the United States, where they could use their existing plants to manufacture it into paper, but the embargos on pulpwood exports from Crown lands prevented them from doing so in the quantities that they required. Ultimately, the best practical option for manufacturers was to build plants in Canada to use the pulpwood there.[27]

The actions of IP, the leading US newsprint producer, are emblematic of this strategy. Throughout the 1920s, IP had chafed at the provincial embargoes on pulpwood exports of timber cut from the Crown lands that the company had leased. In March 1920, IP president Philip Dodge announced that, since it had been unable "to obtain wood from our holdings in Canada, the International Company is now erecting its first paper mill in Canada, being forced to do so in order to obtain a return from its timber investment. This will probably be followed by other mills. In brief, the American manufacturers are being forced to take American capital out of the country and invest it abroad in order to protect themselves." The following year, IP formed the Canadian International Paper Company and invested $20 million in the new venture.[28]

With IP as the vanguard, North American newsprint production continued to expand in Canada as it stagnated in the United States. By 1924, as one commentator wrote in the *North American Review*, "consumers of newsprint in the United States are becoming more dependent upon the forest wealth of the Dominion." The following year, Canadian production exceeded US production for the first time, and the spread between the two countries would continue to expand. From 1925 to 1935, US production would decline from 1.5 million to 912,000 tons annually, while Canadian production increased from 1.6 million to 3.1 million tons.[29]

Some publishers and policymakers remained concerned about the growing US dependence on Canadian newsprint, but, as the decade progressed, few bemoaned the fact that prices had declined dramatically. By 1929, prices were down to $62 per ton, a figure not seen in over a decade, and publishers were pleased. *Editor & Publisher* noted that Canadians now had "modern operations, cheap water supply, and abundant convenient timber" that allowed them to "quote much lower rates and still return a profit." Newsprint prices, the magazine claimed, "have returned to the control of the buyer for the first time since 1916." As the industry boomed in the 1920s, one historian notes, an "arms race was set in motion, as rival producers modernized their facilities with new and even larger machines" in the hopes of producing more paper for export to the United States.[30]

While US publishers enjoyed the declining newsprint prices during the 1920s, some Canadians were concerned that the industry was entering into a cycle of overproduction. As Sir John Aird, the president of the Canadian Bank of Commerce, remarked in January 1929, the "dark spot on the horizon of the business situation in this country is the over-production which has occurred in the pulp and paper industry. That an industry in which over half a billion dollars of capital is invested, which employs many thousands of people and the products of which rank second in value among our principal exports, should be in an unprofitable condition through overproduction . . . seems inexcusable." These concerns had circulated throughout the decade, and they were acknowledgements of the additional risks to having a commodity like newsprint become such a large part of a national economy. With much Canadian production done for export purposes, a significant part of the domestic economy was in effect at the whims of the global market. During the Great Depression, these concerns would become heightened as the "world uncertainties affecting all business" trickled down to the newspaper business and thus the newsprint industry.[31]

When the slide into the Depression began in late 1929, it thus caught the Canadian newsprint industry at a particularly vulnerable moment. Canada had become the world's leading producer of newsprint following a period of heavy capital investment, and much of that newsprint was going to the United States to be used in newspapers that were reliant on using it to print advertisements in order to generate revenues. The onset of the Depression created significant disruptions across this supply chain. In 1929, US newspapers sold $1.2 billion in advertisements. This declined to $926 million by 1931, $735 million in 1932, and $665 million in 1933. Newspaper ad revenues would not return to 1929 levels until 1947. As the Depression caused newspaper advertising to plunge, American publishers found themselves needing less newsprint, which in turn affected the Canadian newsprint industry. As one trade journal noted, the "direct result" of the drop in newspaper advertising was a "reduction in the number of pages," which caused "reduced newsprint consumption." From 1929 to 1933, US newsprint consumption declined from 3.7 million to 2.7 million tons, and the sudden loss of more than one-quarter of the demand dealt a punishing blow to the Canadian newsprint industry and many of those involved in it. As the Depression deepened in the United States, this in turn had direct effects on Canadian newsprint production. "The pulp and paper industry has fallen on evil days since 1929," one Canadian economist noted in 1935.[32]

In response to these dire economic conditions, Canadian policymakers began taking steps to regulate production. In late 1934, Quebec premier Louis-Alexandre Taschereau claimed that it was the "duty of the government to try and take measures to see to it that the mills survive and carry on. They are fed by natural resources of this province and we are not going to allow our forests to be bared without profit to those exploiting them." At the same time, Ontario premier Mitchell Hepburn pledged to cooperate in regulating production so as to keep as many mills as possible open across Canada. Rather than having some mills at full capacity and others having to shut down, Quebec and Ontario officials jointly moved to regulate production in the two provinces so as to most effectively manage the decline in US demand. US publishers were vexed by these possible regulations and alarmed at their inability to influence Canadian policy in the same ways that they could at home. As *Editor & Publisher* noted, the price of newsprint was now "based not on supply and demand, but fixed by a politico-financial dictatorship beyond the reach of American law. The American newspaper industry is slow to move, but the start it has already made toward a domestic newsprint industry may be accelerated."[33]

There is a significant irony in all of this American lamentation. US publishers had fought against what they called the "Paper Trust" for much of the first decade of the twentieth century, and they succeeded in getting duty-free Canadian newsprint starting in 1911. What they soon found was that they were now dependent on foreign production, and they found themselves less able to use their political influence to shape that production to fit their needs. US publishers would find themselves dependent on Canadian newsprint production for decades after, and they would often struggle to adapt to the unintended consequences of market conditions that they had helped create through vigorous lobbying in favor of North American free trade in newsprint.[34]

Alaska, the American South, and the Hope for Newsprint Independence

While North American newsprint production was shifting from the United States to Canada, some US policymakers, paper manufacturers, and publishers expressed concerns about the loss of US newsprint independence which the United States was experiencing. In response, some lobbied to change tariff policy and reenact policies protecting American manufacturing. Others pursued conservationist policies aimed at reforestation and at replenishing spruce stands in the northeastern United States. For some proponents of US newsprint manufacturing, the goal was to find an alternative source of domestic pulpwood, and the two main locations that seemed feasible were Alaska and the South. Each locale presented different sorts of opportunities and challenges, but they promised the possibility of weaning the United States off supplies of foreign newsprint.

Significant federal support for developing newsprint production in Alaska began as early as 1917, as the FTC and the DOJ were in the process of investigating and prosecuting US newsprint manufacturers for alleged price fixing. To its proponents, the primary advantage that Alaska offered was that it had the same kind of abundant spruce stands as Canadian forests, but the significant impediment to developing production there was the great distance between those forests and the major users of newsprint in the northeastern and midwestern US cities. Some thought that a plan for Alaskan production was imminently feasible despite the problem of distance, and in March 1917 Secretary of Agriculture David Houston claimed that it would be possible to manufacture newsprint in Alaska, ship it through the Panama Canal to the East Coast, and sell it there at $35 per ton, a figure well below the prevailing market price of $63 per ton. The United States Department of Agriculture (USDA) remained an enthusiastic booster of an Alaskan newsprint industry as an antidote to Canadian imports. "A condition of dependence upon foreign supplies of newsprint," the USDA noted, "carries with it serious possibilities not only for consumers of newsprint (chiefly our newspapers) but also for other business interests and the public generally. It would afford a dangerous opening for covert interference with the freedom of the press. . . . That such a danger is not imaginary has been evidenced abroad. A permanent domestic industry is therefore a matter of public importance."[35]

Despite some encouragement, the development of an Alaskan paper industry did not take off in the 1910s and 1920s, partly because of the war and partly because cheap Canadian paper began flooding the US market. However, Alaska continued to offer an appealing alternative to some industry observers. As one noted in 1930, "With American newspapers practically at the mercy of the Canadian manufacturers, and the demand for newsprint constantly growing in the United States, the necessities of the situation are forcing newsprint consumers to take a lively interest in the Alaskan possibilities." Over the following decades, the possibility of Alaskan newsprint manufacturing continued to remain marginal in the industry, and one newspaper executive remarked in 1947 that "Alaska, I am afraid, will be more or less a will of the wisp for many years" when it came to newsprint production.[36]

Southern newsprint production proved to be a more successful solution to creating a domestic alternative to Canadian imports. Publishers in the South held planning meetings as early as 1916 to discuss the possibility of building a mill in Florida or Georgia. As planning began, the primary challenge facing those wanting to build southern newsprint mills was the type of regional wood supply. The South had vast stands of pine but almost no spruce, and this created problems for production given existing industry knowledge. Pine had a high pitch content that could cause problems with machinery and make a sheet of newsprint that was too dark without excessive bleaching. Some experiments began in the early 1920s for advancing the process of making newsprint from pine, though paper manufacturers remained skeptical throughout the 1920s about the possibilities of making a quality sheet from this type of wood. Charles Holmes Herty, a Georgia chemist, was the most important of these experimenters, and in 1932 he received funding from the Georgia legislature to begin experiments in how to make southern pine useful for the mass production of newsprint. Soon after, Herty began touting this on the national level as a potential economic boon to the South, something that earned him the enmity of newsprint manufacturers in the northern United States and in Canada.[37]

Southern newsprint production also drew support from leading experts in chemical engineering. Francis P. Garvan, the president of the Chemical Foundation and one of the founders of the National Institutes of Health, urged US publishers to support the project. Newsprint made from southern pine, he claimed, would help in "diminishing unemployment" and in "utilizing waste lands" in the South. Southern newsprint, he added, promised "continuous domestic supply, greatly reduced prices, without tariff, bonus or artificial aids. We consume eighty percent of world consumption. . . . Tomorrow one hundred percent of our consumption necessarily imported from Canada, Sweden, Norway, Finland, Germany. Day after to-morrow these five countries will form cartel and dictate to you, to our school book publishers, to all users, price of paper—in other words, your margin of profit. What will then become of freedom of press on which all our liberties depend?" Herty's experiments also drew increasing attention from newspaper publishers in the region. James Stahlman, publisher of the *Nashville Banner*, became a strong supporter, and by mid-1934 the Southern Newspaper Publishers Association began discussing plans to fund and build a newsprint mill. This, Stahlman claimed, would free the United States from the "continuous meddling" of Canadian policymakers in the supplies and prices of newsprint available to US publishers. Despite significant support for pine-based southern newsprint manufacturing, funding the mill proved impossible during the Depression, and construction was delayed.[38]

By early 1937, however, a collective of southern banking and industrial interests announced the construction of the South's first newsprint mill, a planned $5 million facility to be built near Dallas. James Stahlman claimed that this would mean "better business for everybody—the banks, the railroads, the merchants, the newspapers and especially the pauperized farmers in the pine-belt sections of the South, who will be able to cultivate pine for commercial use on land heretofore going to waste and practically worthless. A new day is dawning for the South and the freedom of American publishers from the domination of Canadian and foreign newsprint interests will shortly be at hand." Funding still proved elusive in the Depression, and in late 1938 the Reconstruction Finance Committee provided a $3.4 million loan to help build the mill, which would be owned and operated by a new company called Southland Paper Mills. Production began in early 1940, and the industry saw a steady expansion in subsequent years.[39]

Even with expanded production in the southern United States starting in the 1940s, many publishers and policymakers were still anxious about US reliance on Canadian imports. US demand for newsprint simply remained too robust for southern production to make up the difference for a decline of newsprint manufacturing in the northern states. Throughout the interwar period and into the post–World War II period, Canadian production continued to be an absolute necessity for the US newspaper business.

Conclusion

To some observers, the travails of US publishers that were dependent on foreign supplies of newsprint offered a cautionary tale to advocates of global free trade in other commodities. As Senator Homer Capehart, an Indiana Republican, remarked in April 1947, the "newsprint situation in the United States should be a warning to the free traders and to those who yell the loudest for one world. . . . Twenty-seven years ago, when newsprint was made tariff-free, 80% of the newsprint used in the United States was produced here while only 20% was imported." By 1947 those proportions were almost completely reversed, and Capehart warned that "this is what will happen to many other industries if the policy of free trade and reciprocal trade treaties is continued and our own producers continue at a disadvantage by reason of foreign competition." Robert L. Smith, the general manager of the *Los Angeles Daily News*, similarly remarked in a 1950 congressional hearing that this dependence on foreign newsprint supplies was a potential threat to a free press. "Let us consider the plight of American industry," he suggested, "if we were dependent upon foreign sources for steel, coal, iron, rubber, oil, and so forth. How then can we expect to have a healthy free press unless we have an adequate domestic source of raw material for this vital industry?"[40]

Canadians were both puzzled and offended by these US concerns about being too reliant on Canadian newsprint. In 1953, R. M. Fowler, the president of the Canadian Pulp and Paper Association, told a US audience, "We have met your demands with speed and efficiency" in supplying newsprint, and thus "it seems strange that in the last year or two there have been frequent statements by a number of your politicians, supported by some of your publishers, which protest against what they call the 'dependence' of the United States on a 'foreign' source of supply." As Fowler saw it, these concerns reflected a broader American anxiety about participating in the global economy. "It seems to us that newsprint is only an example of a more general discomfort in this country over your growing need to depend on 'foreign' supplies of materials you require. If this attitude is widespread or should grow, it has some undesirable political implications; no one likes to be regarded as a 'foreigner' for this is a word that seems inappropriate for neighbours and allies."[41]

By the early 1950s, US concerns about the dependence on Canadian newsprint supplies remained significant, and indeed these anxieties had been widely expressed throughout the previous decades, especially during wartime. In the first half of the twentieth century, newspaper publishers around the world faced challenges getting adequate supplies of newsprint. These challenges were occasionally the result of market forces, and they were at times the result of wars. To many publishers, state officials, and international observers, the promotion and maintenance of a free press involved finding ways of meeting these challenges, and this often involved cultivating international trade relationships. To many in the United States, it proved perpetually troubling that ensuring the practical functioning of a free press required managing a dependency on foreign newsprint. As Canada became the global leader in newsprint production, a key geopolitical challenge for newspaper publishers around the world was getting access to that newsprint.

PART TWO

EXTENDING CHICAGOLAND

CHAPTER FOUR

The Local Newspaper
as International Corporation

When Robert McCormick took over the *Chicago Tribune* in 1911, he continued a family control over the company which stretched back more than half a century to when his grandfather, Joseph Medill, bought a share of the paper. Born in 1823 in Saint John, New Brunswick, Medill immigrated to Ohio in 1832 and by the early 1850s was in the newspaper business there. In 1855, he moved to Chicago and purchased a share of the *Tribune*. As would be the case with his grandson, Medill used his position as a publisher as a platform for his staunch political support of the Republican Party. From 1871 to 1873, Medill was the mayor of Chicago, and he guided the city's rebuilding process after the Great Chicago Fire. Finding himself frustrated with the machinations of urban politics, Medill returned to newspaper publishing in 1874, when he purchased a controlling interest in the *Tribune*. One of the pillars of Joseph Medill's business strategy was a commitment to operating his newspaper as an innovative manufacturing enterprise. Prior to the Great Chicago Fire of 1871, Medill and his partners had invested $250,000 in one of the most advanced printing operations in the country, and Medill advised other publishers to be "very particular with the mechanical execution of your paper." He extended this commitment to efficient manufacturing when he rebuilt the destroyed printing plant.[1] After the fire, one of his oft-quoted maxims became "First, the machinery must work."[2]

After Joseph Medill died in 1899, the marriages of his two daughters, Elinor and Katherine, continued the family's commitments to both newspapers and politics. Elinor married *Tribune* reporter Robert Patterson, and their children, Joseph Patterson and Eleanor "Cissy" Patterson, both would later enter the publishing business, Joseph as the publisher of the *New York Daily News* and Cissy as the editor and publisher of the *Washington Herald* and the *Washington Times*. Katherine married Robert Sanderson McCormick, the nephew of Cyrus McCormick of McCormick Reaper fame. The couple's older son, Medill McCormick, was briefly the publisher of the *Chicago Tribune* prior to serving in both houses of Congress, before committing suicide in 1925 at the age of 47. Their younger son, Robert Rutherford McCormick, would prove to have an extraordinarily successful career, extending his grandfather's commitments to publishing and politics from the time he became the *Tribune*'s publisher in 1911 until his death in 1955.[3]

When Robert McCormick took over the *Tribune* in 1911, it was a successful morning newspaper in a highly competitive market. In the following decades, he built the *Chicago Tribune* into not only the highest-circulating newspaper in Chicago but also the second-highest-circulating newspaper in the United States. The only newspaper to have a higher circulation was the *New York Daily News*, the hugely successful tabloid that his cousin Joseph Patterson started in 1919 as a wholly owned subsidiary of the Tribune Company.

Through his newspaper corporation, Robert McCormick came to exert a tremendous influence over the city of Chicago, as well as areas quite distant from it. A 1941 *Saturday Evening Post* portrait remarked, "If there were such a title as Duke of Chicago, it would go without serious opposition to Colonel McCormick." Journalist Jack Alexander found that the *Tribune* dominated much of the public sphere of the middle of the central United States. "As a public force," he wrote, the *Tribune* "spans the Midwestern plain area like a colossus. . . . In circulation, the Tribune's shadow sprawls beyond Chicago and its suburbs into half a dozen neighboring states, cutting across the spheres of influence of other large-city dailies, which are numerous in the Great Lakes section. Political power follows circulation." One of the reasons that McCormick was able to achieve this position was because of a strategy of vertical integration in manufacturing his newspaper.[4]

As a publisher, Robert McCormick brought to his job not only a strong desire for social and political influence but also a builder's sensibility. Two of the most significant lingering monuments of this trait are the Tribune Tower, which was completed in 1925 and remains an iconic building in a city filled with such structures, and the Canadian city of Baie Comeau, which in the following decade McCormick helped to plan from scratch on the rugged North Shore of the St. Lawrence River in Quebec to house workers for the state-of-the-art newsprint mill that he was building. In between these two sites, McCormick also built another technologically advanced newsprint mill at Thorold, Ontario. Pulpwood and newsprint were shuttled between the forests, Canadian newsprint mills, and the US printing plants on a fleet of boats that the Tribune Company owned, including some that had been specially designed to carry newsprint as cargo. As an industrial enterprise, the Tribune operated across the US-Canada border with a high degree of vertical integration, and McCormick sought every opportunity to make his company self-sufficient in its manufacturing operations. As a trade journal noted in 1947, the *Chicago Tribune* was "no journalistic accident or phenomenon. It is the product of farsighted management which has provided mechanical facilities second to none and established its own newsprint supply 35 years ago."[5]

This and the following two chapters will narrate the creation and operation of this international manufacturing enterprise, paying particular attention to the various political contexts of its development. Though the *Tribune* would like to tout its editorial independence and campaigns against government regulation of the press, it was, as a manufacturing firm, enabled by and dependent on state policies related to trade, regional economic development, and natural resource management.

Though as a publisher he was a peer of such prominent figures as Adolph Ochs, Joseph Pulitzer, and William Randolph Hearst, Robert McCormick in many respects also had as much in common with some of the other aggressively independent industrialists of his era, for example, Andrew Carnegie, John D. Rockefeller, and Henry Ford. Like them, McCormick wanted control over all aspects of his firm's operations, and in pursuit of that goal he was willing to use any commercial and political means available to him. From practically the moment that he took over as *Tribune* publisher in 1911 until his death in 1955, Robert McCormick sought to expand the commercial success and political influence of his publishing company, and his primary strategy to achieve those goals was the vertical integration of his industrial operations. The success of this strategy would reshape the newspaper markets of Chicago and later New York, and the industrial operation providing paper for the *Chicago Tribune* and *New York Daily News* would in turn have significant effects on the economies, labor markets, and landscapes of both Ontario and Quebec.

The Industrial Newspaper in Chicago

Chicago is often called the "second city," and as with many facets of American cultural history, the New York newspaper market has drawn more attention from historians. In some ways, though, Chicago offers an equally compelling site for considering the development and growth of the modern industrial newspaper, given the chronology of its development.[6] Settled in the 1830s but rebuilt in the 1870s after a devastating fire, Chicago is a distinctly modern metropolis in its planning and design, and its urban development was intimately related to the growth of the industrial newspaper. In the twentieth century, the most successful of these industrial newspapers was the *Chicago Tribune*, and a consideration of its growth within the context of the city's development can demonstrate a number of important themes in both media and urban history.[7]

As a manufacturing enterprise, the *Chicago Tribune* increased dramatically in size and scope as the city grew. One can get a rough sense of this by considering the expansion in the daily production requirements for printing the growing newspaper. The first issue in 1847, as one account later described it, "consisted of 400 copies pulled on a Washington hand press by one of the editors." In 1873, the four-page *Tribune* had a circulation of 28,500, and its daily newsprint requirements amounted to three rolls. By 1900, the paper had a weekday circulation of 84,000, and editions ranged from twelve to sixteen pages. On Sundays, the paper ran as long as sixty-eight pages, and it estimated that its newsprint requirements had grown to three hundred rolls per week. This newsprint was brought "11 rolls to the wagon" to the printing plant, and "unloading was an arduous task, requiring the help of many men in addition to the mechanical aids such as trucks, hoisters, trolleys, and dollies that had been devised." Paper was a bulky and heavy material, and storage space was at a premium. In 1900, the *Tribune* estimated, it had room at the printing plant for about seventy rolls of newsprint and space off-site for some four hundred rolls. The paper's increasing needs for a physical plant led it to construct a new $1 million building in downtown Chicago, where it installed state-of-the-art Hoe printing presses.[8]

The newspaper market that the *Tribune* operated in was one of the most vibrant and competitive in the country. In 1900, Chicago had a total of thirty-seven daily newspapers, with sixteen circulating in the morning and twenty-one in the evening. Only New York had more newspapers, with twenty-nine in both the morning and the evening. By 1911, as table 5 shows, Chicago's two highest-circulating papers were evening papers, with the *Daily News* selling 327,634 daily copies and William Randolph Hearst's *Evening American* 315,335 copies. In the morning market, the *Tribune* circulation of 226,000 was closely followed by the *Record-Herald* at 200,000 and Hearst's *Examiner* at 187,021.[9]

In Chicago as in all urban newspaper markets, competition for circulation and advertising was fierce. One of the most important elements of newspaper competition in Chicago involved the race to build more efficient and productive factories to print newspapers. The *Chicago Daily News*, for example, planned and built a massive new structure to be both its corporate headquarters and its manufacturing plant. The building was situated on local river and railroad networks so as to best facilitate the circulation of newsprint rolls into the factory and printed newspapers out of it. As publisher Walter Strong noted, among the paper's primary considerations in choosing a site and designing its building were the availability of "transportation lines and streets for the distribution of newspapers." This was because, Strong claimed, the "economical development of a building to house a manufacturing plant required, especially by evening newspapers," that it be situated at a "central point of distribution because of the shortness of time between the receiving of news and its publication . . . [and] the accessibility to water and

TABLE 5
Daily circulation of market-leading Chicago newspapers, 1906–29

	Morning				Evening	
Year	Tribune	Record-Herald (1906–14); Herald (1915–17)	Examiner	Herald-Examiner	Daily News	American
1906	151,661	145,780	147,410		319,539	328,218
1907	160,859	141,560	182,222		320,030	351,960
1908	162,338	152,056	171,428		321,195	351,960
1909	162,330	143,241	160,571		326,796	350,000
1910	150,000	141,471	177,889		325,214	298,447
1911	226,000	200,000	187,021		327,634	315,335
1912	220,000	204,678	215,052		323,100	324,025
1913	220,000	184,464	204,289		346,361	386,139
1914	253,000	155,319	240,000		350,550	350,000
1915	303,316	167,602	243,655		379,108	350,000
1916	354,520	191,534	232,015		410,000	378,941
1917	392,488	201,759	225,297		437,315	400,031
1918	410,818			260,777	373,112	325,017
1919	424,588			311,831	377,769	339,721
1920	437,158			343,515	388,406	364,769
1921	483,272			359,386	395,665	395,427
1922	517,184			355,127	371,078	387,573
1923	567,628			335,370	377,861	388,352
1924	608,130			349,024	392,731	458,376
1925	658,948			363,162	387,189	441,227
1926	741,493			392,340	395,086	525,771
1927	778,768			434,165	435,580	552,041
1928	809,165			423,623	413,187	538,797
1929	857,595			420,903	430,204	540,516

Sources: N. W. Ayer & Sons, *American Newspaper Annual*, various years; Chicago Tribune Company, *Book of Facts, 1930* (Chicago: Chicago Tribune Company, 1930), 7.

Note: In 1914, *Tribune* managing editor James Keeley purchased the *Record-Herald* and merged it with the *Inter-Ocean* to form the *Herald*. In 1918, William Randolph Hearst purchased the *Herald* and merged it with his *Examiner* to form the *Herald-Examiner*.

rail transportation for receiving the more than 200 tons of raw materials used each day in its manufacturing process." The *Daily News* plant was, he claimed, "in reality a great manufacturing plant turning out each day a half a million separate parcels for immediate sale."[10]

In the course of meeting this kind of competition, the *Tribune* continually expanded its own industrial plant, and by 1920 the company opened a new printing plant on Michigan Avenue. Built at a cost of $1.5 million, the boxy and functional-looking five-story building would eventually be overshadowed by the much grander neo-Gothic Tribune Tower, construction of which would begin several years later. In 1921, however, the new industrial plant was, one account noted, "built from the standpoint of factory production. As everyone knows the ideal factory receives its raw material at as few entrances as possible, delivers it to the various departments, and finally the assembling room . . . without any of the finished material having interfered with the progress of manufacturing. This has been done as far as possible in a newspaper way in the new Tribune plant."[11]

Over time, the *Tribune* used this physical plant to produce an increasingly successful newspaper, and circulation continued to increase throughout the 1910s. By 1915, the paper had a circulation of 303,316. In 1918, the paper attained the highest daily circulation of any newspaper in Chicago, morning or evening, at 410,818. By 1929, the *Tribune*'s circulation had more than doubled to 857,595, a figure more than twice that of its main morning competition, the *Herald-Examiner*. This was also dramatically higher than the circulation of the most successful evening papers, the *Daily News* (430,204) and the *Evening American* (540,516).

This massive circulation growth was the result of Robert McCormick's aim of making the *Tribune* into much more of a regional paper than its rivals by reaching readers outside of Chicago. He succeeded, as by 1920 the *Daily News* sold only 6 percent of its copies outside of Chicago and its suburbs, while the *Tribune* had some 38 percent of its circulation in these areas.[12] If the *Daily News* aimed to be a local paper, the *Tribune*'s ambition was for regional domination. In 1920, copies of the *Tribune* accounted for approximately one in four morning papers circulating in Illinois, Indiana, Iowa, Michigan, and Wisconsin, and the paper called this territory "Tribune Land." In the 1920s, the paper began calling this area "Chicagoland," which it described as an area encompassing "a 200-mile radius from Chicago, north, south, east, and west." This term lingers today as an everyday description of the region that the city dominates.[13]

Beyond the number of newspapers sold, one of the key elements of the *Tribune*'s success was the amount of advertising printed in each paper. The *Tribune*'s classified business proved particularly lucrative, and it boasted that in 1911 it sold 985,174 want ads. This was, the paper claimed, nearly three times the number sold by the nearest morning competitor, and more than the combined totals of all the morning papers in the city. Even when the paper's circulation trailed its competitors, its advertising business boomed, as table 6 shows. In 1906, for example, the morning market was highly competitive in circulation, with the *Tribune* selling 151,661 daily copies, the *Examiner* 147,410 copies, and the *Record-Herald* 145,780 copies. In advertising, however, the *Tribune* dominated, selling 9,781,356 lines of advertising, a sum far greater than the *Record-Herald*'s 7,191,840 and the *Examiner*'s 3,813,312 lines. The *Tribune* printed more advertising than any paper in the city, morning or evening, including the *Daily News*, despite the fact that the *News* had more than double the *Tribune*'s circulation. This gap in advertising only wid-

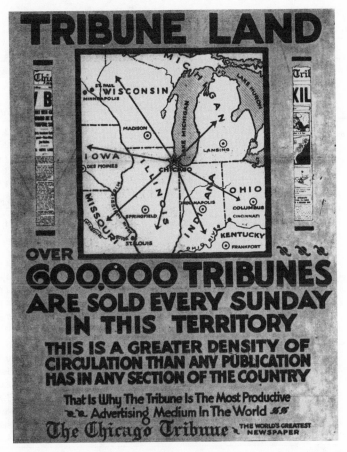

In 1916, the *Chicago Tribune* began describing the region of the country where it circulated as "Tribune Land." In the 1920s, this would evolve into "Chicagoland," a term that remains in usage today to describe Chicago's broader environs. Advertisement, Chicago Tribune Company, *Editor & Publisher* 48, no. 44 (Apr. 8, 1916): cover.

ened as time went on. By 1914, the *Tribune* was selling almost as many ads as the *Record-Herald* and *Examiner* combined, and by 1929 it sold 31,907,811 lines of advertising, more than double the amount of its nearest morning competitor.[14]

Having a newspaper with both an increasing circulation and increasing advertising meant that increasing amounts of paper were needed, and in this sense Chicago newspaper competition became a battle measured in tonnage. The *Chicago Daily News*, for example, used roughly 30,000 tons of newsprint per year from 1915 to 1917, and this had increased to 38,000 tons in 1919 and 48,000 tons in 1920. In 1921, the *Chicago Tribune* claimed to have used 64,524 tons of newsprint. The need to move massive amounts of paper around an industrial plant presented

TABLE 6

Annual advertising linage of market-leading Chicago newspapers, 1906–29

		Morning			Evening	
Year	Tribune	Record-Herald (1906–14); Herald (1915–17)	Examiner	Herald-Examiner	Daily News	American
1906	9,781,356	7,191,840	3,813,312		8,776,035	3,640,515
1907	9,932,109	6,421,923	5,920,530		8,257,227	2,735,859
1908	10,097,058	5,568,858	5,033,484		8,206,647	2,676,093
1909	11,344,455	6,336,000	6,456,780		9,148,479	3,174,927
1910	11,803,623	6,716,832	6,976,791		9,577,296	3,533,046
1911	11,424,765	7,344,204	7,398,456		9,218,874	3,736,584
1912	11,921,985	6,706,107	7,142,271		8,767,524	3,550,512
1913	13,102,881	5,789,667	7,382,505		10,587,045	4,372,470
1914	13,050,777	6,444,216	7,140,417		10,750,203	4,797,099
1915	13,765,965	6,883,446	6,482,439		11,229,939	4,481,811
1916	16,492,398	7,532,049	7,697,199		12,609,087	4,823,070
1917	16,467,450	7,104,261	6,999,435		12,596,526	4,403,580
1918	15,566,430			6,836,022	12,041,841	4,021,971
1919	23,333,184			10,165,539	16,019,139	6,866,358
1920	25,726,071			10,302,765	20,396,781	7,436,838
1921	23,010,993			9,035,691	17,501,673	8,218,764
1922	26,212,563			10,536,561	18,224,817	8,816,595
1923	28,041,477			12,054,003	20,090,682	10,544,646
1924	28,181,121			12,242,742	19,583,370	11,242,890
1925	31,068,561			12,179,577	20,483,166	12,694,074
1926	32,715,561			12,987,654	21,811,503	13,726,488
1927	31,834,173			12,849,684	21,160,335	14,276,163
1928	30,512,112			12,795,117	20,861,232	14,529,489
1929	31,907,811			13,790,532	21,158,202	14,558,133

Sources: Chicago Tribune Company, *Book of Facts, 1921* (Chicago: Chicago Tribune Company, 1921), 68; Chicago Tribune Company, *Book of Facts, 1930* (Chicago: Chicago Tribune Company, 1930), 261.

Note: In 1914, *Tribune* managing editor James Keeley purchased the *Record-Herald* and merged it with the *Inter-Ocean* to form the *Herald.* In 1918, William Randolph Hearst purchased the *Herald* and merged it with his *Examiner* to form the *Herald-Examiner.*

a problem to all publishers, and designers factored this into the construction of the new facilities that the company built in the 1920s. The company allocated room for some nineteen hundred rolls of newsprint, and it also began leasing warehouse space nearby to hold additional supplies. To coordinate moving massive quantities of newsprint from these facilities to the printing plant, the company's longtime warehouse manager "built himself a railroad" that ran through tunnels dug below the city's streets. By 1942, the Tribune had the very prominent Tribune Tower above ground, and it also had, as the paper reported, downtown properties that "number millions of square feet of floor space." The operations above and below ground formed, as one correspondent reported, a massive industrial concern. Correspondent Charles Leavelle wrote,

You stand at the forks of a narrow tunnel. On your right the passage is dark, but the left fork is lighted and it climbs swiftly into the distance. Gradually from the far end comes sound; a steady rattling that rises harshly as steel wheels clash against the tracks that line the floor. A white cylinder six feet long and more than three feet across is hurtling toward you down the tunnel, the noise of its coming swelling swiftly into a reverberating roar. You leap back involuntarily as it thunders past. You are just below the intersection of Hubbard and St. Clair streets, in the tunnel that connects The Tribune's Michigan avenue plant with its vast paper warehouses along the Chicago river. Before sunrise the 1,800 pound juggernaut that just roared by will have been printed, cut, and folded into more than 10,000 *Tribunes*. . . . Every half minute one of these big spools, containing 18,000 feet of white newsprint— paper—will boom down the tunnel to the pressroom. Six nights a week 340 of them go past. On a Saturday night—when the big Sunday *Tribune* is on the presses—you can count 832.

Considering this expanding scope of the *Tribune's* operations, one can get a sense of the increasing magnitude of the industrial operation. "There are many ways of tracing *The Tribune's* rise," the paper noted, but "its mushrooming need for white paper is as dramatic and probably the most graphic medium, now that the reader has visualized the immensity of one week's consumption of newsprint and the vast amount of white paper contained in one of those 1,800 pound rolls."[15]

Throughout its existence, the Chicago Tribune had been a newspaper firm defined by a commitment to building innovative and efficient manufacturing operations. Under Robert McCormick's leadership, this commitment became even stronger, and the paper invested significant sums in a plant that helped it become the dominant newspaper in a competitive market. Newsprint was the basis for this, in terms of both allowing the Tribune to print and sell more papers and providing the space to print more advertisements in each issue. Independent of its content, the *Chicago Tribune* had become, through investments in factories and infrastructure, the most successful industrial newspaper in Chicago, and one of the most successful in the United States.

The Vertically Integrated Newspaper

Robert McCormick's path to becoming the publisher of a major metropolitan newspaper was based on the calculated exploitation of an exceptionally privileged background and upbringing. His family was wealthy and powerful at the time of his birth, and he had ample opportunity to exploit his origins. While his mother had been the daughter of *Chicago Tribune* publisher Joseph Medill, his father, Robert Sanderson McCormick, was a US diplomat whose career included stints as ambassador to Austria-Hungary, Russia, and France between 1901 and 1907. While his father was moving between ambassadorships in Europe, Robert McCormick began to get involved in Chicago politics. In 1904, at the age of twenty-four, McCormick won an election to the Chicago City Council as an alderman, where he was regarded as an active and influential member. A lifelong Republican (his grandfather Joseph Medill had been a strong supporter of Abraham Lincoln back to Lincoln's days in state office), McCormick clashed with Democratic mayor Edward F. Dunne over the management of local public transportation franchises, which Dunne wanted run by the city and McCormick by regulated private franchises. After a year on the City Council, McCormick in 1905 ran for and was elected president of the Chicago Sanitary District. During his five years in that position, McCormick supervised a variety of municipal development projects (this episode in his career is discussed in chap. 5) and was considered by many in the city to be a strong mayoral candidate. His career hit a snag in 1910 when he, like many Republicans in the city, was turned out of office in local elections, and the Sanitary District would prove to be his last stint in elected office.[16]

Finding himself out of a job, Robert McCormick experienced a personally unfortunate but professionally timely death in the family, as his uncle Robert Patterson, then the publisher of the *Tribune*, died in April 1910. Robert McCormick's older brother Medill had served in various executive positions at the *Tribune* and would have been an obvious candidate to take over as publisher, but according to most accounts, he had never been happy working at the newspaper. He also suffered from mental health issues, and at the time of Robert Patterson's death he was in Zurich undergoing treatment at Carl Jung's sanitarium. (He would return to the United States to a career in politics prior to his suicide in 1925.) When Robert Patterson died, Victor Lawson, the publisher of the *Chicago Daily News* and *Chicago Record-Herald*, proposed to buy the paper from the Medill family, but Robert McCormick and his cousin Joseph Patterson strongly opposed the sale and succeeded in keeping the paper in the family. Once he became involved with publishing the *Chicago Tribune*, McCormick transferred his political ambitions to the paper. It became one of the leading Republican newspapers in the United States, and its politics regularly tilted into territory that many rightly considered reactionary.[17]

Besides bringing his politics into his work as a publisher, Robert McCormick also applied his organizational acumen to the paper. Under McCormick's leadership, the *Chicago Tribune* was technically and operationally one of the most innovative newspapers in the United States. At every turn as a manufacturer, the company sought to be independent. When Robert McCormick took over as publisher of the *Tribune* in 1911, one of the first things that he believed that he needed to do if he wanted to expand his business was to find a source for cheap and plentiful newsprint, as this was among his most significant and costly inputs. As his paper would later remark in an editorial, "Since newsprint is the principal raw material of a newspaper and represents the largest single cost of its production," the *Tribune*'s management "foresaw that a newspaper unable to obtain a supply of newsprint at a fair price might be forced into bankruptcy, however able and prudent the conduct of its other affairs."[18]

One of the biggest challenges to getting paper at what was considered a "fair price" was the fact that William Randolph Hearst was such a large purchaser of newsprint that he could get it at prices substantially lower than other publishers. Around the turn of the twentieth century, Hearst was using an estimated 120,000 tons of newsprint per year for production across the newspapers in his chain. In 1904, he entered into an agreement with the International Paper Company (IP) to purchase some 400 tons of paper per day, a deal that one report described as the "largest order for news stock in the history of the paper trade." This was roughly one-third of IP's total production, and purchasing such a substantial amount of paper gave Hearst a degree of leverage in negotiating a price which other publishers did not have. The contract that Hearst got from IP was to run for ten years, and he was to get roughly 146,000 tons per year at an average price of $37.60 per ton for paper delivered to his printing plants.[19]

Hearst's price was substantially cheaper than what the *Tribune* was paying at the time. *Tribune* publisher Medill McCormick had also negotiated a deal with IP in 1904, but this was for about 15,000 tons per year, or roughly 10 percent of the amount that Hearst bought. Medill McCormick had not been able to negotiate the long-term price that Hearst had, and by 1908 his price, which had started at about the same as Hearst's, had increased to $44 per ton. Robert McCormick knew he had to find a way to compete more effectively with Hearst in obtaining newsprint. As he later recalled, by 1911 "Hearst, with his enormous buying power, was getting paper for $5 a ton less than we had to pay. This discrepancy, if continued, would finally have starved us to death. With my experience in electricity and canals and a general instinct for machinery, I decided that we should build our own paper mill."[20]

Most publishers, including William Randolph Hearst, preferred simply to negotiate with manufacturers rather than develop their own newsprint manufacturing. As John Norris of the American Newspaper Publishers Association noted in 1909, the typical publisher "had troubles enough when he ran a newspaper without attempting to operate a paper mill," and for most "it was wiser" simply "to make contracts with others" for paper. When McCormick decided to lead the *Chicago Tribune* into newsprint manufacturing, he was thus pursuing a bold course of action. The *Tribune* was not alone in employing this strategy, but it was by far the most active and successful newspaper to pursue vertical integration in order to gain more control over its own operations and carve out more independence from newsprint manufacturers.[21]

Besides the *Chicago Tribune*, the *New York Times* was the only other major US newspaper that engaged in sustained and successful efforts at creating an international and vertically integrated operation. After an initial failure operating a Brooklyn newsprint plant from 1918 to 1920, the *Times* returned to newsprint manufacturing in the late 1920s when it partnered with the Kimberly-Clark Company to form the Spruce Falls Power and Paper Company. Kimberly-Clark owned 51 percent of the new company, which soon built a mill at Kapuskasing, Ontario, able to produce an estimated 550 tons per day of newsprint, enough for all of the *Times*'s own needs and a surplus to sell on the market.[22]

The *New York Times* and *Chicago Tribune* were the outliers, however, and the experience of many newspapers was that owning a mill created insurmountable financial and operational challenges, and the few who tried eventually abandoned their efforts. When Robert McCormick initiated plans to build his own newsprint mill in Ontario, he was employing a strategy shared by few other publishers willing to tolerate the cost and risk involved. McCormick's 1911 decision to vertically integrate would define the company's operations for the next eighty years and would help make the Tribune Company into one of the most powerful newspaper corporations in North America.

Building the Mill at Thorold

Robert McCormick's 1911 decision to vertically integrate across the US-Canada border was in many ways a bold and strategic move, but it was a decision enabled by not only business acumen but also state policy. This was the exact moment when the US Congress was approving the reciprocity pact with Canada which would allow newsprint to move across the border duty-free. The House passed the bill on April 21, and the Senate followed on July 22. McCormick began advocating the construction of a Canadian newsprint mill at a Tribune Company Board of Directors meeting on July 17, 1911, and President Taft signed the reciprocity bill on July 26. As a newspaper publisher, McCormick stood to receive immediate and enormous savings in operational costs as a result of this change in tariff policy. Like many publishers, McCormick was dismayed by Canada's rejection of reciprocity later that year, but he was also well aware of the loophole by which the American law would allow conditional duty-free newsprint imports from Canada even if the rest of the reciprocity agreement did not pass.[23]

Enabled by the tariff change, McCormick sprang into action. Working with former Chicago Sanitary District engineer George Wisner, he began scouting locations in Canada and making plans for a newsprint mill. He eventually settled on a site on the Welland Canal in Thorold, Ontario. He also invited engineer Warren Curtis Jr. (the son of the IP president of the same name) to be involved in the planning, and he convinced the *Tribune*'s board to commit $1 million to build a mill, a risky proposition at the time. In 1938, while speaking in Ontario near Niagara Falls, Robert McCormick would remark that, when his company decided to vertically integrate, it was "not difficult to come to Canada to erect a paper mill . . . because the founder of the Tribune, my grandfather, was born in Canada." In reality, the initial choice to go to Canada had more to do with basic commercial and operational imperatives than it did with familial ties. The mill was, as one observer noted, "midway between the source of raw supply and the point of consumption." The company could get wood from Ontario and Quebec, much of which could be moved via waterway, thus allowing the company to avoid railroad freight charges in both getting raw materials to the mill and then sending the finished newsprint to Chicago. The company also took advantage of Thorold's proximity to Niagara Falls, which was roughly 10 miles from the mill site and which could provide the mill with a continuous source of affordable hydroelectric power. In February 1912, the Tribune Company created a wholly owned Canadian newsprint manufacturing subsidiary, initially called the Ontario Paper Company. On March 15, the Tribune Company's board formally authorized construction of the Thorold mill, which then began in June.[24]

The Thorold mill was revolutionary in its planning, and one of its major design innovations was that it combined a pulp mill and a paper mill into one manufacturing facility. At the time, the overwhelming majority of mills throughout the world made just pulp or paper, as they either converted trees into pulp or used pulp made or purchased elsewhere to manufacture paper. The Thorold mill would do both. The Tribune Company was able to purchase power from Niagara Falls at exceptionally low rates, as the mill did most of its wood grinding and pulp manufacturing at night, when power rates were cheaper.[25]

The mill's first two machines commenced operation on September 5, 1913, roughly a month before the US Congress passed the Underwood Tariff in October and made all Canadian newsprint imports duty-free. By the end of its first year of operation, the mill had produced 5,713 tons of newsprint. This increased to 31,707 tons in 1914, and production continued to increase annually after that. By 1920, the Thorold mill produced 69,970 tons, a figure that had increased to 112,580 tons in 1930. By 1940, the mill was producing over 150,000 tons of newsprint annually. The Chicago Tribune Company used all this newsprint and did not sell it on the open market. Indeed, the company's lead executive in Canada described the Thorold facility as a "non-commercial mill acting as a manufacturing department for the Chicago Tribune."[26]

As the *Chicago Tribune* began manufacturing newsprint after 1913, it extended control over its industrial operations from the pressroom all the way to the manufacturing of the newsprint that it needed. In exerting control over his newspaper's paper supplies, Robert McCormick aimed to better rationalize his daily production and to make his firm increasingly independent of the market forces that could cause significant difficulties for other newspaper firms. Taking control of his supply of newsprint, he understood, gave him a greater degree of independence than many of his competitors.

Controlling Raw Materials in Ontario and Quebec

As purveyors of industrially produced commodities, newspaper publishers were in many ways similar to automakers: both groups used factory production and wanted steady, predictable, and affordable supplies of essential raw materials. This motivated Henry Ford to try to build a rubber plantation on the Amazon River in Brazil, where he was given control of 2.5 million acres of land to build the plantation and the town of Fordlandia, and this is what motivated Robert McCormick to seek timber lands in Canada. The Tribune Company's aim was the same as Ford's: control the production process as much as possible through vertical integration down to the basic elements needed in production.[27]

In the early years of operating the Thorold mill, the company "bought its pulpwood in the open market," but it soon sought to purchase available forest tracts and to lease Crown lands in Ontario and Quebec in order to control its own pulpwood supplies. Robert McCormick himself made some of the earliest scouting trips in Canada, where he both surveyed timberlands by canoe and met with provincial policymakers. The company's initial land acquisition was the 1915 purchase of 314 square miles of land along the Rocky River on Quebec's North Shore for $350,000. The company would add gradually but significantly to this modest initial acquisition, and by 1974 it would own or have exclusive leases on 11,254 square miles of forest across Canada, an area that was in the aggregate larger than the entire state of Massachusetts.[28]

In Ontario, the company's primary landholdings were located at Heron Bay, on the northeastern shore of Lake Superior. The initial timber limits acquired in 1937 comprised just over 1,000 square miles, and subsequent additions brought the total in the region to 3,083 square miles by 1962. To process the wood in preparation for shipment to the Thorold mill, the company began building a barking

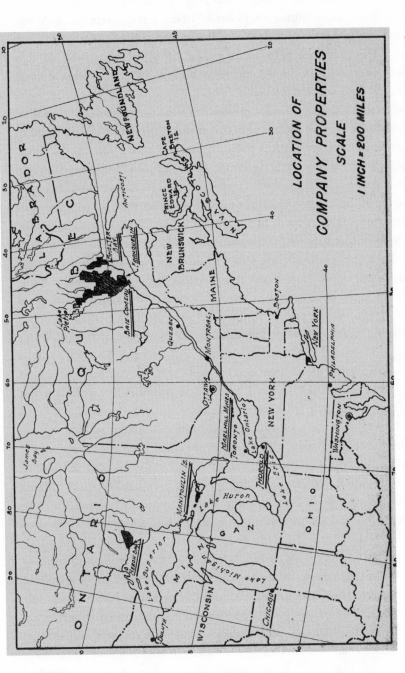

This map shows the size and scope of the Tribune Company's Canadian timber leases and production network as of 1948. "The Ontario Paper Company, Limited and Quebec North Shore Paper Company: The Company's Properties and Production Activities in Relation to the Papers' Newsprint Requirements," 1948, vol. 25316 (acquisition no. 2006-00392-5, box 27), folder Baie Comeau 10th Anniversary, pt. 9, QOPC, LAC. Used with permission.

plant soon after acquiring the Heron Bay timber limits. The plant, completed in 1939 at a cost of nearly $1 million, allowed the company to exploit what one account described as an "almost virgin forest." In this forest, lumberjacks cut logs from the company limits into 4-foot lengths and then used the Black River as a means to transport them to the barking plant. After processing there, the logs traveled down a 3.5-mile flume to the wharf on Lake Superior, where they were loaded onto ships and sent to Thorold.[29]

The Tribune Company's land acquisition in Quebec would prove to be the basis for even more expansive operations in that province. Beginning with the acquisition of the Rocky River limits in 1915, the company added tracts in the region along the North Shore of the St. Lawrence. These included 264 square miles purchased or leased in the 1920s around a settlement that became known as Franquelin and the 552 square miles around what would become the town of Shelter Bay. The company would secure its first major North Shore land concession in 1923, when it obtained a lease on 2,000 square miles of forest along the Manicouagan River after a competitive auction. One Canadian observer of that deal called the land the "most valuable and largest area of timber properties ever put up at auction in the Province of Quebec in one block," and, along with the holdings at Shelter Bay, this would provide the basis for what would be a major industrial undertaking over the following three decades. The company paid an $870,000 fee to initiate the lease, and it also committed to paying roughly $30,000 in annual fees for water rights, ground rent, and fire protection. The company also agreed to build within seven years either a pulp or a newsprint mill at which it would manufacture either pulp or newsprint from wood cut on the limits.[30]

One condition of the lease which would create tremendous political difficulties later was the agreement to pay a royalty fee called "stumpage" based on the amount of wood cut. This included a standard fee of $1.35 per cord. In addition, the company had to pay what was called "premium stumpage" as part of winning the auction for the land. These different kinds of stumpage fees were revenue-generating mechanisms instituted by the Quebec government when it licensed portions of public Crown lands to private corporations for cutting. These private licensees paid annual ground rent on their land, and the government charged the standard stumpage fee on all cutting. The premium stumpage fee was determined at auction, when all parties interested in a particular piece of forest submitted bids for the premium fee that they were willing to pay on top of the standard stumpage fee. The highest bidder would be awarded the exclusive license to cut on the particular piece of land. In the case of the Tribune Company's 1923 North Shore lease, the premium stumpage fee that it agreed to pay beyond the standard fee was $1.675 per cord. In addition to the exclusive right to cut pulpwood on the land, the company was also granted the privilege of developing hydroelectric power on the nearby Outardes River. As it did through the stumpage fees, the company would also pay royalties to the province for the power it generated. The province benefitted from the lease by way of not only the revenue from these royalties but also the fact that it obliged the company to build a hydroelectric dam on the Outardes River, which at the time was undeveloped, and a pulp mill with a 100-ton daily capacity.[31]

In essence, provincial officials licensed the power rights to a river and the cutting rights to trees in a prime piece of forest in exchange for the private capital necessary to develop these resources, and they used policy mechanisms to draw foreign investment to an area that had no industrial development and a small population that was "very scattered," as Honoré Mercier, the Quebec minister of lands and forests, claimed. The North Shore had "abundant natural forest and hydroelectric resources," Mercier noted, "the exploitation of which suggested a future full of promise." Back to the 1910 embargo on pulpwood exports from public lands, Mercier noted, Quebec policy had sought to "encourage the establishment of provincial forest industries," and the massive land concession given to the Tribune Company and the "privilege to exploit the forest" were in keeping with that. Indeed, Mercier argued, the company would be in the vanguard of developing the North Shore, and its operations offered "many advantages" to Quebec and would help bring "material prosperity" to a region that had not enjoyed it as much as the rest of the province.[32]

This was in many respects a representative example of Quebec resource and development policy under Premier Louis-Alexandre Taschereau, who served from 1920 to 1936 as a member of the Liberal Party. Taschereau was a modernizer having much in common with fellow Liberal and former prime minister Wilfrid Laurier, and like Laurier, he welcomed US foreign investment in Canada, especially in hydroelectric development and resource exploitation. In Taschereau's vision for a modern Quebec, the province's forests were perhaps the most important of these resources, and this disposition proved particularly helpful for corporations like the Tribune Company which manufactured newsprint for export to the United States. Taschereau's understanding of economic development being generated from public-private partnership harmonized well with Robert McCormick's Republican philosophy, and Taschereau's views in general were also in many ways aligned with the US Republican Party of the period, including President Herbert Hoover, when it came to understanding the benefits of partnerships between state and industry.[33]

With its land concession and the support of the Quebec government, the Tribune Company set to work immediately in developing operations in the North

Shore region, and the key figure in Canada was Arthur Schmon, who became the president of the Ontario Paper Company and the most important Tribune Company executive in Canada. Schmon, born in New Jersey in 1895, was a battery officer in McCormick's First Division in France during World War I, and he impressed McCormick with his service. When the war ended, McCormick offered Schmon a job managing the North Shore pulpwood cutting operations, and over the next four decades Schmon worked his way up the company ladder, eventually becoming one of the top Tribune Company executives in North America. Schmon was the quintessential company man, personally and professionally. Besides his service to the company, he named his son Robert McCormick Schmon in honor of his boss, and in November 1963, just four months before Arthur Schmon died, he was remarried (his first wife had died earlier in 1963) to the widow of former *Tribune* publisher Chesser Campbell, who had died in 1960. Robert McCormick Schmon would himself go to work for the Tribune Company, and he would serve as the chairman and chief executive of the Canadian subsidiary from 1979 to 1984. Arthur Schmon was essential to the Tribune's success in Canada, and virtually all the company's operations there went through him.[34]

With McCormick directing from Chicago and Schmon operating in Quebec, the Tribune Company continued its development programs after 1923. The company began planning construction of a dam on the Outardes River in Quebec with the aim of providing power for its paper mill. From the outset, McCormick looked at the Tribune's North Shore operations as part of a long-term commitment, remarking in 1924 that "with the vast forest up the river we should construct works of the utmost permanence, and reduce repairs, renewals, and re-construction to the minimum." The company began construction on the Outardes dam in 1926 and would spend $2.25 million on the dam and power facilities on the river over the following decade. The dam was supposed to be operational by the spring of 1931, part of a total investment in the region which the company estimated to be $6.25 million by 1936.[35]

The Tribune Company's North Shore operations brought them into a controversial partnership with the paper manufacturer Frank Clarke, whose company had built what was at the time only the second pulp mill in Quebec in 1908. The mill, near the small North Shore port town of Sept-Îles, was supplemented by the company town of Clarke City and was one of the only industrial developments along the North Shore prior to the Tribune Company's arrival. Over the following decade, the Clarke-owned Gulf Pulp & Paper Company became a successful manufacturing concern, and it attracted the attention of the Harmsworth brothers, Alfred and Harold, also known as Lord Northcliffe and Lord Rothermere. The Harmsworths owned one of the most powerful newspaper corporations in the

In the 1930s, the Chicago Tribune Company built a dam on the Outardes River to fulfill one of the conditions of its North Shore forest concession and to generate power for its operations in the region. This aerial photo shows the operational dam in 1937. Vol. 25198, folder Outardes (1 of 2), QOPC, LAC. Used with permission.

world, and they had previously formed the Anglo-Newfoundland Development Company to manufacture pulp and paper from their timber holdings in Newfoundland to serve their growing British newspaper empire. In 1920, Lord Rothermere purchased Clarke's Gulf Pulp & Paper Company.[36]

Frank Clarke soon began expanding operations in the interest of serving his new British owner, and a North Shore forest concession that he obtained in late 1923 created a great deal of controversy in Quebec's Legislative Assembly. In the process, the *Tribune* was dragged into allegations that corrupt practices by the Liberal Taschereau government had led to land concessions being granted as political favors at great cost to the province. The problems stemmed from a roughly 2,300-square-mile timber lease granted to Clarke in November 1923, several months after the Tribune Company had obtained its own lease in the area. The lease was located just across the Manicouagan River from the large concession that the Tribune had obtained earlier that year. In a heated exchange in Quebec's

Legislative Assembly, Conservative member of Parliament (MP) Charles Smart fumed that Clarke's lease had been "made to benefit friends of the Government," and he claimed that Clarke and the Tribune Company had "worked in collusion" to make the auction price artificially low. The public notice of the auction, Conservative MP Arthur Sauvé added, had been made only thirty days before the date of the actual auction, leaving no time for other potentially interested parties to scout the land and assess its value. Since the Tribune Company and Clarke were the only parties already operating in and familiar with the region, each had a clear inside track to a successful bid, and they were ultimately the only two bidders. It was, Sauvé remarked, suspicious that "there had been but two tenderers" for such a valuable forest concession, and he claimed that the fair market value of the land was sacrificed to cronyism because the auction was not more competitive. Smart alleged that, during the bidding process, Clarke had made a trip to Chicago to meet with McCormick, and he claimed that the pair agreed not to drive the price up through a more competitive bidding process. Smart declined to produce any actual evidence of collusion or to give the source of his information, which provoked Taschereau to dismiss the matter as "old maid's gossip," but the financial difference between this auction and the one that the Tribune had won earlier in the year was significant. In order to win its own earlier competitive auction, the Tribune Company had been forced to pay $1.675 per cord in premium stumpage. This time, with the Tribune and Clarke the only parties to the auction, and with both aware of that fact, the bids were dramatically lower. The Tribune Company offered a mere 2.5 cents per cord, and Clarke won the auction by doubling that bid to five cents. Given that this was $1.625 less per cord than the Tribune had paid, Smart claimed that this collusion to keep the price down would cost the province $20 to $25 million over the life of the concession, and Suavé added that there were two "forces at work in the destruction of Quebec forests, the first being forest fires and the second the American interests." Criticisms of the concession would follow Clarke and the Tribune Company for years in the province.[37]

Lease in hand, Clarke began expanding his operations throughout Quebec to serve Rothermere's growing newsprint needs, and these partners formed a new corporate entity, Anglo-Canadian Pulp & Paper Mills, to supply Rothermere with newsprint for his UK newspapers. Despite facing significant public criticism of their timber leases on the North Shore, the Tribune Company and Clarke continued cooperating to exploit the North Shore timber resources that they controlled. On October 4, 1926, the Tribune Company and Clarke's Anglo-Canadian Pulp & Paper created a joint venture called the Quebec Logging Corporation, which would cut pulpwood on both properties on the Manicouagan. The arrangement would give little comfort to those who had accused the two companies of collusion. In essence, the two companies created a shell company to cut wood on the combined limits that spread out the respective premium stumpage costs between the partners. As Arthur Schmon noted of the Quebec Logging Corporation, it was "based on the principle of share and share alike. All of the preamble and preliminary agreements on the subject—and what is more important between partners, the discussions—stated that the purpose of the organization was to ensure that the Ontario Paper Company would receive 50% of the wood on the Anglo Canadian Company's limits and the Anglo Company would receive 50% of the wood on our limits." The agreement had the aim and effect of "providing for the joint operation of the two Manicouagan properties" and "averaging the premium stumpage on both limits," thus lowering timber costs for both parties. Had Clarke been forced to pay at auction the same premium stumpage as the Tribune, a fee of $1.675 per cord, the companies would have paid a combined $3.35 per cord in premium stumpage. Instead, with Clarke paying five cents per cord after the alleged collusion, the total premium stumpage cost across both limits would be only $1.725 per cord. The Quebec Logging Corporation created an entity through which each partner was "entitled to one half of all the pulpwood on the respective limits of each of the Companies." Agreeing to log across both properties and split the premium stumpage, each company only paid $0.8625 per cord in premium stumpage fees to cut pulpwood.[38]

Though the arrangement promised to dramatically lower its long-term pulpwood costs, Tribune Company officials were often frustrated by the pact. Early in the partnership, it became clear that the Tribune Company had grander designs for the North Shore than did Clarke. As one Quebec historian notes, Clarke and his internal partners were "un peu plus timides" in their plans for development in the region, and the Tribune Company would drive the development much more. To provide pulp and paper for Rothermere, Anglo-Canadian in 1925 built a mill near Quebec City. To supply that mill, Clarke chose to use pulpwood from other forest holdings that the company had in Quebec, leaving the Tribune Company operating mostly alone on the North Shore on the shared limits. Clarke essentially treated the Manicouagan limits as a hedge for future development, with the annual ground rent being his only cost. He indicated to the Tribune Company in 1928 that he was "not planning to obtain any wood from the Quebec Logging Corporation for an indefinite period, which might be ten years."[39]

The Tribune Company, on the other hand, was committing significant capital to industrial and infrastructural development in the region, and it was considering building a company town in order to provide greater stability for its operations. To Arthur Schmon, Clarke began to seem like a freeloader. "I have always been suspicious" about Clarke's goals for and statements about North Shore development, Schmon remarked in 1929. "The advantage, which he is trying to get out of this situation, is first to get us to develop the Manicouagan and then later for him to come into a settled and developed townsite; add a few touches, and operate with a lower capital investment than he could if he developed simultaneously with us, and with the benefit of our experience, which under the best conditions will cost us money."[40]

Despite the uneasy relationship with Frank Clarke, the Tribune Company's Canadian operations were robust and successful. In the three decades after building its newsprint mill at Thorold, the Chicago Tribune expanded its operations into the forests of both Ontario and Quebec, and the company eventually controlled vast tracts of forest that it used in a vertically integrated process of newsprint manufacturing. As the company expanded the scope and reach of its operations, it continued to rely on political connections in both the United States and Canada. As trade policies had allowed the company to bring Canadian newsprint to the United States duty-free, provincial resource policy allowed the company to control the raw materials that it needed to maintain production.

The Network of Shipping

Manufacturing newsprint involved a host of logistical challenges for publishers. The raw materials (the logs) were heavy and bulky, and the finished product (the newsprint) was not only similarly heavy and bulky but also fragile. Paper storage was costly in urban areas, and the newsprint rolls required careful handling, as the paper could be damaged easily in the shipping process and thus rendered unsuitable for use in industrial printing. To run a successful vertically integrated newspaper, Robert McCormick had to manage a perpetually moving supply chain of materials and products, and the tonnages involved in doing so could be massive. In the nine years after completing the Heron Bay barking plant in 1939, for example, the Tribune Company shipped 366,822 cords of pulpwood to Thorold. Between 1920 and 1947, the company shipped an additional 2,421,402 cords of pulpwood from Quebec's North Shore to the Thorold mill.[41]

To move this material, the Tribune Company took advantage of the inland network of canals built in North America starting in the nineteenth century, particularly the Welland Canal, which cut across the Niagara Peninsula and connected Lake Ontario from the north with Lake Erie to the south. This allowed for more local commerce and traffic between the United States and Canada, but perhaps more importantly, like the Erie Canal, it linked cities along the Great Lakes to the Atlantic, in effect transforming a number of otherwise-landlocked midwestern cities like Thorold, Ontario, into extensions of the Atlantic economy. Throughout the nineteenth century, the Welland Canal was a boon for the Niagara

Peninsula, and by the 1920s it was being described as the "principal industrial gateway from the United States to Canada." It was certainly of benefit to the Tribune Company, as it offered not just an easy passage for ships to and from the St. Lawrence River but also proximity to cheap hydroelectric power, with Niagara Falls only 10 miles to the east.[42]

To take advantage of these ocean and inland waterways, the Tribune Company pioneered in the development of a system of boats to move its newsprint. Its efforts to vertically integrate encompassed not only forests but also a shipping fleet, and it created a wholly owned subsidiary shipping company initially called the Ontario Transportation and Pulp Company in January 1914. This allowed it to essentially build a fluidly functioning network of newsprint production, transportation, and consumption which was entirely under its own control. For its first two years, the company leased ships to make the trips from Quebec and Ontario to Chicago. In 1916, however, the company purchased its own ships, though it found them "ill adapted" to its needs. In response, the company itself began commissioning the construction of ships that were "specially designed" to handle its cargos of bulky and fragile newsprint. By 1928, the renamed Quebec and Ontario Transportation Company was using these boats to make regular in-season trips carrying pulpwood on the St. Lawrence from Shelter Bay to the mill at Thorold, a journey of roughly 900 miles. The subsequent 800-mile trip from Thorold to Chicago took an estimated 106 hours in 1932.[43] The *Tribune* consistently ran brief articles about its paper shipping operations, often in the spring when the first ships began arriving after the thaw.[44]

As the company expanded its fleet and shipping operations, its practices of ship design remained attuned to the particular needs of the bodies of water on which the ships were traveling. The Thorold mill was situated at a waterway juncture that had on either side different ideal ships for the conditions. Those coming up the St. Lawrence, usually laden with pulpwood or pulp, were vessels more suited to canal transport, as they were "limited as to length, beam and draft by the dimensions of the lock chambers" along the system. Those heading to Chicago with finished newsprint were the larger cargo ships that moved through the deep but relatively calm waters of the Great Lakes. By 1949, the Tribune Company's fleet was up to nine ships, and at its maximum size in 1978, the shipping company owned sixteen vessels moving materials between the various stages of operations.[45]

The shipping operations undergirded a continuously operating and routinized supply chain. As one industry profile noted, "An interesting feature in the shipping angle is the fact that boats that transport the pulpwood make a complete circle tour. Leaving Thorold with a cargo of paper from the Ontario Paper Company's plant, freighters ply to Chicago, where is situated the Chicago Tribune. Unloading the paper there the boats proceed to Heron Bay light and take on a full load of pulpwood for Thorold." The North Shore was a similar transportation node, as ships would return from Thorold with, among other cargoes, the "limestone, coal and foodstuffs" necessary for the Quebec operations. This transportation network connected all these far-flung elements of the Tribune Company's production network. Through it, the company moved raw materials and finished goods in a continuously flowing supply chain from the forest to the pressroom. Though operating across an international border, the Tribune Company operated with a high degree of independence and in many respects functioned as an almost entirely self-sufficient manufacturing firm.[46]

Building the First Company Towns at Heron Bay and Shelter Bay

The parts of Ontario and Quebec in which the Tribune Company operated were remote and initially largely undeveloped. Given the significant capital investments in industrial infrastructure, as well as the desire to provide amenities for its workers, the company began building company towns adjoining its timberlands. In Heron Bay, Ontario, the company had some six hundred employees in its woods operations in the region, and the barking plant promised to provide seasonal work for up to two hundred people and year-round work for roughly thirty-five. The company established a townsite to house the workers, and in addition to building its own hydroelectric power facilities for the plant and town, it also erected municipal electrical, sewer, and water systems. To observers, Heron Bay was a "thriving little town," as one visitor noted in 1949, "and if the plans of the company are carried out it will be a bustling little community for many years to come." By the early 1950s, according to reports, the town of Heron Bay was well apportioned. The living conditions, as the *Tribune* described them, were "as comfortable as in a large city. Recreation facilities are centered in a community club building operated by the employes. Dances, card parties, concerts, and social gatherings are held in the main auditorium of the building. Adjoining the auditorium is a lounge and reading room. In the basement are facilities for billiards, ping pong, and similar games. A rink for hockey and curling is popular in the winter."[47]

In Quebec, Shelter Bay became the company's first attempt to build more than a rough logging camp in that province. Initially, the town provided a base for the lumberjacks who would head into the forest in the winter to cut down trees. The company built a sawmill to process those trees into smaller logs for shipping, and the dock facilities were in perpetual use in spring, summer, and fall as pulpwood was loaded onto ships for the journey to Thorold. The town began in 1916 as a group of "rude log cabins" that initially were simply built from the trees that had

By 1940, the settlement at Shelter Bay had come to encompass a cluster of modest homes on the shore of the St. Lawrence River. Vol. 25177, Scrapbook 1926–34, QOPC, LAC. Used with permission.

been cut down to clear land for the site. The company soon realized that it wanted this company town to be more appealing and hospitable to workers. "Foresight," the *Tribune* later remarked, "went far beyond these expedient cabins. Looking into the future, the Tribune saw a time when hundreds of families would regard this place as home. People would marry and die there, and children would be born. Foresight called for a town; the Tribune built one. It was not a mere congeries of shacks, but a town modern in all its ways, scientifically planned and rigidly zoned." By 1930, Shelter Bay had a population of nearly eight hundred people, and the Tribune Company had begun a more extensive program to encourage home construction by employees. The Tribune offered building sites to prospective homeowners, and it offered them design consultation on home construction. After this process, the company provided free "rough lumber" and all "other materials and labor at cost." Through the building of a factory and townsite, the *Tribune* claimed, the Shelter Bay enterprise had grown to be a "monument to the enterprise of the newspaper maker," according to the *Tribune*, and it was complete with "many modern homes, electric lights, and a central heating plant." The town was, the paper claimed, "far more modern and livable than many a larger and older one."[48]

As the town of Shelter Bay grew, the Tribune Company's North Shore operations began to orient themselves not only up the St. Lawrence River to Chicago but down the river to New York City via the Atlantic. One of the primary causes was the 1919 founding of the *New York Daily News*, a paper that had by 1930 become the highest-circulating newspaper in the United States. Thorold might supply the *Tribune* with newsprint, company officials surmised, while a new mill built on the North Shore could supply the *Daily News*.

The *New York Daily News* and the New Metropolitan Daily

In McCormick's early years running the Tribune Company, he shared editorial responsibilities with his cousin, Joseph Patterson. Compared to the aristocratic and rather rigid temperament that McCormick displayed, Patterson exhibited more mercurial tendencies. McCormick had taken immediately to the business side of the operations, while the personally eclectic Patterson exhibited a "knack for conceiving and developing newspaper features," including the comic strips Moon Mullins and Little Orphan Annie. By 1911, Patterson had established himself as being, in addition to an aspiring publisher, both a competent playwright and an ideological eccentric who flirted with socialism. Despite the significant personal and political differences between Patterson and McCormick, the two initially worked well together running the newspaper. Within a couple of years, however, the cousins began to have some interpersonal problems sharing editorial responsibility, and Patterson wanted to find a way out of the arrangement. When "trouble started in Veracruz in 1914," one account later noted, Patterson "seized the chance to get away from the *Tribune* by going to Mexico as its correspondent." Immediately following that, he went to Belgium to cover World War I.[49]

The cousins both joined the military once the United States entered the war in 1917, with Patterson becoming a captain and McCormick a colonel. (This was the origin of the military titles that both men would use for the rest of their lives.) The *New York Daily News* was born of this wartime experience. In 1918, both men were in France, and McCormick drove to Mareuil-en-Dôle, near where Patterson was stationed. According to one version of the origin story, "the pair tried to talk in a farmhouse" that was "temporarily in use as a field headquarters." There was too much noise and activity for the pair to converse, so they "climbed out a window and sat on a manure pile in the back yard, from where they could hear the rumble of the guns and see shells bursting. In this pungent setting the tabloid New York News was conceived." The germ of the idea was a previous meeting that Patterson had had in London with *London Daily Mirror* publisher Lord Northcliffe, after which Patterson was convinced that the tabloid model could be successfully imported to New York. McCormick had been considering expanding into the New York market already, and before the war he had unsuccessfully engaged James Gordon Bennett Jr. in discussions about the possibility of buying the *New York*

Herald. Patterson's idea made commercial sense, and it also appealed to McCormick because it would get his cousin out of Chicago and allow McCormick to assume complete control of the *Tribune*.[50]

When the *New York Daily News* started publishing in 1919, it inaugurated a new style of tabloid journalism in the United States. Previously, the *New York Daily Graphic* had a sixteen-year run in the 1870s and 1880s in the tabloid format, but otherwise the United States had not had the kind of tabloid newspapers that had been hugely popular in Britain. As the first of its kind in the twentieth-century United States, the *Daily News* was, like the British tabloids, not aimed at providing deeply incisive journalism. As one report described it, "Like the London publications it resembles, it will not print news events in detail. It will not attempt to give its readers complete news reports, but only that which is of remarkable interest and which lends itself readily to illustration." Instead, the *New York Daily News* was "designed along sprightly lines that appeal to the tired business man and the tired housewife," and its content would be "snappy stuff" that could be "seen and read quickly. If anyone wants a complete compendium of world events, he or she can read other newspapers. To be immediately entertained or interested," in contrast, one could turn to the *New York Daily News*.[51]

The *Daily News* published its first edition on June 26, 1919, entering a crowded New York market that included seven major morning and ten major afternoon papers. Despite the competition, the *News* was a quick and spectacular success. The explosion of the paper's appeal was somewhat controversial in New York journalism circles. Many considered it to be a "bizarre experiment," and one critic remarked, "No newspaperman returning from the war considered applying for a job on it until he had exhausted all other possibilities and had spent his discharge bonus. To accept its pay meant to lose caste professionally." To some, it was the presentation of pictures that both created mass appeal and elicited elite derision. As Philip Payne, the paper's managing editor, noted in 1924, "Certainly pictures are the chief attraction, for they are the very essence of tabloidism. . . . A large part of the lure is based on the feeling it gives the readers of being eye-witness to the big news of the day." To critics, the appeal of the *Daily News* was not just the pictures but also its content, which many considered deplorable. "Probably no paper in the world has been so criticized," one journalist wrote in 1934. The *Daily News* was widely "denounced and pilloried, not only by the lay public but by the 'professional' publishers. Its daily performance, the critics say, has only been successful because it catered to mass cravings, because it dished out the news in the same raw condition that it received it, and without diluting it for public consumption." Roy C. Holliss, the paper's general manager, defended the paper's editorial philosophy against what he derided as the "academic criticism" of its allegedly lowbrow content. "The 'intelligent adult' goes to prize fights where he enjoys a violent brawl in the same manner as does a 14-year-old," Holliss noted. "He revels in a display of brawn and mastery. That is human nature. Perhaps that same man will come home from the prize fight and pick up Shakespeare and read it and enjoy it. It is our feeling that the 14-year-old mind . . . is a part of the mental equipment of the average human being. A part of everyone's mind is 14 years old. . . . We try to edit the News to meet the varied requirements of the average person's mental equipment." The *Daily News* itself touted its populism in an advertisement in asking, "Can a million people be wrong in their choice of a newspaper?" Assessing the overall rise of the tabloids and their wide appeal, press critic Silas Bent in 1927 called them, in terms that well predated those used by cultural historian Lawrence Levine, the "printed daily folklore of the factory age."[52]

Whatever the appeal, the *Daily News* was perhaps the most stunning success in American journalism in the 1920s. In October 1921, the paper became the second-highest-circulating morning paper in the United States, behind only the *Chicago Tribune*. "Now watch us catch momma!" the *Daily News* subsequently promised. Daily circulation continued to rise throughout the 1920s, going from 186,000 in 1920 to 482,000 in 1922, and by 1924 it had the highest circulation of any newspaper in the United States. The circulation hit a million in 1925 and 1.6 million by 1935. Sunday circulation rose at a similarly meteoric rate after publication began in 1921, starting at 504,000 in 1923, reaching a million in 1925, and then passing two million in 1934 and three million in 1937. Advertising linage rose with circulation, from 1.7 million lines in 1920 to 20.3 million in 1937, when it had the "largest display lineage in the U.S." By the mid-1930s, after fifteen years in existence, the paper was, as one observer noted, the "greatest revenue producing paper in the country," and by the end of the decade only three other newspapers in the entire world (the *London Daily Express*, *London Daily Herald*, and *News of the World*) sold more copies on a daily basis.[53]

Though the *New York Daily News* had appealing content, one of the reasons for its continued success was that it followed the pattern of the *Chicago Tribune* in being an innovator in manufacturing. It had a staff of only nine in the beginning, and they worked in rented office space in the Evening Mail Building on City Hall Place. The paper was also printed using the *Evening Mail*'s equipment at the outset. In 1921, the *Daily News* moved to a new facility at 23–25 Park Place which it equipped with a "complete modern newspaper plant." There were "three high-speed octuple presses" in the building, two of which were Goss presses that were "especially designed for picture printing and have superior ink distribution and other features adapting them to the finest quality of half-tone work." This new facility helped the paper immensely, but by 1927 the paper's growing circulation meant that it needed even larger production facilities, and the company built a new six-story printing plant in Brooklyn at a cost of $2.25 million. "Its location emphasizes the factory purpose of the building," one account noted. "Easy inflow and outflow of raw materials and finished products are insured, since the building stands close to the main arteries of Brooklyn and Long Island traffic. Directly opposite the main entrance run the Long Island railroad tracks and within a minute's run are main highways to all parts of Brooklyn." This was, as another account described it, a "newspaper factory, not a newspaper, for all it contained was presses and stereotyping equipment. The section was made up in Park Place and mats of the pages were sent—with mats for the regular paper—by car over the Brooklyn Bridge."[54]

At the end of the decade, the company announced its success in a new physical way, building a skyscraper in Midtown Manhattan on 42nd Street between Second and Third Avenues. The architect was Raymond Hood, the same person who had designed the Tribune Tower, but the Daily News Building was a very different sort of structure. Whereas the Tribune Tower attempted to communicate pedigree through its neo-Gothic styling, the Daily News Building communicated modernity. The building was both monumental and sleek, and it meant to announce itself as a physical vanguard of Jazz Age New York City. The real estate and construction ultimately cost over $10 million, and the building aimed not only to project a particular public image but also to sit at a transportation node. "The new building will give the Daily News a strategic location with regard to distribution facilities," a report noted. "Within two blocks of Grand Central Terminal, it will have means of out-of-town delivery almost at its doorstep." This would allow it "quicker access to the Manhattan, Bronx and Westchester residential districts," an essential thing for a morning paper like the *Daily News* in a large city like New York. Many afternoon papers were purchased downtown, which favored a downtown printing plant, but the morning paper often was delivered to subscribers at home prior to the ride into the city on mass transit, and the publisher had to get his printed newspaper there before the commuters got on the train.[55]

Within a decade of its founding, the *New York Daily News* had evolved from an experimental example of British-inspired tabloid journalism into the highest-circulating newspaper in the United States. In the process, the paper had moved from rented quarters and into a building so iconic that it would be the model for the *Daily Planet*, where the fictional character Clark Kent worked. The success of the paper was to some degree a reflection of its popularly appealing content, but it was also due to its being a part of one of the most innovative newspaper firms in the United States, and in the following decade the Tribune Company would dramatically expand its pattern of international vertical integration to serve both the *Tribune* and the *Daily News*.[56]

The Politics of Vertical Integration

Robert McCormick's use of vertical integration made the Tribune Company one of the most powerful media companies in the United States. The strategy gave the company a significant amount of control over its industrial production, but it also meant that the company was perpetually negotiating with policymakers in the United States and Canada in order to maintain operations as it saw fit. This was especially apparent during World War I and its immediate aftermath, when significant increases in newsprint prices created problems for publishers across the United States. In Chicago, for example, the *Herald* claimed that its newsprint costs had increased $400,000 between 1916 and 1917, and the *Chicago Daily News* claimed a $1.4 million increase in its paper costs. In May 1917, this prompted publisher Victor Lawson to call it an "impossible business situation," and he doubled the sale price of his *Daily News* to two cents. The *Tribune* also doubled its sale price to two cents, but it was able to weather the conditions better than its competitors because it had vertically integrated. Frank Glass, the publisher of the *Birmingham News*, remarked enviously that the *Tribune* "owns its paper mill and is more independent than most of us." In 1921, the *Tribune* boasted that its Canadian operations allowed it to maintain a great degree of self-sufficiency, and it claimed that this helped its journalism. "The economic independence" of the *Chicago Tribune*, the company claimed, "parallels independence of editorial stand, of newsgathering facilities, and of advertising policy, which have combined to establish this great newspaper in its unique position."[57]

The independence that the Tribune believed it had carved out was at times at odds with the degree to which policymakers understood that independence. During World War I, for example, when R. A. Pringle was trying to manage newsprint supplies in Canada at the behest of his government, the Tribune claimed that, since its Thorold mill provided newsprint only for its internal use and was not a commercial mill, it should be exempt from any wartime rationing or redistribution programs. The Tribune, McCormick wrote to R. A. Pringle in May 1917, was "dependent for its supply of paper upon this mill, which was built for no other purpose," and he claimed that "our private property, which is nothing but a part of our own business, should not be commandeered for other private businesses." The Tribune Company in effect wanted its Canadian operations treated as an extraterritorial appendage of its American operations, and it wanted freedom from Canadian regulations on wartime newsprint production.[58]

Despite McCormick's protests, the Tribune did accede to Pringle's requests and sold newsprint to a number of newspapers across Canada, though it perpetually sought ways out of doing so. In late 1918, the company pleaded to Pringle that the Thorold mill was not "producing enough paper" for the *Tribune*'s needs. In providing newsprint to other Canadian newspapers, the Tribune Company claimed, it was doing so "at great sacrifice," and the company made what it called an "honest request that no further calls be made on us."[59] By 1920, the Tribune was much less polite in response to these Canadian requests for assistance in providing newsprint to domestic customers. In May 1920, McCormick in a speech called the Canadian government's newsprint allocation programs "not robbery, but downright bolshevism." Tariff policy had allowed McCormick to commence operations in Canada, but his practical exploitation of the privilege to import newsprint to the United States free of any duties showed him to be a selectively cooperative continentalist.[60]

One of the starkest examples of McCormick's strategic exploitation of North American tariff policy came soon after the end of World War I when he began building another mill in the United States at Tonawanda, New York, to make paper for *Liberty* magazine, a national weekly that his company would begin publishing in 1924, and for a special rotogravure section of the *New York Daily News*. Because of the specifics of North American tariff policy, newsprint crossed the border duty-free while other grades of paper did not. Wanting to print on a higher-quality sheet of paper than regular newsprint, the Tribune Company believed that it would be cheaper to do so in the United States than in Canada, and thus it built a new mill at Tonawanda. In November 1923, the company laid the cornerstone for the mill, which it claimed had a daily capacity of 60 tons. One of the ironies of the mill was that it drew electricity from Niagara Falls, the same power source that the Tribune Company's mill used in Thorold. But given the fact that the company would have to pay a tariff on the mill's paper if it manufactured it in Canada, it chose to do the manufacturing on the US side of the border while importing much of its raw material from Canada. One Canadian journalist noted of the arrangement that the tariff was essentially "taking competent labor from this country" by encouraging the Tribune Company to manufacture in the United States. The overall result of this arrangement, he claimed, contributed to the "dual menace of forest and population depletion!" Ultimately, however, *Liberty* never performed well for the Tribune Company, and in 1931 the company sold the magazine to publisher Bernarr Macfadden and the mill to IP.[61]

If *Liberty* and the Tonawanda mill were unsuccessful, the rest of the Tribune Company's vertically integrated operations were a stunning success overall, and by the end of the 1920s the company sought to expand them further on the North Shore. In doing so, the company's actions revealed the degree to which the firm's operations were enabled by public policy and by connections to public officials. As of the late 1920s, the *Daily News*, unlike its corporate parent the *Chicago Tribune*, remained firmly linked to IP. By 1928, the *Daily News* was using roughly 100,000 tons of paper per year, most of which it purchased from IP. As had been the case with the *Chicago Tribune* since 1911, Robert McCormick wanted to break away from the arrangement with the paper conglomerate and control the newsprint supply within the company. Throughout the 1920s, the Tribune Company explored various ways of trying to do this and also to fulfill the condition that came with its initial Quebec forest concession requiring it to build a newsprint or pulp mill in the province by 1930. The company considered a series of scenarios for developing operations in Quebec, including one that involved building an integrated pulp and paper mill on the North Shore and another involving a paper mill in Quebec City or New York which would use pulp made on the North Shore.[62]

By early 1929, the company's plans had crystallized around a pulp mill on the North Shore. The company engaged Leonard Schlemm, a member of the Montreal Planning Commission, to help them in the early stages of planning a townsite. Knowing that they would not meet the terms of the 1923 lease and have a mill operational by 1930, the company sought and received an extension to July 1934, and it began to consider building a newsprint mill rather than just a pulp mill on the North Shore. As work continued, the Depression hit and the Canadian newsprint industry collapsed. In 1931, the Quebec government asked the Tribune Company to "for the time being at least, abandon the project by reason of the alleged over-capacity and under-consumption existing during the depression." Despite having invested what it claimed was $6.25 million on North Shore infrastructural development, the company agreed to delay construction of a pulp or newsprint mill there in order to see whether the newsprint market would stabilize.[63]

Conclusion

The Tribune Company's Quebec development projects would expand dramatically in the second half of the 1930s, as the next chapter will discuss, but they were hamstrung as of 1931 because of the politics of Depression-era Quebec. Whatever decisions it made as matters of business strategy, the company was always limited by the political contexts in which it operated in the United States and Canada. Pulpwood cutting would continue for the Thorold mill after 1931, but the pausing of industrial development in Quebec meant that the Tribune Company would be sitting on a significant but wholly unproductive capital investment on the North Shore in the form of the dormant dam that it had built on the Outardes River in order to generate hydroelectric power for its planned but unbuilt mill.

The Tribune Company would return to this development with great vigor later in the 1930s. In those activities and in all of the projects undertaken in Canada after 1912, the Tribune Company sought to expand the vertically integrated manufacturing operations that enabled it to produce and distribute its newspapers as efficiently and widely as possible. Though Robert McCormick and other Tribune officials liked to present the company's operations in terms of business strategy, these operations were always structured by state policy in both the United States and Canada, as tariffs, timber leases, hydroelectric generation privileges, and canal construction all provided it with significant public subsidies and infrastructural support. Ultimately, the two highest-circulating newspapers in the United States in the interwar period—the *Chicago Tribune* and the *New York Daily News*—achieved their success because they were enabled by public policy across North America.

CHAPTER FIVE

Robert McCormick
and the Politics of Planning

Robert McCormick hated the New Deal and, with the platform available to him as the publisher of the *Chicago Tribune*, relished telling the public why. Franklin Roosevelt's administration, *Tribune* editorials repeatedly and consistently claimed, was transforming the country into an "economic and political dictatorship." In myriad ways, the paper asserted in 1936, the New Deal infringed on private rights while expanding and centralizing public power. "Each part in the program fits in its place with the others. The pieces join. When they are all in place they will make a picture. It will be the picture of a country under a socialistic dictatorship."[1]

In the 1930s, McCormick was a member of a loosely connected group of anti–New Dealers hostile to the expanded size and reach of the federal government. In addition to McCormick, individuals like the industrialist du Pont brothers and groups like the Liberty League and National Association of Manufacturers spent large amounts of money and energy fighting the New Deal at the grassroots and electoral levels. Most supported Alf Landon's presidential run against Roosevelt in 1936, McCormick so strongly that he provided free office space in the Tribune Tower for Landon's campaign volunteers. This enthusiastic support of conservative presidential candidates would continue throughout the New Deal period, and it would lead McCormick to back a line of losing candidates and later to generate a famous example of comically inaccurate reporting. It was McCormick's *Tribune*, after all, that boldly, wishfully, and incorrectly declared on its front page that Thomas Dewey was the winner of the 1948 presidential election over Harry Truman.[2]

McCormick saw nefarious purposes everywhere in the New Deal, but his attitude toward federal planning and spending on public works elicited particularly angry and shrill remarks. During the New Deal, federal spending on public works increased dramatically, and during Roosevelt's first two terms some two-thirds of federal emergency spending was on programs like the Public Works Administration and Works Progress Administration. The results of this massive expansion of spending on public works were tens of thousands of structures around the country, including such major projects as the Hoover Dam and New

York City's Triborough Bridge. McCormick was aghast at these developments, with the *Tribune* editorializing that they represented a "vast reaching out for political and economic control by a centralized authority, the destruction of private enterprise, the reckless spending of money exacted by increasing taxation, and the descent to public and private bankruptcy through the foolish experimentation of doctrinaires and demagogs."[3]

Among these loathsome public works programs, Robert McCormick held particular contempt for the Tennessee Valley Authority (TVA), which his paper claimed should be regarded as "revealing the motive of the New Deal program" and an "extension of government controls which would give the United States a planned society no different from that attempted in Russia."[4] The TVA was one of the signature programs of the early New Deal and represented an extraordinarily ambitious effort to use federal power to transform an entire region of the country. The basis for the program was the construction of hydroelectric dams along the Tennessee River, which at the time only had the Hales Bar Dam near Chattanooga and the Wilson Dam at Muscle Shoals, Alabama, and which defined a region encompassing parts of seven states and some 40,000 square miles of land. Taming the Tennessee River would make it more navigable, and the dams would provide electric power for new industries to develop, all the while putting thousands of Americans back to work. Many assessments of the TVA in action were glowing. Sociologist Philip Selznick surveyed the project in 1942 and 1943 and found that, for many, the "TVA has become not merely an administrative model and prototype, but a symbol of the positive, benevolent intervention of government for the general welfare."[5]

Robert McCormick did not see it this way, and his paper editorialized that the program was "madness in the Tennessee Valley" perpetrated by dictators masquerading as bureaucrats. In agencies like the TVA, the "sovietized radicals are scrambling for the highest places. The secretaries of the cabinet have visions of themselves as dictators, each a Hitler in his own right. Under such impulsion the government would soon be in full flight from the constitution, the representative order and the republican form."[6] The town of Norris, Tennessee, became in many ways a focused point of hatred for McCormick.[7] Norris was needed to house workers on the Norris Dam, the first major construction project undertaken by the TVA, but it became through the work of engineer Arthur Morgan and urban planner Earle Draper an attempt to demonstrate how New Deal principles could be reflected in the built environment. The town was to be integrated into its natural surroundings, and housing and services were to be available to residents affordably and efficiently. Built in 1933, Norris became a concrete and prominent example of how the federal government could improve Americans' material lives. The TVA at the regional level transformed the economy, and to individuals during the Great Depression it provided not only work but also housing.[8]

Ultimately, the TVA combined three elements that McCormick most deeply despised about the New Deal as he understood it: the profligate use of public funds, the unwarranted intrusion into what should be private markets, and the centralized planning of public and private affairs. In spending, it demonstrated an abdication of "clarity and financial rectitude,"[9] and the entry into the electricity market was the sort of "burocratic management . . . favored by theorists who wish to destroy private enterprise and by politicians who wish to spend public money to their own advantage" and evidence that Roosevelt was "earnestly . . . following the lead given him from Russia,"[10] where Lenin used the state to control electrical power generation and distribution. Assessing the ways that the TVA structured the lives of the citizens in planned communities like Norris led the *Tribune* to conclude that there was "no longer room to doubt that . . . the communism of Lenin and Stalin has taken root in the United States." As the *Tribune* editorialized in late 1934, "*No man in Norris may engage in private business. No man in Norris may work for wages except for the government. No man in Norris may worship in church or build a church to worship in according to his conscience.* Can any one doubt that in the inaccessible mountains of Tennessee is being grown the germ culture that is intended to infect America?" Created and allegedly controlled by the federal government, the town of Norris came to symbolize everything that Robert McCormick claimed to despise about the New Deal.[11]

It is somewhat ironic, then, that in 1937 the vocal antistatist McCormick himself founded a town in Canada for workers at a new Tribune Company newsprint mill that brought industrial development to a region that previously had very little of it. The town was named Baie Comeau, and as the head of a major US newspaper corporation, Robert McCormick oversaw it as a distant and, in his mind, benevolent patriarch. Located on a remote stretch of the North Shore of the St. Lawrence River in Quebec (250 miles downriver from Quebec City, and almost 420 miles from Montreal), Baie Comeau housed workers whom the Tribune Company employed to extract logs from Canadian forests and process them in a newsprint mill built to be the most advanced in all of Canada. The finished paper was then brought via waterways to Chicago and New York, where his company used it to print the *Chicago Tribune* and the *New York Daily News*, the wholly owned subsidiary of the Tribune Company. Quebec's North Shore had little settlement and virtually no industry at the time of McCormick's initial investment there in the 1910s, but over the following decades it became a prosperous and environmentally transformed region through the exploitation of the raw materials needed to produce two of the leading daily newspapers in the United States. Building Baie Comeau and exploiting the region involved the Tribune Company privately developing hydroelectric power on the Outardes and Manicouagan Rivers through the construction of massive dams, and the town's construction in the 1930s represented, according to one estimate, the "largest single construction undertaking" in all of Canada since the start of the Depression in 1929. If Roosevelt was to have his TVA, McCormick wanted his as well.[12]

The history of Robert McCormick's involvement in Canada is filled with irony. On one level, it demonstrates that, in Quebec, one of American's leading defenders of free enterprise and critics of the New Deal state was at the same time engaged in a municipal development project driven by a level of centralized planning that was in line with and even impressive by New Deal standards. This municipal project was the first step in a broader post–World War II pattern of development on the North Shore of the St. Lawrence River, most importantly through the exploitation of hydroelectric power on the region's previously untapped rivers. Moreover, it is deeply ironic that the basic fact that the antistatist McCormick was able to involve his company in Canada at all had everything to do with state policy on both sides of the US-Canada border. It was through arrangements with Canadian policymakers that McCormick received favorable leases and concessions on timberlands and water rights to be used to generate the hydroelectricity that powered his Quebec newsprint mill. McCormick also reaped the benefits of a fortuitous shift in tariff policy beginning in 1911, after which newsprint could cross the US-Canada border without a duty. Ultimately, through the help of policymakers in two countries, Robert McCormick created an international and vertically integrated newspaper corporation with operations that eventually encompassed over 11,000 square miles of timber lands under its control, two state-of-the-art newsprint mills, a fleet of ships to transport the newsprint to the United States, one of the largest hydroelectric dams in the entire country of Canada (immodestly but unsurprisingly named the McCormick Dam), and the modern company town of Baie Comeau. And yet, ironically, readers of the *Tribune* were warned that government projects like the TVA were threats to liberty.[13]

The Publisher as City Builder

McCormick's hostility toward the New Deal at times shaded into the pathological, and the animosity he directed toward Franklin Roosevelt seemed to many commentators odd and overly aggressive. Some biographers have speculated that the two had a dislike stretching back to their shared prep school days at Groton, but there is scant evidence in the record for what the specific reasons for this might be. But one thing is clear: McCormick saw himself as being in direct personal and political competition with Roosevelt. In some respects, if a handful of events in McCormick's life had been different, he might have had a career as an elected official rather than as a publisher.[14]

In the decade before he took over the *Tribune*, that is in fact what he was. In 1904, he was elected as an alderman to the Chicago City Council representing the Twenty-First Ward, and he displayed the characteristics of a diligent public servant. He was always in attendance at city council meetings, and his general aldermanic stance was in line with many of the city's progressives. In 1905, Mc-Cormick ran for and won the office of president of the Chicago Sanitary District. This was a heady and fortuitous time to be in that office. In 1909, Daniel Burnham's Plan of Chicago was published, and city officials undertook one of the most expansive and significant policy-driven urban redevelopment plans in the country to date.[15]

As president of the Sanitary District, McCormick supervised a range of major infrastructural projects, one of the most important of which involved the extension of the Chicago Sanitary Canal. This had been one of the major national infrastructure projects of the era, which included the reversal of the flow of the Chicago River so that it no longer emptied into Lake Michigan. In his work on the canal, McCormick displayed the kind of hands-on work that he would later show with his timber scouting in Quebec. He was eager to put his canoe in the water, as it were, and to tell the tale. When describing the canal in 1910 as "today the greatest artificial water way the world has ever seen," McCormick related in his published annual message how he had obtained firsthand knowledge of how clean the canal's water had become. That summer, he had taken to the canal in a "small boat" with a traveling party of four. At Joliet, where the canal had been extended in 1907 under his supervision, he marveled at how clean the water was. "Its oily, sooty appearance has begun to depart, and the gasy [sic] smell is absent." While marveling at the work he had done on the canal's development, McCormick ran into trouble as night approached. He later acknowledged that he had not paid attention to the "sound of rushing water" coming from a dam at Dresden Heights. This, he soon realized, "made a veritable cataract of the stream, and our first knowledge of this fact was in the capsizing of our frail vessel and the submerging of ourselves." McCormick and his traveling companions passed safely through the rapids, and after about thirty minutes they were able to get their boat ashore. At that point, McCormick wrote, he became aware of the fact that he had been floating in a drainage canal for sewage. "During this time none of us had noticed any peculiarity either in the taste or smell of the water, which forty miles farther north would be considered poison. There was a certain uneasiness of mind among us during the two weeks which followed, but no one developed any kind of intestinal disorder."[16]

McCormick's work on projects like the Sanitary Canal looked in some ways like the start of a promising political career. However, when he ran for reelection as Sanitary District president in 1910, he was defeated, as were most Republican candidates for city offices that year. That would be McCormick's last campaign as a candidate for public office. It is unclear whether the fact that he never ran again had to do with the sting of losing, a dissatisfaction with the rhythms of public sector employment, or the fact that the chain of events that led him to take over the *Tribune* happened just after. Whatever the case, the events were fortuitous, and being a publisher allowed him to continue to have a public presence. When the possibility of building Baie Comeau emerged, it allowed McCormick to indulge some of his older municipal technocratic tendencies on a blank canvas.

The early stages of development of what would become Baie Comeau began in the 1920s while the Tribune Company was planning a more general expansion of North Shore operations. The company had agreed to the Quebec government's 1931 request that it halt these expansion plans so as not to add additional newsprint production capacity to an already-oversaturated market, a move that was on the Tribune's part commercially inconvenient but politically necessary. The company continued internal planning discussions despite the forced delay in its industrial expansion, and the basic plans for what would become Baie Comeau took shape. By 1934, the company was pleading with Canadian officials to help it restart its Quebec development projects and commence construction of a new newsprint mill. The company claimed to be "in a position at the moment to aid unemployment in this section of the country" through this new capital investment. In essence, the company claimed that whatever problems it caused by adding excess newsprint production capacity to a depressed market would be offset by the economic activity it would generate on the North Shore.[17]

By this point in the 1930s, Tribune Company officials were becoming increasingly anxious about restarting the construction of a newsprint mill, not only because its significant capital investment in the Outardes Dam was essentially sitting idle but also because the company's newsprint needs had increased during the period. The *Chicago Tribune*, along with the *New York Daily News*, formed a two-paper combination that was by 1934 the second-largest chain in the United States in terms of papers sold, with a combined daily circulation of 2,200,098 and Sunday circulation of 2,643,219. Only the Hearst Newspapers chain had a higher total circulation, but it took that company twenty-four dailies and sixteen Sunday papers to sell 3,951,852 and 4,686,214 copies, respectively. The third-place chain, Scripps-Howard, had twenty-four daily newspapers and six Sunday newspapers, with a combined circulation of 1,705,234 and 275,804, respectively. So, with just the *Chicago Tribune* and *New York Daily News*, the Tribune Company had created one of the most successful newspaper corporations in the United States by the middle of the 1930s. Because of the delay in expanding its North Shore papermaking operations, the Tribune Company claimed that it had to purchase newsprint elsewhere when it could have been manufacturing its own. In 1936, the company noted that between 1933 and 1935 the *Daily News* had purchased 348,999 tons of paper from the International Paper Company (IP), and only 22,243 tons of that had been manufactured in Quebec. Were the Baie Comeau mill open, the company argued, Quebec would have reaped the benefits of this industrial activity. Overall, the Tribune Company claimed, it spent some $67 million buying paper on the open market between 1920 and 1935 to print the *Chicago Tribune* and *New York Daily News*, and it aimed to have a second mill at Baie Comeau to supplement the production in Thorold, Ontario, and to end these large expenditures to third parties.[18]

While the Tribune Company was waiting to recommence its Quebec expansion, company officials also considered expanding to the southern United States, where newsprint inventor and entrepreneur Charles Holmes Herty had been trying to attract Tribune Company investment for some time. At one point, Arthur Schmon went so far as to threaten Canadian policymakers in saying that if they did not allow companies like the Tribune to continue expanding in Canada, "Southern newsprint will become a reality." Despite the fact that a Tribune mill in the South would have both provided a boost to the American rather than Canadian economy and also freed the company from dealing with foreign policymakers, Tribune Company officials had resisted Herty's overtures. On March 6, 1935, however, several Tribune Company executives visited Herty's lab. Ten days later McCormick himself received a tour, during which part of Herty's pitch was that, in the middle of the Depression, American papermakers could be providing work for the unemployed in the South rather than relying on foreign labor. Tribune Company officials remained unpersuaded, and Arthur Schmon remained skeptical of pine as a raw material that could make high enough quality newsprint for the Tribune Company to use. An "inferior wood can only make an inferior product," he remarked, and the high resin content in southern pine made the material undesirable for newsprint manufacturing.[19]

With leading executives believing that the South lacked potential as a site for newsprint manufacture as of 1935, the Tribune Company remained focused on building its new mill in Quebec. When the company began planning to restart

industrial expansion in the mid-1930s, it reconsidered projects that it had initiated in the province in the late 1920s. One of these potential developments involved building a pulp mill on the North Shore and a paper mill in Quebec City,
from which it could ship paper to either Chicago or New York. For a variety of
reasons, as Arthur Schmon pointed out, that plan would be a better option than
building a North Shore newsprint mill and an accompanying company town,
as it not only "eliminate[d] the risk of winter navigation" but also ensured that
"better personnel can be obtained and retained in a centre of population." On
the isolated North Shore, Schmon believed, the overall development needed by the
company would be much grander and costlier. Ultimately, Schmon claimed, the
North Shore "proposition has possibilities of improving as time goes on, but
the Quebec one is more certain, simpler and less radical. Therefore even were the
two propositions equal, I would still recommend the Quebec one, because of its
immediate advantages."[20]

Though he valued Schmon's opinions, McCormick ignored all of this advice,
and ultimately the symbolic advantage of the North Shore in the era of the TVA
proved too difficult for him to resist. As the TVA continued its work in the southern United States, McCormick seized the opportunity to undertake a similar
project in Quebec, and he chose to build the mill and the company town of Baie
Comeau on the North Shore and to begin developing that region. In so doing, he
not only discounted Arthur Schmon's advice but also worked to overcome resistance from his cousin and *New York Daily News* publisher Joseph Patterson. As
Patterson remarked of the project's risks, the "idea of a fifteen or twenty million
dollar further investment up in Canada . . . with possible tariffs and so on, kind
of frightens me. If the bonds . . . failed to pay their interest, it might wipe out my
children's fortunes." McCormick, while sensitive to Patterson's financial concerns,
pointed out that the company had already invested over $13 million in Canada,
and he noted that "we will have to do something if we wish to save these investments." Patterson did participate to some degree in the planning process, and
McCormick made it clear to Schmon that "Captain Patterson's approval" was
essential to any course of action, but ultimately this was McCormick's project,
and Patterson was only a marginal figure in the planning and building of the North
Shore newsprint mill and Baie Comeau.[21] One longtime Baie Comeau resident
told me in 2013 of Patterson that, while "I know the name," she did not have "any
recollection" of him being a presence in Baie Comeau. Of McCormick, on the
other had, she recalled, "He owned us . . . he owned the town. He was a multi-
millionaire, and when he said 'jump,' you jumped. You said 'how high?' "[22]

Though McCormick in subsequent years would love to talk about the Baie
Comeau undertaking as the result of private initiative, it was in fact the result of

a close partnership between his company and Canadian policymakers, and one that had been enabled by international trade agreements between the United States and Canada. In building Baie Comeau, the Tribune Company was also aided by its close connection to provincial officials, especially Maurice Duplessis, whose party and personality dominated Quebec from the mid-1930s to the mid-1950s.

Newsprint Politics and the Rise of Maurice Duplessis

Though many in Quebec praised Baie Comeau and the Tribune Company's role in developing it, there was a consistent and persistent line of criticism from provincial observers of forest and resource policies. In the early 1920s, this had involved allegations that the Tribune Company colluded with Anglo-Canadian Pulp & Paper Mills to get favorable conditions on timber leases. These sorts of criticisms continued in some circles for years after, and they were often supplemented by allegations that the Tribune Company was aided by corrupt practices among Quebec officials. Whether they were ethical or not, the Tribune Company's relationships with federal and provincial policymakers were always close, and throughout Baie Comeau's planning and development, Robert McCormick and his executives continued to cultivate them in order to advance the company's interests.

In Quebec, the Tribune Company worked closely and amicably with Liberal premier Louis-Alexandre Taschereau, who served from 1920 to 1936, and with members of his government. In one instance of this friendly relationship, Arthur Schmon remarked in 1931 that Minister of Lands and Forests Honoré Mercier "gave me a sympathetic ear on working out our land requirements." In hopes of getting favorable treatment from the government, Schmon arranged for another company executive to see Mercier "some evening at his Quebec apartment where he can go over the whole matter quietly to have it fixed in his mind. Mr. Mercier will then write me his suggestions, at which time we will decide what our next step will be. In the meantime none of his officials are to be consulted, in order to get away from the conventional attitude toward these problems." This public-private partnership defined the relationship between the Tribune Company and the ruling Liberal government throughout the 1920s and into the 1930s, and the relationship always remained fruitfully symbiotic.[23]

To many conservatives in Quebec, on the other hand, this close relationship had long shaded into outright cronyism. Members of Quebec's Legislative Assembly routinely denounced Taschereau's relationships with American industrialists generally and with the Tribune Company specifically. In the period leading immediately up to Baie Comeau's construction, the friendly relationship between the Tribune Company and the Taschereau government resulted in a particularly controversial episode when the government proposed to cancel the premium stumpage fees that the Tribune Company was required to pay on trees cut on the North Shore limits in December 1935. Under the terms of its 1923 forest lease, the Tribune Company was to have paid a premium stumpage of $1.675 per cord on wood that it cut, and officials claimed that this made the operations economically impossible given the fact that newsprint prices had declined over time. In 1923, when the company agreed to the lease, newsprint in the United States was selling at $81.80 per ton, and it had sold for as much as $112.60 in 1920. By 1930, the price had declined to $62 per ton, and in 1935 the price hit the historic low of $40 per ton. Given this lower market price, the Tribune claimed that it could purchase newsprint from a third-party provider like IP at a lower price than it could manufacture its own with the premium stumpage fee attached. Given how much money the company was investing in the region, officials told the Taschereau government, it would be a benefit to all parties if the premium stumpage fees were removed so that the company would be able to operate economically and to continue its work in the province. The Tribune Company's request was extraordinarily self-serving, as the company at that point was well into the plans to build a North Shore mill using wood that it cut from its limits there. These plans were nearly certain to continue, with or without the premium stumpage fees. Asking for the province to remove those fees was tantamount to the Tribune Company requesting a direct and preemptive rebate of costs that it had agreed to pay to the province.[24]

The Taschereau government ultimately acceded to the request and signed an agreement with the Tribune Company to remove the premium stumpage fees on December 21, 1935, just prior to an election that returned the Taschereau government to power with what one company official described as a "majority of only six votes. There was no opportunity during the following six months, with so small a majority, for any legislation to be passed by the Taschereau government. As a consequence, a special Act confirming the premium stumpage order-in-council and contract could not be enacted." Despite the fact that the agreement was not finalized, the Tribune Company forged ahead with the Baie Comeau development, all the while expecting that it would be able to secure provincial legislative approval as planned. As they did so, major developments in provincial politics were taking shape around them, and their relationship with the Taschereau government became a topic of significant public and legislative debate.[25]

As the Depression worsened during the 1930s, Taschereau's government increasingly came under attack for its ineffective response, and since the newsprint industry was such a major factor in Quebec's economy, the government's relationship with companies in this sector came under more intense scrutiny. By 1933, because of mill closures and decreased production, some Quebec cities were in economic crisis. In Chicoutimi, for example, paper mill closures had put more than half of the residents on direct government relief. Conservatives began assailing Taschereau's ineffective response to the Depression, and many increasingly added strident claims about corruption within the government. Maurice Duplessis, the leader of the Conservative opposition, formed the Union Nationale (UN) party in 1935 and began attacking Taschereau in the Quebec's Legislative Assembly and in public. Duplessis began claiming publicly that he had evidence of massive and widespread malfeasance by Taschereau and members of his government. In the legislature, Duplessis used his official position as opposition leader to initiate hearings, which became, as historian Bernard Vigod describes the situation, akin to a "spectacular criminal trial. The government stood accused of systematic corruption on an unimaginable scale. . . . Day by day the evidence mounted: sinecures and inflated salaries for relatives and retired candidates; distribution of statutory grants, especially for colonization, at the whim of Liberal deputies; fraudulent invoicing; huge printing contracts for government newspapers." In total, Vigod remarks, "it was the enormity, the pervasiveness, and the brazenness of abuse which first staggered, then enraged the people of Quebec. In the midst of unprecedented poverty and suffering, while leaders preached the virtues of self-reliance, family responsibility, private charity, and 'fiscal responsibility,' this behavior was regarded as positively obscene. No civilized society could freely tolerate such a regime while preserving its self-respect. The government must go, and it was irrelevant who the opposition were and what they stood for."[26]

As Duplessis mounted these attacks, the Tribune Company became a part of this discourse because of its close relationship with the Taschereau government. In February 1936, a Montreal attorney named Noel Dorion sent a letter to Maurice Duplessis in which he confided that he had "fairly accurate information" from a "serious source" that members of Taschereau's own party were angered by the allegations of corruption. One Liberal member, Edgar Rochette, was apparently "tired of being fooled" by Taschereau's allegedly hypocritical and duplicitous style of governing. One of Rochette's most significant concerns had to do with what he claimed were the "extravagant" concessions given to the Ontario Paper Company, the Tribune Company's wholly owned Canadian subsidiary, just before the previous provincial election in the form of the removal of premium stumpage fees, which Rochette had opposed. Arthur Schmon began hearing similar sorts of rumors around the province, remarking to Robert McCormick that he had received a confidential letter from an acquaintance who had just spoken to a French news writer, who writes special articles for a French syndicate. This man says that he is planning to write a series of articles on the scandal of the Ontario Paper Co. steal. This man did not know that this particular person knew me. He said that there would be a caucus in the Opposition over the week end at Montreal and that purpose is for a concerted campaign against our company, pretending that the government of Taschereau has bled the treasury of a total sum of about $10,000,000 through lowered charges and other concessions, and through this agitation to bring about the downfall of the Government, making the Ontario Paper Co. the goat. Of course the fat is now in the fire and the coming weeks will be anxious.[27]

At the same time, some of these previously private accusations about the corrupt relationship between the government and the Tribune Company began filtering into public discourse. In 1936, the Quebec Catholic Farmers' Union's bulletin *La Terre de Chez Nous* published a lengthy article about the removal of the Tribune Company's premium stumpage fees, in which it calculated that Taschereau, through a "simple stroke of the pen," signed away some $42,752,000 of potential revenues, and for no good reason other than to benefit the company. In doing this, the "disgraceful regime wanted to treat its favorites with unsurpassed and unsurpassable generosity." This was "swindling on a grand scale," the author noted, and he remarked that it would be "interesting to know the secret considerations" agreed on by the "actors in this national tragedy." Whether or not the premium stumpage cancellation was achieved through corrupt means, it created substantial savings for the Tribune Company. According to the company's private internal estimates, the cancellation of premium stumpage would save it $20,000,000 over the life of the lease, a sum that itself would cover a significant portion of the cost of Baie Comeau's construction. In essence, the major construction project of a newsprint mill and company town, which was being touted widely by Robert McCormick and the Tribune Company as a pioneering private development, was funded substantially by this direct public subsidy.[28]

The result of these attacks on the Taschereau government was that an election held in August 1936 elevated opposition leader Maurice Duplessis to the office of premier as his UN party won a large majority in the Legislative Assembly. As one Tribune executive noted, throughout his campaign Duplessis, "in seeking ammunition with which to attack the Taschereau Government, seized upon the Ontario Paper Company order-in-council and agreement relating to premium stumpage as a political football. Throughout most of the hustings in the province the Taschereau government was charged both on the platform and over the radio with having given away to the Ontario Paper Company natural resources in the province in fabulous sums amounting to thirty or forty millions of dollars."[29]

After this rhetoric succeeded at the ballot box, Duplessis's election created significant anxiety among Tribune officials. Robert McCormick noted that a "radical and possibly revolutionary party has assumed power. It is determined to prosecute and possibly persecute the old ministers. We may be a handy agency for this purpose. . . . We have certain investments at Comeau. They may retain their full value or may be lost. The essential thing is not to incur any more obligations than we have to until we find out what is going to happen." In hopes of navigating these newly turbulent political circumstances, Tribune Company officials immediately sprang into action to lobby Duplessis, whom they found wanted both economic development in the North Shore region and to communicate to foreign investors that Quebec was a place where "capital is safer than any part of the world." Upholding Taschereau's controversial concession was expedient, Duplessis believed, because "the Government needed revenue" and the Tribune Company was working "in an isolated part of the country" where there was tremendous economic potential but no industrial activity. Still, the company faced significant challenges in winning over Duplessis's favor. As one of Schmon's Montreal contacts told him after the election, Duplessis had "received information from a source which I have not yet been able to discover of rather a damaging nature to your Company. In particular, it is intimated that the last Government accorded your Company favours at considerable loss to the Province."[30]

The Tribune Company's relationship with the Liberal government, which had benefited the company greatly for nearly two decades, now appeared to be a significant liability. Shortly after Duplessis took office, however, Schmon reported that he had lunch with him and several of his ministers in an effort to cultivate a closer working relationship that would allow the company to continue operating in Quebec as it had under the Taschereau government. Schmon reported that he "had no lawyers present with me, feeling that this was a matter that we should deal with on principles and not on legal points." Schmon and Duplessis were quick to find common ground, and the relationship between the company and the new government blossomed. Soon, Schmon wrote to Robert McCormick that "with the explanations given and with judicious personal contact, it will not be long before our company will be in just as good graces, if not better, than we were with the previous Government."[31]

Despite his campaign rhetoric promising radical economic reform, Duplessis in government exhibited a disposition toward the economy which in practice was continuous with that of the Taschereau government. Under Duplessis, the provincial government would extend Taschereau's encouragement of both private enterprise and foreign investment, and US corporations found that the business climate in Quebec remained the same if not friendlier for them. In the specific case of the Tribune Company's operations, Duplessis expressed a clear and strong desire to work with it and sought a politically expedient way of doing so. As a Tribune Company executive noted, Duplessis "found himself under great pressure in dealing with all problems" related to the Tribune Company, "by reason of the charges made during the campaign by himself and members of his party." In office, Duplessis was

> anxious to settle the issue in a practical way which will protect his own position and at the same time avoid embarrassment to the Ontario Paper Company interests through any admission of its being a party to an illegal transaction. Mr. Duplessis frankly stated to us that he recognizes that we were dealing with the existing government which had a mandate from the people and that he does not blame us, as businessmen, for making as good a deal as possible with the government in power, irrespective of his views concerning the moral or legal right of the Taschereau government to remove the premium stumpage, which was imposed by bidding at a public auction.

The terms that Duplessis proposed to complete the cancellation of the premium stumpage fees were in many respects as conciliatory as those that Taschereau had offered, though the new agreement came with a payment that was somewhere between bribe and tribute. Duplessis suggested that, in exchange for the premium stumpage fees being canceled, the Tribune Company would pay a $250,000 fee to the government that would "transfer . . . all of our Quebec limits to a Quebec corporation" to be formed in order to domicile the Tribune Company's Quebec operations in the province. As one company executive noted, Duplessis justified the $250,000 payment as necessary "for political reasons" and claimed that this "arbitrary transfer fee" and the ensuing legislation that he would promote would "remove for all time any question concerning the title of our limits which is under attack and might always be made a political football."[32]

Though the $250,000 fee was significant, Tribune Company officials believed that it was necessary to pay so that operations could continue and remain stable. "Since the new government has been in power," one noted, "we have been hamstrung in every quarter in getting important questions settled, orders-in-council approved and signed, subsidies for roads, bridges, etc., granted, and the proposed arrangement, in addition to settling the legality of the renewal of the premium stumpage, will eliminate these difficulties and without doubt enable us, through added concessions which we seek, to absorb the substantial outlay indicated in a relatively short time." There was little the company could do but cooperate, given that "Mr. Duplessis is now approaching the peak of his power and can exercise almost dictatorial independence in dealing with governmental problems." As a result, the "legislative charter confirming titles and giving assurances of regular stumpage dues will be a Duplessis-sponsored bill, the terms of which will be prepared and agreed upon promptly and in advance of the time when we make the initial payment on the transfer fees." The company's agreement with Duplessis over the cancellation of the premium stumpage which Taschereau had agreed to was formalized by the two sides in 1938, and the Tribune Company formally created a separate corporate entity called Quebec North Shore Paper Company as a wholly owned subsidiary. McCormick was not necessarily eager to do this, but he clearly understood the politics of the matter. Duplessis had "insisted that our Quebec properties be put into a Quebec corporation," McCormick stated flatly, and he had "ample powers to compel us to do it, so the only thing to do is to go along pleasantly."[33]

This agreement demonstrated the shared strategic maneuvering by both the Tribune Company and Maurice Duplessis which helped them navigate the changing political climate of 1930s Quebec. As Arthur Schmon noted to McCormick, Duplessis had to occasionally say critical things in public about the Tribune Company

> to satisfy the criticisms within his own party by making it appear that he has imposed onerous conditions on the company, and secondly to take the credit now for his Government for the development and to put the previous government on the defensive. Finally he wishes to establish an atmosphere in which he can be openly friendly to our company. I know that you will have the same feeling that I have of indignation and resentment, that we have to be the object of such political maneuvering. However, I am convinced, as a result of my recent visit to Montreal that Duplessis has had a real political problem within his own Government in dealing with us and that he is preparing for a new election and does not want to have it thrown against him that he criticized the former Government for its friendly attitude towards us and upon assumption of power he has swallowed his words and has shown the same friendly attitude. As I have said before I am convinced that the results are what will count and not the methods.[34]

Maurice Duplessis, as historians have shown, ruled the province of Quebec with a centralized authority far greater than any of Robert McCormick's most caricatured New Deal autocrats. McCormick and Duplessis, while never close personal friends, shared a staunch anticommunism and developed a stable and consistently amicable working relationship that each used to his advantage. Ultimately, McCormick got his town and the ability to expand his newsprint operations in

Quebec, and Duplessis was able to consistently tout this development as an example of how public-private partnerships between his province and American companies worked in Quebec's interest. In many ways, the Tribune Company's activities in Canada fit with Duplessis's economic goals of developing provincial industries with the help of foreign capital. They also worked well personally with the new premier, whose style of governing relied heavily on patronage and personal relations.[35]

Baie Comeau and the Promise of the Company Town

In spite of the Depression and its effects on Canada and the United States, Robert McCormick decided to begin building his new mill and the accompanying townsite. The timing was in many ways an attempt by the Tribune to show by example what could be done by private enterprise as the economy in North America flagged, and company officials consistently promoted their role in stimulating the economy in Canada. In touting its massive spending on construction materials, for example, the company claimed that it had by 1936 agreed to purchase "5000 tons of structural steel from the Dominion Bridge Company . . . one million bags of cement from the Canada Cement Company . . . one million bricks, and . . . several million feet of lumber." All of this, company officials stated, was aimed at "contributing additional employment all along the line which leads to economic recovery and the lessening of expenditures for provincial and federal relief." Indeed, according to company estimates, in 1936 and 1937 it alone had spent more than one and a half times what the Canadian federal government had spent on public works in Quebec in the period from 1932 to 1937. These expenditures went to both materials and wages, as the company estimated that it would employ at least 1,700 workers in the construction project.[36]

On one level, the company needed to build a mill and town to support its operations as an industrial newspaper, but Robert McCormick also was able to demonstrate that major infrastructural projects could be undertaken successfully through the private sector. In many respects, the Tribune Company's development on the North Shore was a corporate variant of the "high modernist ideology" that political scientist James Scott shows was deployed by twentieth-century planners around the world. In the case of the TVA, Scott argues, its founders "hoped that it would become a model for regional development that would eventually be applied throughout the nation," and Robert McCormick thought the same about his own projects in Canada. As a planned community, Baie Comeau was meant to be both a philosophical and physical rebuttal to the TVA. In its execution, the town was also meant to show the degree to which the Tribune Company was committed to private development in Canada.[37]

This 1936 aerial photo shows the clearing that had been made to accommodate the Tribune Company's new Baie Comeau mill. Vol. 25177, Scrapbook 1926–34, QOPC, LAC. Used with permission.

The intellectual roots of Baie Comeau's planning go back decades and included several international undertakings in city and regional planning. The example of Henry Ford's Brazilian development, Fordlandia, provided some evidence of the challenges confronting US corporations trying to obtain raw material supplies in the international market. In Ford's case, his company wanted stable supplies of rubber (like McCormick, Ford needed trees, in this case *Hevea brasiliensis* rather than spruce) which it could control outside of the nexus of plantations in Southeast Asia dominated by British, Dutch, and French interests. In 1927, Ford received a land concession of about 2.5 million acres in the Brazilian state of Pará, and he set about building a rubber plantation and company town for workers. The undertaking was a spectacular failure. Planners misunderstood the dynamics

GENERAL VIEW   24/8/37

By late August 1937, the Tribune Company had made significant progress on its Baie Comeau mill construction. Vol. 25178, Scrapbook 1934–41, QOPC, LAC. Used with permission.

of planting rubber trees, which worked better when allowed to grow under natural conditions where pests could not travel easily between trees planted closely together, as Ford's managers did. The town's design exhibited a similar lack of concrete experience with the regional environment, and neither the rubber plantation nor the workers' housing was well conceived or executed to be successful in local conditions. By 1930, the area around Fordlandia had a population of about five thousand, though it was marked by chronic labor instability and turnover. This culminated in a December riot that destroyed a great deal of the plantation's property. Ford struggled throughout the decade to make the project work, but he was unsuccessful at both rubber cultivation and corporate colonialism. In 1945, the company abruptly abandoned the project.[38]

McCormick's approach to industrial and urban planning in Canada was in some ways a comment on Ford and his Brazilian projects. In size and scope, the two corporate developments were quite similar, though McCormick in Quebec had the advantage of a climate slightly more familiar to a midwestern urban planner. McCormick also had the kind of adversarial relationship with Ford which he did with Franklin Roosevelt, though the former relationship warmed over the years. The roots of the McCormick-Ford conflict went back to 1919, when Ford filed a libel lawsuit against the *Chicago Tribune* for $1,000,000 after the paper called him an anarchist. After a protracted court case, the *Tribune* was found guilty, though Ford was awarded only six cents in damages. McCormick refused to pay, in effect daring the automaker to sue him again to recover the award. The relationship thawed over the years, and in a 1953 *Tribune* column McCormick remarked of Ford, "Later I came to know and esteem Henry Ford, then known largely for his political peculiarities." During the time that Ford worked on building Fordlandia, the *Tribune* occasionally covered the construction and rarely avoided the opportunity to report on its problems. In 1939, after the Baie Comeau mill and townsite had begun operating successfully, the *Tribune* reported of Fordlandia that it was still struggling mightily. "The investment in the Ford venture to date is about 8 ½ million dollars. . . . Whether or not the money will be invested or whether the whole venture is successful or not is problematical." A 1941 report acerbically noted that "Ford experts had learned that the Amazon valley is no picnic ground."[39]

When McCormick set about building Baie Comeau, he found himself engaged in a project that was occupying some of the most important urban theorists of the period. Lewis Mumford, for example, wrote in 1938 that "today our world faces a crisis" because of the fact that cities had become physical settings that were ill suited to structuring healthy lives for their populations. Cities, Mumford believed, had become inhospitable to what he understood to be a good life. "The standard-ization of the factory-slum," Mumford believed, "was the chief urban achieve-ment of the nineteenth century." One possible solution to this problem, Mumford believed, was the kind of industrial town designed by Robert Owen, who "pro-posed, in order to enable the new industrial workers to rise out of the squalid state in which they lived under the new factory system, to build small balanced commu-nities in the open country." This ethic animated New Deal planners, and to some degree it animated Robert McCormick in his desire to show that his firm could engage in industrial capitalist manufacturing while building a well-functioning community for his workers.[40]

Baie Comeau in many respects was also a comment on two cities that were parts of Robert McCormick's own particular history as a Chicagoan: Pullman, Illinois, and Gary, Indiana, the former the revolutionary but ultimately unsuc-cessful company town built just south of Chicago's city limits in the 1890s, and the latter the town built just outside Chicago and across the state line on Lake Michigan by US Steel. To its early admirers in the Progressive Era, Pullman seemed to be a solution to some of the most pressing problems of industrial capi-talism and urban development. Pullman sought both to rationalize production of his railroad cars and to create a built environment that would motivate his work-ers to be happily productive on the job and then moral and thrifty in their off-hours. By building a town with open boulevards and municipal amenities such as a library and churches, Pullman believed that he would have lower worker turnover and less labor unrest and that he could demonstrate the benefits of wel-fare capitalism. By 1884, four years after George Pullman announced his plans to build his eponymous town, it had some 8,500 residents, and by 1893 the Pullman Palace Car Company had some 14,000 employees, more than one-third of whom lived in the town. The town received "hundreds of thousands of visitors" in its first decade, historian Stanley Buder remarks, "and an overwhelming majority left im-pressed." Press coverage was laudatory in tone. In 1894, however, Pullman's designs went awry, and his town was the site of one of America's major labor disputes, as workers organized under Eugene V. Debs's American Railway Union and went on strike in May. The strike turned violent and stretched into the summer, when federal troops were called in to quell it. The town, as well as Pullman's company, suffered greatly as a result of the actions, and it was widely regarded in the early twentieth century as an urban planning failure. Pullman had wanted to use a planned built environment to create a stable and happy workforce, but his error was to exert too heavy a hand of paternalism. Workers eventually rebelled against the fact that they could not own homes and had to shop in the company store.

Partaking in what Pullman understood to be amenities seemed to workers to be onerous and unfair impositions of corporate discipline. Pullman succeeded in rationalizing industrial production of railroad cars but not in urban planning.[41]

If Pullman represented the dangers of a company town built with too much control of the built and social environment, Gary, Indiana, represented the danger of a company town built with insufficient attention to these matters. US Steel began planning Gary construction in 1904 but declined to hire professional urban planners. The company instead elected to simply lay out a grid of streets, build basic infrastructure, and then try to induce workers to build homes on residential lots that it sold to them. Workers were slow to do this, and an ad hoc collection of poorly constructed rental units built by speculators began springing up. Still, the city was faced with a chronic housing shortage, and much of the housing that was built was of poor quality. Along with its poor housing stock, general living conditions in Gary were grim, and in 1925 its murder rate "encouraged comparisons to Al Capone's Chicago." For a publisher in Chicago planning to build a city in Quebec, Gary offered powerful lessons of what not to do.[42]

Robert McCormick understood the mistakes of Pullman and Gary, and from the beginning Baie Comeau was planned to communicate to workers and residents that it was not a "company town" in the traditional sense. The Tribune Company also found useful models for Baie Comeau away from Chicago, especially in the examples of the private developments in Vandergrift and Hershey, two towns built in Pennsylvania by the Apollo Iron and Steel Company and the chocolate manufacturer Hershey, respectively. As one Tribune executive remarked, Hershey was "established and is being maintained along lines exactly the same as those proposed" for Baie Comeau, and he suggested that "much could be learned from what the Hershey people are doing." In each of these cases, company owners created towns in which homes were privately owned, and the towns were built into the surrounding landscapes in harmonious manners. Company discipline was engineered to be subtle, and the urban spaces were designed to make workers happy to live and work there. For American corporations in the period between the Haymarket riot, Pullman, and the creation of the Congress of Industrial Organizations, a well-designed company town offered to many an intriguing and in some ways ideal solution to the problem of labor relations.[43]

In addition to trying to demonstrate that Baie Comeau could show the possibilities of privately funded municipal and regional development, McCormick also wanted to show his company's superiority to the New York Times, which had at Kapuskasing, Ontario, engaged in the only sort of newspaper-driven development that rivaled the Tribune Company's. Initially just a "watering point on the Transcontinental Railroad," as *Times* publisher Arthur Sulzberger remarked, the town took shape during World War I when a "trainload of alien prisoners was unloaded there" and assigned to begin clearing the area for the settlement that became Kapuskasing. The pace of the town's development picked up momentum after the war, as the Ontario government added a "large number of buildings, including saw mill, planing mills, store, administration building, school, boarding houses . . . [and] twenty-one houses." In 1918, the Government of Ontario issued S. A. Mundy and Elihu Stewart a lease on 1,740 square miles of timberlands near Kapuskasing, and the pair operated under the name Spruce Falls Pulp and Paper. Kimberly-Clark bought that company in 1920, and it soon built a sulfite pulp mill that went into operation in late 1922. In 1926, the Times and Kimberly-Clark formed a partnership (of which Kimberly-Clark owned 51% to the Times's 49%) to build an expanded mill and townsite.[44] The Times's newsprint operation, while a tremendous help to its business, was never as ambitious as the Tribune's in regional development. Robert McCormick even remarked derisively of the operations that the Times "went into the paper mill in imitation of us and had to take a paper mill man in partnership. Ours is the only newspaper that has the paper mill know-how." The Times and Kimberly-Clark had also let the province of Ontario do much of the planning for Kapuskasing, a course of action that provincial officials thought would be a way "to prevent the Town becoming a 'closed' or 'Company' Town.'"[45]

With these examples in the background, McCormick and his executives had extensive debates about the merits and shortcomings of building Baie Comeau as a "closed town" or an "open town."[46] In the former case, the advantage would be that the company would have "full control . . . of all activities" in the town, as was the case in Shelter Bay, which the Tribune Company had established on the North Shore in the early 1920s. This model, however, threatened to create problems in the 1930s, as pointed criticism had been emerging in Canada of closed towns like Dolbeau, Arvida, and Noranda. In order to promote itself through the new Baie Comeau development as a more benevolent corporation, as executive Arthur Schmon suggested, the open town might be the better option. In this model, the company "generally owns all of the land and therefore really controls the development of it. The incorporators of the town, which are really named and trusted employees of the company, are authorized to name for a period of two years, the Mayor, Alderman and such other officers required for the town government." In many ways, the open town was the best of both worlds, as it offered the company a degree of control that was operationally close to total but in appearance much less so. The closed town "always creates impression and belief that the company controls workers," Schmon believed, "not only when they are on the job but also when they are not, hence many employees not only feel but declare that there is no personal freedom. Men, especially intelligent labor, do not do their best work in such an atmosphere." In the closed town, "most of the initiative, interest and feeling of responsibility that flourishes in a community of home owners is absent and building up civic pride is very difficult, if not impossible." The open town, in contrast, not only communicated corporate benevolence to workers and the larger public but also had the advantage of getting "government assistance under Quebec Municipal Act for water supply, fire protection, schools, hospitals, etc."[47]

Robert McCormick initially sought to have it both ways in thinking about how to build Baie Comeau. He suggested to Arthur Schmon,

> I wonder if we cannot have both a closed town and an open town; with such houses as the company has to build on its own property with private streets and adjoining the mill, and the open town beyond where people can own their homes and conduct stores. I have in mind the possibility of eventual labor trouble. The men that stay by us then would be on private property and not subject to attack going to and from work. The hotels would be on this property. It might be unwise to extend the fence around it, although if the fence is up at first, it might never be noticed.

Schmon remained a strong proponent of the "open town" model, noting that it would probably be "impossible or inadvisable to carry out your suggestion of both an open and a closed town." He did, however, agree that the company should consider how to use the built environment as a means of social control, and he remarked that the company could "obtain the same result you want in planning the town layout, having in mind eventual labor troubles."[48]

The internal debates about whether Baie Comeau should be an "open town" or a "closed town" also had to do with the politics of Quebec. One newspaper report in April 1937 noted "strenuous opposition" to the initial bill incorporating the town, which some claimed would create a "closed municipality" and "give the company dictatorial powers." The initial plans for Baie Comeau created significant legislative debates in the province, as critics were highly skeptical of the company's professed commitment to building an open town. Philippe Hamel, a member of Duplessis's ruling UN party, claimed that Baie Comeau would be in all significant operational matters a closed city and that the Tribune Company would be the "king and master of the region" since it built the town and would manage all aspects of life there during its early existence. "If a doctor wants to establish a practice in town," Hamel remarked, "he has to get the king's permission. If the company does not want the doctor there, he cannot go." Likewise, UN member Oscar Drouin remarked that the original bill was "tout simplement un bill atroce," and he and fellow UN member Joseph Bilodeau remarked that there had to be "necessary precautions in place to avoid repeating the mistakes of Arvida." To proponents of Baie Comeau like UN member Arthur Leclerc, it was "la clé du développement" on the North Shore, a line of argument that fit with the prerogatives of party leader and premier Duplessis.[49]

After vigorous debate, the legislation incorporating Baie Comeau passed on May 20, 1937. The bill made "special provisions derogatory to the Cities and Towns' Act" in order to spur the project along by waiving provisions relating to the election of local officials. Instead of requiring a popularly elected government from the outset, Baie Comeau was allowed initially to be governed by an appointed municipal council. This five-person council was composed entirely of company employees, and it was charged with selecting a mayor. The legislation mandated that the town hold elections in March 1940 and March 1942, with two and then three of the original council members set to be replaced in each contest. Baie Comeau would, the Legislative Assembly stated, "create a considerable influx of people to the territory," and it was incorporated as an open town and given a clear process and timeline to transition from being controlled by the company for expediency's sake into functioning as a normal city.[50]

In the twentieth century, many business leaders, policymakers, and reformers became intrigued by the promise of the company town. To some, the planned town in a remote area could offer a way for a company engaged in resource extraction to create an environment conducive to a stable and satisfied workforce less likely to engage in unrest that would be bad for business. To others, the company town offered a solution to important and perpetual problems created by industrial capitalism. The areas around plants were often unhealthy and undesirable, and workers and their families suffered accordingly as a result of these conditions. If a company could provide a better environment inside and outside the factory, many believed, that might create significant benefits for both workers and shareholders. This, in many ways, was the ethic guiding the Tribune Company's actions in Baie Comeau. For the workers who moved there, the bargain was that obtaining jobs and houses involved submitting to a variety of carefully designed procedures and policies disciplining daily life.

The Well-Ordered City in the Wilderness

When the construction of the Baie Comeau mill commenced in 1937, it was a major and costly undertaking. The Tribune Company budgeted for 25,000,000 hours of labor at a cost of $12,000,000, and it estimated that this meant full employment for five thousand people for a period of two years. As the town construction commenced, the company began trying to attract workers and residents.

Even during the Depression this was no small task, given Baie Comeau's remote location. In order to cast a wide net, the company searched for both Francophone and Anglophone workers and made an effort to ensure that potential Protestant workers would be comfortable in an isolated town located deep in Catholic Quebec. A promotional brochure entitled "Baie Comeau: A Modern Model City" promised in bilingual English and French text that workers would enjoy life in a "modern, planned community—planned for comfort, convenience and attractive appearance." Workers would have the "opportunity to build their own homes in delightful surroundings in residential areas," and they and their families would enjoy a "healthy climate in picturesque surroundings" and "employment at good wages." Businesses were promised a "constant market of employed workers receiving good salaries and wages" and a "planned business section maintaining high standards of efficiency and convenience." This was the exact opposite of the popular understanding of what a "company town" was, the brochure claimed. "Baie Comeau will be an 'open town.' It will be governed by its own citizens through a duly elected town council and town manager. . . . The Company will be pleased to co-operate with merchants in planning suitable buildings and will supply architectural designs. Merchants establishing businesses at Baie Comeau will have a permanent, assured market from the start. They will live in a model community where fair wages, fair prices and mutual consideration in all business dealings will be accepted facts." Given Robert McCormick's opposition to central planning and federal relief in the United States, it is rather ironic that his enterprise in Canada promoted a well-regulated market and full employment.[51]

With this company inducement, home construction by employees began quickly, and in 1938 some forty-two employees built homes in Baie Comeau. Though this seemed to promote the kind of development the company wanted to encourage, some officials were concerned about what sorts of workers were doing this. As one executive remarked, a "danger exists in having too many unskilled build their own homes, particularly at the values that have existed this year." Though the company promised full employment to workers who moved to Baie Comeau, it remained anxious about its ability to deliver. The company might also, some believed, "have some difficulty in employing all the possible workers in the families," and managers did not want to take responsibility for housing these people. "We believe that eventually a shack town will develop somewhere near Baie Comeau," and that would be a perfectly acceptable place for the company to draw its "casual labor." The company should, the executive concluded, "therefore strive to allow no one but our skilled and semi-skilled workers to build their own homes." In so doing, company officials would ensure that unskilled labor did not get a

By May 1937, home construction in Baie Comeau was well under way, as shown by these newly built houses on what was called Champlain Avenue. Vol. 25178, Scrapbook 1934–41, QOPC, LAC. Used with permission.

foothold in Baie Comeau and thus could be more easily encouraged to move away if there were not sufficient jobs.[52]

The Baie Comeau housing market represented a hybrid of patronage and social engineering, as company officials doled out the privilege to build homes to favored employees. In one instance, an executive noted that there was a "large waiting list" of employees in the Woods Department, the group involved in forest work, who wanted to build homes. The workers

> have been seeking permission to build for some time now but we have been holding them off because it was felt that the Mill Department should be given first consideration. Now, however, we have reached the point where some of these men have become discontented and are dissatisfied because they are away from their families practically all year and wish to have them at Baie Comeau and thus reduce living expenses. Some of these men hold positions which are important in the Woods Department and we would not like to lose them. All of them are key men and we would have difficulty in adequately replacing them now.[53]

The ultimate result of all of this private planning was a housing market and a town that were far less "open" than the company claimed publicly. Schmon believed that nearby towns would provide housing for the "young and ambitious man" who was hungry for work during the Depression and who would "be willing to be close to it under almost any kind of conditions until he does get some sort of a permanent opportunity." Thus, he believed, "we must not try to meet the problem of housing the lower wage earner until time gives us a better conception of how to tackle the problem." For those already there but not yet owning a home, Schmon planned a "careful study of each laborer who now has got a permanent job with us so that we will make a good selection of men from the point of view of qualifications of citizenship as well as value and stability." The way that the company controlled the housing market represented a strong structuring of the local market. "There must be," Schmon claimed, "a very deliberate attempt to maintain a scarcity of living quarters of all kinds and classes. This will maintain property values. . . . We must never permit too many people to build or to own homes to such an extent that there will be vacant houses, with consequent competition from private owners to sell their homes at bargain prices or to rent them below cost." As the Tribune Company planned Baie Comeau, it clearly understood its workers' living conditions to have direct connections to its overall manufacturing operations. As Schmon noted, the "problem of dealing with our housing is a complex one, that really reaches back into the manufacturing of newsprint. We must maintain discipline and a certain amount of contentment among our employees. Discipline is important." For Schmon, a key part of this overall disciplinary ethic was the aim of preventing the unskilled labor in the company's town from owning homes. "We must not allow any other people to build than members of the staff or skilled workmen; even then we must hold it down to a low number, possibly 20 houses." From the beginning, the company printing a newspaper aggressively promoting free enterprise and private initiative was itself planning housing shortages so as to ensure that prices stayed high and that the town did not attract settlers or residents who were not company employees. "After 1940," Schmon stated bluntly, "we must plan at all times to maintain an artificial shortage of about ten to fifteen percent; this in order to maintain property values and to prevent an unemployment problem."[54]

In practice, this was what the company did, despite complaints from employees. In late 1940, Schmon reported, there was "considerable pressure to build some homes next year, but I believe we should continue again next year our policy of creating an artificial shortage." This policy remained in effect for several years after Baie Comeau had been open to residents. According to one study, the town had 1.36 occupants for each available room in 1943, up from 0.97 residents per room in 1938. "There are several instances" in late 1943, the report found, "where three families are living in one house. Recently and possibly at the present time

29 people resided in a 14 room house of which 8 were normal bedrooms. There are 9 families living in cellars at the present time."[55]

The Tribune Company's control over Baie Comeau covered not only houses but also people, and the company was quick and relentless in trying to drive away anyone who came to the city without a job. In May 1938, Baie Comeau mayor H. A. Sewell wrote to Maurice Duplessis that the situation was a "dangerous one and unless some drastic steps are taken immediately may get beyond our control. Could permission be obtained to forbid the sale of tickets to men who have not a written authorization from The Ontario Paper Company that work will be found for them upon arrival here?" As the unemployed in the region continued arriving in town, Sewell remarked that the "question of accommodation is very serious, because we have not available the beds in which these men can be placed to sleep. At the present time we have a number who are sleeping on the floors in the camps, which is contrary to Provincial Government regulations and is an unhealthy situation." This continued despite the fact that the company had "posted notices at various points on the South Shore of the St. Lawrence, advising men that there is no work available and that they should not travel over here." Robert McCormick suggested other ways of preventing unemployed and undesirable desperate laborers from overrunning Baie Comeau. "I think the priests might be able to help in keeping the unemployed away, by letting other priests know that we have more citizens than jobs. Perhaps a sign on the dock would help dissuade people from landing."[56]

In the winter of 1938/39, with the Depression still ongoing, town and company officials believed they were being overrun by the unemployed. As Mayor Sewell claimed, "vagrants" were arriving in Baie Comeau and taking residence in empty and decaying logging camp buildings outside of town. "Many undesirable characters" had been arriving of late, and some were "given jail sentences varying from two weeks to three months." Not wanting to pay for their confinement, the town offered to suspend the sentences so long as the prisoners left and promised never to return. Eventually, the company just tore down many of its decrepit buildings in which these desperate job seekers had stayed on arrival.[57]

As town construction went on, the Baie Comeau newsprint mill officially opened in May 1938. It was national news in both the United States and Canada, with the *New York Times* giving note. Maurice Duplessis remarked at the opening ceremony that Quebec "should co-operate with the United States with the right sort of men." Many provincial supporters of US foreign investment agreed, with the *Montreal Gazette* editorializing that "co-operation in industry spells progress for all concerned, hence the great importance of the newsprint and power development of the Ontario Paper Company at Baie Comeau. . . . A new community has been created in Quebec, capital has been put into service and employment provided for many workers and their families."[58]

Many visitors to Baie Comeau marveled at what they perceived as the quality and efficiency of planning, and almost as soon it was built, the town elicited wonder from observers. One Quebec newspaper, *Le Progrès du Golfe*, remarked in June 1938 that "from a bay around which there was, just a year and a half ago, such emptiness and desolation, to now a beautiful little city with a huge paper mill . . . it seemed that a fairy had waved her magic wand." A *Montreal Gazette* correspondent was similarly impressed, remarking,

> Today most of the construction bunkhouses and work sheds have disappeared. In their place are stores and brick buildings and long rows of residences. The great mill and the storehouse are completed. Walking along the main street one passes the brand-new bank and the town hall and the church. There are women's dresses in the windows of the shops—the latest from Paris for the ladies of Comeau—and you can go for a stroll through the residential district, past neat little houses in fresh coats of paint. . . . Amazing changes have taken place in a year. The wilderness has been tamed.[59]

Others were impressed by how well planned the development had been and how well workers and residents were provided for by the local amenities. A journalist for *L'Echo du Bas St-Laurent* remarked that Baie Comeau was nothing like the "cities of mining camps of the American West that were created overnight and disappeared just as quickly. It is a real city . . . complete, well organized, providing all the necessary services to the demands of modern life, but it has not gone through the normal stages of childhood and adolescence before reaching maturity. . . . I would say that it is a fantastic dream materialized in steel and cement. The sense of strength and power that one feels in seeing it is stunning." The *St. Maurice Valley Chronicle* editorialized that the town did a great job of providing for the Tribune Company's workers:

> Colonel McCormick and his enlightened aides evidently attach as much importance to the social welfare of the workers as they do to the industrial efficiency of the mill. . . . Quite obviously, Colonel McCormick and his associates appreciate the fact that their new-found industry incurred definite obligations to its workpeople and to the province when it was given a license to exploit our natural resources. Instead of trying to evade these responsibilities they have definitely embraced them, and that fact must surely have influenced Premier Duplessis and his government in his dealings with the company. Canada needs men like Colonel McCormick, not merely because they are capitalists, but because they have sufficient enlightenment,

sufficient imagination to see beyond the question of costs and manufacturing economies to the essential social realities.[60]

The Ironies of Planning

By the late 1930s, the Chicago Tribune Company was one of the most powerful and profitable newspaper corporations in the United States. Some of this had to do with the fact that its two newspapers had a wide popular appeal, and their high circulations drove high advertising revenues. But the company was also successful because it had created an industrial infrastructure, making its manufacturing operations highly efficient and profitable. However appealing the informational content of the *Tribune* or *Daily News*, getting that content to readers was based on the industrial manufacturing and physical distribution of paper objects. And here the Tribune Company was singularly successful. At the time of completion of the Baie Comeau mill and townsite by the Tribune Company, its newsprint needs for its two newspapers were massive. By 1940, the company estimated that it was using one-tenth of all the newsprint in the United States to print the *Chicago Tribune* and the *New York Daily News*. In 1942, according to estimates by the US government, the total newsprint consumption in all of South America was 150,000 tons. In the United States that year, the *Daily News* alone used 190,000 tons and the *Tribune* 130,000 tons. By 1947, the Tribune Company newsprint requirements had gone up even higher, as the *Tribune* was using 161,000 tons of paper (86% of which came from Thorold, 12% from Baie Comeau, and 2% from the open market), and the *Daily News* was using more than 250,000 tons of paper, half of which came from Baie Comeau and the other half from IP. Despite relying on IP for some of its newsprint, the Tribune Company had by the end of the 1940s created a manufacturing firm that supported the two highest-circulating and most successful newspapers in the country.[61]

As the Tribune Company's newsprint needs increased in the pressroom, the mills' need for trees increased in kind. In 1943, the company estimated that, given its existing timber leases, it had seventeen years of wood supplies remaining to keep Thorold operating at capacity, after which these forest reserves would be "not sufficient to guarantee any appreciable expansion in production, or, to give security in continuous operation." This would put a strain on the North Shore limits that were now supplying the Baie Comeau mill, and the company began considering ways of adding to its forest holdings in the region. The best immediate option, company officials believed, would be buying out the adjacent limits that had been acquired controversially in 1923 by Anglo-Canadian Pulp & Paper Mills.[62]

By October 1937, the Tribune Company was well into the construction of the massive newsprint machines in the Baie Comeau mill. Vol. 25195, folder Baie Comeau, Mill and Town Construction (1 of 3), QOPC, LAC. Used with permission.

The Tribune's relationship with Anglo-Canadian on the North Shore had been fraught since the 1920s. Anglo-Canadian had built a mill near Quebec City in 1928, which it supplied with timber from its other limits in Quebec, and the company never seemed as enthusiastic as the Tribune about North Shore development. The two companies had formed the Quebec Logging Corporation in 1926 as a way to share premium stumpage costs along the Manicouagan and to partner on industrial and infrastructural development along the North Shore, but Tribune executives were complaining constantly about Anglo-Canadian's "dilatory tactics" in pursuing these projects. Throughout the 1930s, Tribune officials increasingly believed that Anglo-Canadian was simply riding their coattails on the North Shore, and by 1944 these Tribune officials were preparing plans to buy out Anglo-Canadian's North Shore holdings. Negotiations continued over the following year, and the Tribune eventually bought all of Anglo-Canadian's North Shore leases, an area of nearly 2,300 square miles, for $2.75 million. By 1948, the Tribune Company's landholdings on the North Shore amounted to 5,752 square miles, an area slightly larger than Connecticut. By 1950, the landholdings in the region had increased to 6,100 square miles.[63]

Through vertical integration, and aided by public policy in the United States and Canada, the Tribune Company had created a manufacturing infrastructure undergirding a highly successful media corporation. Anchored by the *Chicago Tribune* and the *New York Daily News*, the Tribune Company had created one of the most successful media corporations in the world, and it did so through the creation of a routinized supply chain connecting the forests in two Canadian provinces to the two highest-circulating newspapers in the United States. This supply chain had been enabled by public policy on both sides of the US-Canada border, and much of it remained obscure to US readers of the company's newspapers.[64]

Robert McCormick was proud of his company's achievements in Baie Comeau, and he was rarely hesitant to tout his role in the private development of the town and region. But McCormick was also deeply conscious of the political context of his development, and he always worked to keep the appearance of public humility in Quebec. Prior to Baie Comeau's grand opening, for example, he remarked that he found himself "looking forward to the Comeau celebration with a degree of foreboding. . . . I am afraid that the politicians will see the wealth of our mill and ignore the labor and speculation of the enterprise." As such, he told Arthur Schmon that he had resolved to be "as affable as I can be to the mighty men of Canada. I appreciate that it is necessary for me to be there, although my presence will accentuate the fact that it is foreign ownership. Not only do I not live in Canada, but I am known as an American publisher. However, that cannot be helped and I will do the best I can." And yet, at the same time, McCormick did not want to cede to Quebec officials and Maurice Duplessis the sense that they were responsible for the development. "I would rather not have the Premier open the mill," McCormick remarked. "I do not like the suggestion that he opened it and that in consequence he can close it. I am glad to honor him as the head of the province, but I do not like to endow him with any dictator powers. He has entirely too much leaning in that direction. . . . I cannot help feeling a good deal of anxiety about this official opening. It gives the politicians and others such a great opportunity to march in."[65]

One of the ironies of the Tribune Company's activities was that, for all of McCormick's grousing about Duplessis's dictatorial or domineering tendencies, his company was not operating in a democratic way on the North Shore, and there was operationally far less openness and transparency than his company publicly proclaimed. This was true in the way that company officials structured the housing market, and it was true in the way that they governed Baie Comeau. Though the original legislation had clearly specified that Baie Comeau was to be an "open" town and that governance would be the same as any other Quebec municipality, company officials worked behind the scenes to make sure that those in charge of the town were working in close concert with the company. In 1940, for example, Arthur Schmon wrote to Robert McCormick that he would be "pleased to know that two candidates we put up for election as Councillors at Baie Comeau were elected on March 1st." Both men were company employees, and Schmon claimed that it was a "good thing to have had this election and now everybody seems satisfied." Schmon himself was "particularly pleased because, virtually, the Management of this Company has been running the Town insofar as policies are concerned."[66]

This practice of shadow municipal government continued well into Baie Comeau's operation. In 1948, Schmon began planning for dealing with Fred Duchesneau, the first mayor to serve in the town since it had opened with H. A. Sewell in the position. Sewell, a company employee, had worked in close concert with Tribune executives, and Schmon was trying to continue this practice. The new mayor, Schmon remarked, "will have to keep us informed of what is going on and what has resulted from each Town Council meeting." Were significant decisions "involving policy" to arise on the Town Council, Duchesneau "should be carefully instructed as to how to proceed." In the event that unexpected issues arose, Schmon went on, the new mayor "should stall the decisions, but always in a manner that will not indicate to the councilmen that he needs clearance or approval. He can say he will hold the matter over until the following meeting in order to give everyone time to think about it." This, Schmon believed, "will permit him to discuss it with our officials." To facilitate these discussions and ensure that the town was governed as the company desired, Schmon suggested setting up a company committee that he would chair. That committee would meet with the mayor and have minutes taken that "should not be made available to the Town Council. . . . It is most important that the organization of the Committee and all matters relating to it should be on a confidential basis. In no way should the Mayor refer to it to the members of the Council." In essence, Schmon aimed to set up an unaccountable secret committee to use the mayor as a puppet. Besides the mayor's office, company officials noted the importance of having a "well balanced" town council "with as few radicals as possible." This meant encouraging company allies, or "sound men who are popular," to run for office. Otherwise, the possibility was that, as in any democracy, "there can be elected by the negative element almost any individual who wishes to spend some time and money on his election."[67]

These kinds of activities created significant disjunctures between the Tribune Company's public statements and its private activities. In many instances, there were rich ironies in the contradictions between the paper's denunciations of activist government and celebrations of individual freedom and free enterprise and the daily practice of the company's own corporate and municipal governance. Indeed, many Baie Comeau residents were unaware of these disjunctures. One former resident told me in 2013 that she was "not aware of the political background surrounding Col. McCormick."[68]

Sometimes, however, Canadians did perceive the significant ironies of the Tribune Company's involvement in their country. One of the strongest and most curious articulations of this argument is in an obscure manuscript, a copy of which is held in the *Tribune*'s archives outside of Chicago. In 1944, a Canadian journalist named Corolyn Cox found herself frustrated by the pervasive anti-British sentiments published in the *Tribune*, and she was outraged that Canadian forests and workers were being exploited to enable this. "Col. McCormick here brings the trees down out of our forests . . . has our good habitants make the wood into newsprint, stows it on his own boats, [and] carts it down to Chicago to say 'damn the British' in his *Tribune!*" Cox traveled to Baie Comeau to write a story about the town and the *Tribune*, and she found McCormick's presence towering over the place. In getting a driving tour of Baie Comeau, Cox noted that "we met the shadow of Col. McCormick at every turn. It was 'Col. McCormick this, Col. McCormick that'—all day my curiosity about this man was to be mounting. He personally pervaded the whole territory."[69]

Parts of the town impressed her, especially in the sense of how well planned and built it was. It had been designed to "avoid the evils of the company store and its credit enslavement of employees," the school system seemed well run, and there seemed to be positive relations across the Catholic-Protestant and Francophone-Anglophone divides, no small accomplishment in midcentury Quebec. Cox described the town as a "fascinating social industrial experiment. Baie Comeau is a laboratory experiment in planning, and Canada is a 'natural' for planning. Vast country, with its resources as yet only scratched, no wonder youth looks it over in the light of modern scientific development and modern political philosophies, and dreams Baie Comeaus from coast to coast!" And yet, she wrote, she could not get past her objections to McCormick's politics, as despite the fact that Baie Comeau was "the product of intelligent use of individually owned capital," it was also "handled by an isolationist whom Canada has come to look upon as almost the enemy of all that she holds dearest. . . . I found myself liking this shadow-man because of his imagination in tackling the remote, rugged, wild north shore of the St. Lawrence, and he did it all so well—the while I still boiled over the stupid misrepresentations of British institutions that continually appear in the *Tribune*."[70]

After visiting Baie Comeau, Cox followed the literal paper trail to Thorold, where she toured the Ontario mill, and she then proceeded on to Chicago, where she secured an interview with McCormick himself. As she had with Baie Comeau, Cox found much to like about McCormick. Soon, however, her political viewpoints got the better of her, and she grew angry that the anti-British McCormick refused to recognize the irony of his role in Canada. "Talk about our King— it's YOU who have the kingdom," she railed at him. Cox found herself unable to control her irritation with what she perceived to be McCormick's hypocrisy. "I continued to abandon tact," she wrote, and told McCormick, "You talk so much about getting rid of controls in industry . . . yet you're running a most perfect planned economy in your own personal kingdom." McCormick disagreed, remarking, "Oh, but I am not against planning. . . . I'm all for it. What I don't want is bureaucracy." Given the evidence of autocratic oversight that McCormick established and maintained over Baie Comeau, it would seem as if he succeeded as a town planner in institutionalizing his aversion to the bureaucratic.[71]

Conclusion

McCormick's reply to Corolyn Cox and his defense of his managerial philosophy reveal a great deal about his relationship to Baie Comeau and his understanding of his responsibilities to those living and working there. McCormick's opposition to contemporary federal projects such as the TVA had to do with his particular understanding of the different moral and practical efficacies of the public and private sectors. Federal planning, McCormick believed, was both inefficient and undemocratic. Public officials, many of whom were unelected, used public funds wastefully and unconstitutionally, McCormick thought, in the interest of pursuing projects that were better and more properly left to the private sector. In contrast, he understood his Canadian projects—activities that were in many respects strikingly similar to New Deal public works—to be in keeping with what he perceived to be the tradition of free enterprise, in which individuals staked their own capital on their projects and succeeded or failed because of their talent, ingenuity, and initiative. For Robert McCormick, what Baie Comeau demonstrated was that the private sector did a better job than the public sector in both creating industrial development and employment and looking after the welfare of workers and their families. McCormick found it consistently convenient to overlook and understate the degree to which his private development had been made possible and directly subsidized by public policy and public funding.

The ultimate fate of Corolyn Cox's article is unclear. She wrote to McCormick in early 1945, enclosing the twenty-four-page piece with a note stating that she hoped to get it published in *Harper's* or *Reader's Digest*. It does not seem to have ever appeared in print. McCormick read the piece and found it amusing, writing back to Cox that it was "very readable, if extreme." It does not seem to have done anything to change his disposition toward running a company town in Canada. Baie Comeau continued to provide not only a base for his industrial operations but also a symbolic rebuttal to many things that he thought were objectionable about the New Deal state.[72]

CHAPTER SIX

Work and Culture along the
Newsprint Supply Chain

In August 1937, the *Chicago Tribune* reported on a wedding. Though the paper did this regularly in its local coverage, this was a wedding of particular significance: it was the first one in "Canada's newest and fastest-growing baby city—Baie Comeau." Where one year before "there was no town of Baie Comeau," there were in 1937 the physical signs of industry and settlement. "And now," the paper remarked, a "bride and bridegroom make the town nearly complete." The groom was Kurt George Hayden, a painter employed by the Ontario Paper Company, and the bride was a Montreal native named Sophia Maynan. The artisanal wedding ring had been fashioned from "hammered brass made by the Baie Comeau foundry," and Hayden's coworkers used their paintbrushes to form a wedding arch, through which the bride and groom walked before getting in the back of a pickup truck and being driven away.[1]

In some ways, the *Tribune* report on the Baie Comeau wedding was an effective bit of promotion for the town that it was in the process of building, and it offered the paper a human interest angle from which to tout its industrial and municipal development projects in Quebec. But the article also revealed something about the people who had moved to this new town and started new lives there. For people like Kurt Hayden, the Baie Comeau mill offered the possibility of employment during the Depression. By design, Baie Comeau was intended to be more than just a place to house male workers laboring on the Canadian resource frontier. This was not meant to be a place resembling logging or mining camps, many of which were built to provide little more in the way of amenities for the mostly male workforce than places to eat, drink, and sleep between shifts. Instead, Baie Comeau was meant to be a permanent and attractive town populated by both workers like Kurt Hayden and people like his wife Sophia Maynan and potentially their children. As a planned community, Baie Comeau was meant to be an exemplar of a new kind of welfare capitalism providing not only jobs but also a community for people like Hayden, Maynan, and all the others who participated in the wedding ceremony.

As the Tribune Company created an international industrial newsprint supply chain, it organized space and resources in two countries and gathered together a

After the first wedding ceremony held in Baie Comeau, the newly married Kurt George Hayden and Sophia Maynan were driven away in the back of a pickup. Vol. 25178, Scrapbook 1934–41, QOPC, LAC. Used with permission.

diverse and dispersed collection of workers whose labor was essential to the production of its two daily newspapers. In the United States, the production of the industrial newspaper included not only journalists and editors working for the *Chicago Tribune* and *New York Daily News* but also a host of production and clerical employees in Chicago and New York. In Canada, the Tribune Company relied on a completely different sort of labor force, and the supply chains that led to urban US pressrooms and editorial offices started in the Canadian forest with the labor of lumberjacks. As the Tribune Company's newspapers shaped public life in the regions in which they circulated through their printed content, its industrial operations shaped public life in the areas in which its newsprint was manufactured through its harnessing of natural resources and human labor.

Natives, Lumberjacks, and the Landscape of Newsprint Production

When the Tribune Company began the industrial development projects that would lead it to building the city of Baie Comeau, it encroached on land that had been home to native people for centuries. In Canada as in the United States, the introduction of white settlers and the cultivation of space for industrial capitalism disrupted and destroyed the long-established routines of native inhabitants. On the North Shore of the St. Lawrence River in Quebec, these people referred to themselves as "Innu" but were more commonly known in North America by the term that the French conferred on them, "Montagnais," or "mountain people." Many of them lived for long stretches in the woods hunting and fishing and spent only parts of the year in some of the few settled towns near the St. Lawrence. Estimates by contemporaries and historians vary, but all point to the presence of one to two thousand people living across a large area in the early twentieth century. For the white North Americans in the twentieth century who began to see the North Shore's resource potential, racialized conceptions of the native population made it rather easy to think of the area as devoid of settlement.[2]

When the Tribune Company started acquiring timber limits on the North Shore, it took control of territory that had not only an indigenous population but also merchants operating small camps and trading posts established up the region's rivers. In 1929, Arthur Schmon expressed concern about some of these traders, noting that they "dispense some liquid refreshment, and are therefore a menace in our woods, particularly because of the fire danger in the spring and the fall." Schmon was concerned that natives were "loitering around these trading places," and he began plotting ways of removing them and the traders. "I am afraid that unless we can combat this menace effectively, it is going to be a very serious thing. We know that these men are up there not only to trade legitimately but also against the law. The Indian Department must certainly be interested in this matter. These traders are doing no good to the Indians. The presence of the Indians at this time of the year in the bush . . . is certainly a great menace to the forest." Company officials began investigating ways of seeing that these traders were "expelled from our property and that no more traders are allowed on our limits." This project took place with the full authorization of provincial officials, who assured the company that its timber leases meant that it held the "exclusive right not only to the timber, but also to the land on which the timber is located. . . . In other words we have the exclusive right to the property and no-one else has any right to build anything on any portion of our limits." For the Tribune Company, clearing the land of what it perceived to be undesirable elements was a part of organizing the territory that it controlled in order to manage the forest reserves necessary to maintain its industrial newsprint production. The only people the company wanted in that territory were the loggers whom it employed.[3]

These woods workers formed the foundation of the Tribune Company's newsprint supply chain, and the printing of the *Chicago Tribune* and the *New York Daily News* was impossible without them. As the *Chicago Tribune* noted in 1930, the "swing of the lumberjack's keen-edged axe is the first step taken in the manufacture of paper; the whining rasp of his saw is the second. For all their importance, wharves, loading plants, power plants, ships, and towns, are but intermediate agencies transferring the product of his toil to the mill. Without the lumberjack the forest would never yield its riches of pulpwood. He is as essential to the production of paper as is the miner to that of coal." The size of the company's lumberjack labor force in the woods was massive. In 1946, for example, Arthur Schmon claimed that the Tribune Company relied on the labor of some seven thousand woods workers, most of whom were French Canadian. Schmon noted that "99% of the labor used on the operations was migratory French-Canadian lumberjacks. . . . They were a different breed entirely from the French-Canadians who were raised on farms or in cities. They were illiterate for the most part but they were a simple, honest folk and seemed to me to possess more nobility than their kin who were closer to civilization." Other company accounts of its French Canadian labor force continued in this kind of essentializing, as one promotional book noted of the French Canadian lumberjack that "lumbering seems to be born in his blood, and with it goes a profound, instinctive knowledge of the forest and all its ways. To him the forest is not merely a place wherein he earns his livelihood; it is more than a place, more than an impersonal thing. It is a personality he knows, whose varied methods he senses shrewdly, and with the sympathy that can exist only between living beings."[4]

In Chicago, the *Tribune* occasionally published articles about its chief forester, Paul Provencher, and the newspaper often described Provencher in similar terms. Born in Trois-Rivières, Provencher earned a degree in forestry at Laval University, after which he went to work for the Tribune's Canadian newsprint subsidiary. He was instrumental in performing the woods survey work that led to the company constructing the mill at Baie Comeau, and his skills in the woods made him known as one of Canada's leading experts in the field. His reputation grew to such heights that, as the *Tribune* reported, during World War II he was "selected to teach soldiers how to live and how to fight in the woods, the idea then being that Canada might be invaded and that it would require a sizable military force of woodsmen to repel the enemy." The Tribune "lent him to the Canadian army," and he spent the spring of 1943 teaching Canadian soldiers how to best function in the woods during combat. In 1953, Provencher wrote a book detailing the practicalities of surviving independently in the forest, with Robert McCormick adding a laudatory foreword. "Many of us who have cruised the properties of The Ontario Paper Company Limited and the Quebec North Shore Paper Company have thought we were pretty good," McCormick stated. "I was among those until I read Paul Provencher's book. It goes far beyond my knowledge of woodcraft. It is so valuable that I think it should be made available to everybody entering the woods, not only to save themselves trouble and hardship, but also their lives. This book will serve as a contribution to the welfare of the modern *coureurs de bois*."[5]

In many ways, the labor of Paul Provencher and the thousands of other French Canadian lumberjacks in producing the *Chicago Tribune* showed how, to use political scientist James Scott's terms, industrial high modernism and "practical knowledge," or what he calls "mētis," intermingled. For Scott, these concepts were often oppositional, as high modernist urban planning and systems of industrial organization like Fordism were at odds with the "vernacular character of local knowledge" that people who have working understandings of environmental conditions demonstrate. As the Tribune Company maintained an international industrial supply chain to manufacture its newspaper, it was dependent on lumberjacks cutting down trees to start the process. One of the most technologically and organizationally advanced newspaper corporations in the world was utterly reliant on lightly educated but highly skilled French Canadian lumberjacks at the base of its operations. At its core, the high modern industrial operation that the Tribune Company utilized to manufacture its newspapers rested on the incorporation and mobilization of a labor force in possession of deep practical understandings of the forest.[6]

In addition to their practical knowledge about the forest, the attitudes of these Quebec workers harmonized with Tribune Company imperatives because of their seeming lack of interest in forms of collective action like unionization. As one 1934 account in the newspaper trade press noted,

> Canadian pulpwood has been cheap on account of the French-Canadian's native skill as a woodsman and river-driver, his ability to live in the woods under the roughest conditions, on an extremely simple diet, and his energy and thrift when making logs on contract. Few Americans, except perhaps those who have fished or hunted in the Canadian woods, will realize what an asset the French-Canadian habitant has been to the Canadian paper companies. Until recently he has known practically nothing about unions, wage-scales, the fifty-hour week or other such new fangled ideas. He ordinarily makes logs on contract, by piece work, and when working in the usual zero temperatures of the northern woods is energetic as well as skillful.

It was the great fortune of newsprint manufacturers to have at their disposal a native labor force that was both capable of doing work that was physically very difficult and willing to do so on terms that were good for business.[7]

For the *Tribune*, as for all US newspapers, this woods labor was crucial to the daily newspaper's production. Through midcentury manufacturers' estimates, we can arrive at a rough estimate of the physical work and forest resources involved in producing a daily newspaper. According to the Canadian Pulp and Paper Association, an average lumberjack could produce 1.5 cords of wood per day. A stacked cord of cut logs measures 4 feet wide, 8 feet long, and 4 feet high and contains 128 cubic feet of wood, bark, and air. This amounted to 5,065 pounds of material, 2,570 of which was wood fiber, 1,960 of which was water, and 535 of which was bark. After processing and manufacturing, this cord of wood yielded 2,304 pounds of newsprint, which could then be used to print roughly 10,000 copies of a standard-sized sixteen-page newspaper. To put it another way, the average day of a lumberjack led to 1.5 cords of pulpwood and was the start of a commodity chain resulting in 15,000 standard-sized newspapers circulating in the United States.[8]

As physical labor, woods work was difficult on many levels, perhaps most importantly because of the weather. The newsprint production process started in the winter, when logs were easier to move across paths covered in slippery ice. Depending on the vicissitudes of the weather, the log drive could be delayed significantly and leave lumberjack crews idle in the woods waiting for the weather to change. In one account written on April 13, 1924, just on the cusp of spring, a participant in the log drive reported that his party had been "loafing for a week" in a logging camp waiting for the weather to allow the drive to start. "When we came up we expected to be able to finish in three weeks but the sudden cold weather has delayed the log drive. . . . We may have to wait one week or perhaps two before the ice goes." When the weather cooperated, once the logs were cut down in the woods, they were "hauled to the nearest river-bank. Here they are stacked on the sloping bank and held in place by upright props. When spring comes the props are released and the logs slide into the water. Sometimes logs are also piled on the frozen rivers and lakes to await the thaws which will enable them to float seaward."[9]

As woods workers toiled in winter to start the process of producing the industrial newspaper, their camp accommodations were often primitive, and the almost exclusively male workforce had few comforts or amenities. Many woods camps were rustic, often involving workers simply tramping down snow and pitching a tent made of silk. The tent floor would then be lined with a layer of balsam branches roughly a foot deep to keep the floor warm and dry, and a tarpaulin was then placed on top of that. The material for a structure that would house five men might weigh 30 pounds, but it would provide adequate shelter despite the cold weather.[10]

Well into the twentieth century, the lumberjack remained a key part of popular understandings of the Canadian North. In the decades after World War II, however, some of the romance of lumberjacks' work in the woods began disappearing owing to significant changes in the technologies of logging. Whereas in the 1920s the men would spend winters living "mostly on salt pork, bread, pork and beans, pea soup, molasses, and tea," working with hand tools, and dragging logs with

The beginning of the Tribune's newsprint supply chain started in the woods, where lumberjacks cut trees and stacked them near the river for conveyance to the mill. Here, in May 1929, several workers participated in the spring log drive along the banks of Tibasse Creek. Vol. 25177, Scrapbook 1926–34, QOPC, LAC. Used with permission.

horses, by the 1950s much of the work had been mechanized. As one 1953 description put it, the "Canadian lumberjack, who once worked with only an ax, a hand saw, and a horse, is becoming a skilled operator of mechanical tools and vehicles. . . . Gone are the days of hardship and adventure. The machine age has come to the Canadian woods." This took place across Canada, and the job of a woods worker was transformed from one involving mostly human power and hand tools to one involving the use of machinery. As in many occupations at midcentury, mechanization transformed the physical labor of lumberjacks.[11]

The midcentury mechanization of the lumberjack trade had other significant social implications, as it began disrupting the traditional patterns in the relationship between farm and woods work. Prior to the mass adoption of mechanization in the newsprint manufacturing industry, the typical start to forest work was in November, after the first snows had fallen and felled trees were more easily moved around by simply sliding them on ice. By the mid-twentieth century, advances in cutting and transportation technologies had allowed woods work to begin in September instead. As Arthur Schmon noted in 1958, in the 1920s

> the labor force came in the fall and when the last boat left in the fall the labor force
> had to remain until the opening of navigation the following spring, which might
> have been in those days April 1st to April 15th. Logging operations in those days
> started full-blast when all the personnel were in for the winter with the arrival of
> the last boat. Now with improved navigation and with the airplanes we complete our
> cutting program by Christmas and our handling operations early in January. Now,
> actually, we may complete all of our work in the fall and early winter before the first
> of the year.[12]

This earlier commencement provoked significant concern among church leaders on the North Shore, as Napoléon-Alexandre Labrie, the bishop of the Gulf of St. Lawrence, called for logging companies to reconsider cutting schedules that would promote a "harmonious advance toward another order of things more in conformity with the dignity of free men." The problems were legion, Labrie argued. Farm crops still in need of harvest in September were being left to rot as workers took to the woods early. The disruption of the traditional cycle of farm work and woods work had also begun attracting more seasonal woods labor to the North Shore, creating an unstable population. There was "damage caused," Labrie argued, "to the crops, to autumn work, to cattle and to the family itself," as mechanization altered the long-standing seasonal rhythms of the male workforce.[13]

The disruptions of traditional work and life rhythms caused by the development of the industrial newspaper's operations also affected the North Shore's native population. As Robert McCormick recalled in 1953, his company began offering irregular work to natives, though "at first they would not work steadily." This changed over time, however, which made McCormick a bit wistful. "Little by little, they took to work around the dock. The masters of the forest became common laborers. There is something sad about this." McCormick overstated to some degree the modernizing trends on the North Shore. In 1956, a *New York Times* correspondent visiting the area commented that a "land of isolation" had on the one hand become a "curious old-modern blend of busy pulp towns, quaint fishing villages, tethered sled dogs and mining booms." On the other hand, the correspondent noted, the region was "home, too, of the country's most primitive Indians." By the mid-1950s, across the North Shore, the manufacturing processes that the Tribune Company had initiated had transformed the environment and labor market of the region in significant ways. The industrial production of newsprint for the export market had prompted a reorganization of lives and landscapes along the North Shore.[14]

The Hazards of Newspaper Work

Newspaper work is occasionally dangerous, and reporters have been the subject of intimidation, violence, and even assassination. The war correspondent can be subject to particularly hazardous conditions. But when one considers "newspaper work" as comprising not just reporting but also a range of occupations stretching from woods work to factory work, the matter of workplace safety takes on a different cast. Industrial newspaper production created a range of hazards, some of which could be fatal, across the supply chain from the forest to the paper mill.

Woods labor could be dangerous and even deadly. Laboring as a lumberjack involved working with sharp cutting implements, and, as the *Tribune* noted, "at least once in the lifetime of every woodsman he may become careless and hurt himself." Paul Provencher related a story about how one day, while working in the winter about 2 miles from his camp, "his ax glanced off a tree and its sharp bit sliced into one of his feet. He made a tourniquet with his sash, but it was too painful for him to endure so he removed it." Bleeding in the frozen woods, Provencher decided to try to make it back to the camp, "leaving a trail of blood." After a while, he became weak and sat down on the ice of a frozen lake. Luckily, one of his coworkers spotted him, and "set out with a dog team and sled to rescue him. He brought Paul into camp and administered first aid. A little more bleeding might have proved fatal." Wounded or not, woods workers could also become lost and disoriented in the wilderness, and death by freezing was not uncommon.[15]

Work-related injuries and deaths were unfortunately reasonably regular occurrences in the Tribune Company's newsprint manufacturing operations. In total, according to company estimates, there had been 139 fatalities in Canadian woods and industrial operations between 1912 and 1953, or an average of 3.4 Tribune Company workers killed on the job every year in the service of manufacturing newsprint. In the woods, the company reported, the causes of death included being struck by a falling tree, getting swept away by a log jam, drowning after breaking through thin ice on a lake, being kicked by a horse, being struck by debris during a dynamiting procedure, and getting hit by cut logs falling off sluices. The list of work-related deaths in manufacturing compiled by the company reads like a documentary of the macabre. For example, at the Thorold mill in 1922, "ironworker employee cut in half by cable during construction of No. 3 and No. 4 machine room." In 1936, "employee killed when stone tower elevator cable broke as elevator was being lifted." In 1940, "while pulpwood was being unloaded by measured cord method, chain broke and released pulpwood on head of workman in boat." Workers were killed on the job in similarly grisly fashion on the North Shore. For example, the company noted, in 1936 a crane operator drowned when his crane fell over and off the wharf. In 1937 at the Baie Comeau mill, while one laborer was working in the steam plant, the "large water main was being tested. Plug blew out, and laborer was internally injured as he was swirled around in water in pit." In 1941, a worker was "electrocuted on scaffold in steam plant, when, as he passed between a defective electrical welding machine and a boiler, he touched machine and boiler frame and died instantly." In 1943, "while washing grinder pit in groundwood mill, workman reached into pit to retrieve an article he had dropped. He got caught in machinery and suffered fatal chest and head injuries." In 1947, a "stevedore, loading pulp at wharf, was crushed on ship's deck when hit by bale of pulp that fell out of sling." In 1951, one worker died a particularly tragic death when the "brake on bulldozer blade failed as he crawled under blade to unhook cable attached to another tractor. Crushed skull."[16]

In addition to the fatalities in its woods and mill operations, the shipping operations of the Tribune Company could be deadly as well. In mid-1935, the company took delivery of a new boat called the *Joseph Medill*, which was to haul pulpwood on the St. Lawrence River. Built in England, the ship sailed for North America in early August. Sighted by a Swedish ship on August 17, the *Joseph Medill* was never seen again, and the full crew of eighteen men died. Fourteen other shipping operations workers died between 1939 and 1953, many of them from either falling overboard or being struck by equipment.[17]

One of the most tragic and ultimately well-publicized deaths suffered by an employee came in the winter of 1930, when the company was working on the foundations of the Outardes dam. Peter Trans, a diver from Montreal, "slipped as he was going down the ladder to work on the bed of the river where foundations are being constructed for the big dam. His air tube and safety lines became entangled in the ladder and other obstructions and efforts which have been continuously conducted since then have hitherto failed to release him." Though he was pinned underwater, Trans's "air pipe is not choked off and he is being kept supplied with needful oxygen," and he remained trapped in the freezing waters for over twenty-four hours, long enough for the *Montreal Star* to publish several reports about the situation. "One volunteer descended twice and vainly attempted to release the man but failed in his attempts." The company tried to have "two expert divers" flown in to handle the situation, but snow was "falling heavily along the north shore and flying rendered impossible."[18]

Two days later, the *Star* reported that Trans "stopped signaling early yesterday and although rescuers still pumped air down the air line, little hope remained that Trans could be brought up alive." With "deadly, heart-exasperating snow preventing rescuers from arriving," one of Trans's friends, a "fellow-workman" named Arno Silyala, "who though he had never been inside a diving suit in his life went down twice into the icy waters in a desperate attempt to save his pal." Silyala was able to locate and reach his friend, but he found him dead. Of this "tragic meeting under the frigid waters of Riviere aux Outardes," the *Star* wrote, "anything more pathetic could scarcely be imagined. The first time Silyala could only see Trans at a distance standing straight up. The current prevented his getting closer. During the second descent the amateur diver clutched for one of the hands of the victim and caught it. Repeatedly he shook and shook it, but it gave back no heartening answer, it seemed lifeless." Workers on the surface hatched a plan to "build a crib, sink it around Trans and pump it out. " Before they could finish, rescue workers from Quebec City finally arrived, though they only succeeded in bringing the body of the deceased to the surface. "His body torn and bruised, his diving suit practically torn off his back by the force of terrific currents and with one leg smashed, the lifeless form of Peter Trans was brought to the surface," as the *Star* reported. "The tragedy of Trans . . . is typically modern. Trans, like hundreds more of the human titans, who labor at the raw work of building up a new Canadian Empire in Northern wastes, harnessing of forces of nature for Canada, came to Riviere Aux Outardes by the air. He was used to danger. He was one of hundreds of men who take chances against death every day, without a thought about it, until danger becomes trivially commonplace. But sometimes the elements have their innings."[19]

The Tribune Company was far from being the most dangerous paper company in Canada, but overall industry estimates did rank the company in roughly the lower third in terms of worker safety in Canada in the late 1930s. It is a sobering fact to consider death in the newspaper business. In many cases, this involves journalists killed in the act of reporting the news. In some cases, these deaths come in war zones, as journalists risk their lives to report the news to those far removed from the scene of danger. In some authoritarian societies, these deaths can be the result of targeted political assassinations. But deaths can also arise in the process of newsprint production, and they can also be the result of physical labor in the woods, settings where the work involves its own set of risks and dangers.[20]

Leisure, Religion, and Welfare Capitalism on the North Shore

Before the Tribune Company's arrival to the North Shore, daily life in the region was often a struggle between people and the environment in a distant and disconnected space. Once the Baie Comeau development started, public descriptions of the North Shore operations increasingly focused on the degree to which it had become settled, and workers' lives there were less likely to be represented as demonstrations of personal and physical fitness than they were as being simply comfortable. Corporate capitalism had enabled the North Shore's industrial and municipal development, and workers were the beneficiaries of that. By 1939, the *Tribune* was even touting Baie Comeau as a tourist destination. With a newly initiated commercial cruise from Montreal, vacationers could "visit the miracle town that sprang up almost over night out of the forest" and see the former "frontier post" that was now "Canada's newest and fastest growing city."[21]

Throughout the planning and building of Baie Comeau, the Tribune Company was guided by the paternalistic desire to create a municipal infrastructure that made workers feel that the company cared for them and had provided them with a place to live that was a stable and thriving community rather than a temporary town built solely to facilitate the extraction of natural resources. As Arthur Schmon remarked just after Baie Comeau was built, the goal was to make workers "feel that Baie Comeau is a place that they can be proud of—and not ashamed of—as a permanent home and as a permanent place for working." As the company implemented these plans, it created a community in which workers had many amenities available to them outside of work. In 1939, almost immediately after the mill went into operation, the company established what it claimed was the "only public library east of Quebec City." By 1951, the library's holdings had expanded to some 12,000 volumes, and the town's community center also had "sports facilities as diversified as skiing, bowling, billiards, hockey, curling, archery, badminton, and—in summer—an outdoor swimming pool. The Association's club house provides lounges, a well-stocked library, classes in handicrafts and physical culture, and serves as a musical centre." Baie Comeau residents got movies not long after larger cities, and its theater had a "seating capacity of 420 and is now operating on a schedule of two complete shows nightly, except Monday."[22]

The creation of these amenities happened within the context of a town that had grown quickly and dramatically. The town's population of 1,102 in 1938 had nearly doubled by 1942, and it was over four thousand by 1951. The number of families nearly tripled from 296 to 889 between 1938 and 1954, and the population began to skew younger as the birth rate increased. There had been forty-four births in Baie Comeau in 1938, and this increased to 120 in 1944 and 173 in 1954. By 1952, roughly a third of the town's 4,144 residents were under the age of eighteen. By this point, the former wilderness now had public facilities in the form of a post office, a telegraph office, and a customs office. There were municipal fire, police, and electrical departments. The company had spent $210,000 on school construction, and the town had a Catholic high school and primary school staffed by twenty-four teachers and a Protestant school with five teachers. There was a Catholic cathedral and a Protestant church. There was a public swimming pool and a recreation center. The town had a private dentist, surgeon, and medical doctor, and that was complemented by a company-owned hospital with one doctor

On its main downtown street in 1947, Baie Comeau offered residents a range of stores and services. Vol. 25197, folder Baie Comeau, Townsite (1 of 3), QOPC, LAC. Used with permission.

and six nurses. There were numerous private stores and businesses, and a number of civic associations had been established, including chapters of the Canadian Legion and Knights of Columbus.[23]

Part of the Tribune Company's project of creating a sense of harmony and community in Baie Comeau was grounded in an awareness of the culture of Quebec, as the town's population was mostly Catholic French Canadian. Cultural differences between English- and French-speaking Canadians remained fraught at the time of Baie Comeau's construction. While skewing younger over time, the population was consistently and overwhelmingly Catholic. In 1952, Protestants made up only 358 of a total population of 4,144, and Catholic births outnumbered Protestants 167 to 6. For the Tribune Company in Quebec, one of the most important venues for accommodation with provincial religious culture involved creating a rapprochement with the Catholic Church, which remained a pillar of French Canadian society. On the North Shore in the 1920s and 1930s, Eudists were the dominant order. Founded in France in the mid-seventeenth century, the Eudists had sent missionaries to the North Shore starting in the late eighteenth century. Besides trying to convert native people, Eudist priests also provided a sense of community for the migrant agricultural labor composed almost exclusively of

Catholics. Having established a stable group of missions along the North Shore by the 1930s, the order focused on the emerging town of Baie Comeau as the potential center of church operations in the region. Knowing that Catholicism was central to its workers and their families, the Tribune Company during Baie Comeau's construction gave $10,000 toward the "building of a temporary church."[24]

At the same time, the company also wanted to make sure that the local clergy did not move too quickly to establish a firm foothold in the region by having a bishop take up residence in the new town. As Robert McCormick remarked in 1937 as town construction was getting under way, "I certainly hope that the Bishop does not make his headquarters at Comeau. Irrespective of the personality of the first bishop, sooner or later, there would be a cantankerous one who might well raise the very devil." However, resigned to the fact that the Catholic Church would use Baie Comeau as its base of regional operations, Tribune Company officials worked with leading clergy to finance the construction of a more permanent building. While the company was contributing this financial assistance, Robert McCormick's wife, Amy, died, and he quickly arranged for the church to be named St. Amélie's in her honor. For the new church building, McCormick also paid for a lavish stained glass window and a number of frescoes done by an Italian artist. According to a local clergyman, "in memory of his late wife whom he said was a great artist herself, he wanted something which, were she living, she could admire at Baie Comeau." Ultimately, Robert McCormick built one of the most lavish Catholic churches within hundreds of miles in Quebec, and one longtime Baie Comeau resident remarked of town lore that locals "always say that that's the biggest Catholic church built by a Protestant."[25]

Though the company was enthusiastic in these public demonstrations of its support of the Catholic Church, its executives remained rather more lukewarm behind the scenes, especially as they continuously sought to prevent the church from having too much influence in the town. During World War II, rumors began circulating that the Vatican was considering the creation of a diocese on the North Shore and that Baie Comeau was being considered as the site of the formal residence of the new bishop. Company officials were not happy about the possibility, with one noting, "I believe it would be a serious mistake for the church to establish the head of a diocese in a town like Baie Comeau." In many ways, this corporate opposition aimed to prevent the church from having too much influence on workers' lives off the job. "Our principal problem as I see it," one company official noted, "is how can we prevent the clerical influence from making conditions in Baie Comeau such that our people there will not feel restrained in their social and recreational activities, and that people of other religious beliefs may continue to feel free to live their lives as they see fit, without feeling that they are the subject of criticism by any other group." Arthur Schmon was similarly unenthusiastic, but he noted that it would be "embarrassing and difficult to oppose" publicly the possibility of a greater church presence in Baie Comeau. Still, Schmon privately tried to persuade the bishop to move the seat of his new diocese elsewhere, telling him at one point, "We do not want our people to feel that the ecclesiastical authorities and institutions are more important than our business."[26]

Despite the company's reservations, in early 1946 the Vatican announced the formal establishment of a new diocese on Quebec's North Shore under Bishop Labrie's direction and with Baie Comeau as the seat. "I thought I had persuaded the Bishop that the seat of the Diocese should be either at Seven Islands or Godbout," Schmon noted. "At that time, he told me he would discuss the matter with higher authorities, but the final decision would be made by Rome. The situation now requires the utmost in diplomacy. The majority of the population will consider that an outstanding honour has been bestowed upon the Town." McCormick did not want the bishop meddling in his corporate fiefdom, noting in 1946 that it would be a "very great nuisance to have him in town and we have not got the space for it." But, private concerns aside, by the end of the 1940s the Tribune Company had developed a working agreement with the church which both parties found satisfactory. As Bishop Labrie praised the company in a speech to celebrate the establishment of the new diocese, "You have been the greatest promoter in the material development of the North Shore. You merit great praise and thanks for this, but it is not only for this reason that I wish to praise you, but chiefly because underlying your industrial expansion is a true sense of social advancement. The people who live and work under your direction are a group of happy people."[27]

While the Tribune Company made its accommodations with the Catholic Church, it also initiated measures to appease and support the local Protestant population, members of which remained vital to its operations. Though Protestants were a minority in Baie Comeau, they were a significant one, as they occupied most of the company's managerial positions. As Arthur Schmon remarked, "It seems to me that from a company point of view, the Protestants that are located in Baie Comeau are here because they have a particular contribution of skill or ability to make to the companies who are located here. I don't think these Protestants would just naturally be living in Baie Comeau. Consequently, the companies are directly interested in seeing that these people have proper education and that families are satisfied." As the town developed, the company supported the creation of separate educational and religious institutions. Because of the relative paucity of

Protestants along the North Shore, the various denominations were compelled to unify around the Church of St. Andrew and St. George in Baie Comeau, which by the early 1950s had members from Anglican, Presbyterian, Baptist, and Lutheran denominations. The Church of St. Andrew and St. George held regular Sunday services, and during the week its Reverend Gourley traveled "by boat and plane" around the North Shore to conduct services in the small communities of Forest-ville, Franquelin, Godbout, Trinity Bay, and Shelter Bay.[28]

All of these various efforts at creating a new and functional city in Baie Co-meau formed a matrix of corporate practices that reflected a mixture of welfare capitalism and social control. For many residents, the mixture seemed to work, at least in the early years. While much of the local culture of Baie Comeau was struc-tured by company activities, local people also developed significant voluntary forms of association. Both English and French residents formed community the-ater groups that gave public performances. A group of Anglophones formed the Baie Comeau Community Players, with the aim of providing entertainment to residents and diversions to "those who are interested in amateur theatricals." French residents formed the Petit Théâtre de Baie-Comeau, "with a view to giv-ing our population a healthful pastime," and their members soon began putting on plays of their own. In another case of voluntary community activity, more than one hundred residents paid $30 each to form a farm cooperative. After obtaining 35 acres of land, the group cultivated it and began growing vegetables and raising chickens. Overall, these sorts of practices meant that, as Baie Comeau mayor J. A. Duchesneau claimed in 1952, the town was defined by the "complete tolerance among all religious, social, civic and cultural groups for one another. Nowhere will you see French and English speaking communities working so closely and cordially together in every way."[29]

This was to some degree overstating the degree of cultural integration in Baie Comeau, and the peaceful tolerance of relatively prosperous English and French workers could be mistaken for genuine bonds of friendship and community across lines of language, culture, and religion. At the Community Center, for example, a series of cooking classes were divided into alternating evenings aimed at English speakers and "French ladies." However, at least in its first decades, Baie Comeau was seen by many residents and outside observers to be a well-planned and har-monious community, and the company that built it was understood to have taken a meaningful interest in the well-being of its employees, on and off the job. As the Francophone newspaper *L'Aquilon* noted in 1953, the company had not only "created a true industrial center on the North Shore" but also "shown a true social character in every sphere of human activity. They have encouraged sports through the construction of an ice rink, of a swimming pool, of playgrounds, etc. Nothing has been spared for education and medical expenses." Overall, *L'Aquilon* claimed of Tribune executives' approaches to Baie Comeau, the town's "social, religious, artistic, and educational" activities had "benefitted from their generosity." A Canadian correspondent for the *Times* of London was similarly impressed by the town's overall appearance, noting that "Baie Comeau is far from being a frontier town. . . . The visitor is immediately conscious of a standard of living as high as in the prosperous towns of Ontario. . . . The town itself has almost a Blackpool brilliance at night."[30]

From the company's perspective, the creation of this kind of culture in Baie Comeau was to some degree based on the need to give workers a reason to move to and then settle in what remained at midcentury a remote area, but it was also in line with other contemporary ideas about welfare capitalism. The company, as one executive noted, was "thinking along the same lines as other progressive employers in realizing that the old days of ignorant labour are gone, and that the present generation which mans our mills and woods operations has different standards. The older generation compared its lot with its past on the poverty stricken farms and in the slums of Europe. Today's worker compares his lot with the most fortunate person of his acquaintance. He reads the same newspapers, the same magazines, and often the same books that his boss reads."[31]

The Tribune Company was certainly not the only company engaged in creating and supporting institutions that provided culture for workers off the job, but it was one of the most active and creative. Over the first decades of Baie Comeau's operations, the company created and encouraged a range of community activities across secular and sacred realms, and it sought to create an inviting environment for workers and their families. As the company's internal discussions about the Catholic Church showed, these aims at creating community were not entirely altruistic but were to a great degree about social control. Having built a community, the Tribune Company was hesitant to share authority in shaping its culture.[32]

Unions, Paternalism, and the Hidden Mechanisms of Social Control

Alongside its commitment to these extensive welfare capitalist projects relating to workers' leisure and cultural activities, the Tribune Company was also supportive of its mill workers' attempts at unionization in the interest of maintaining outwardly harmonious industrial relations. Indeed, to some degree, one can see some of the major trends in the labor history of North American paper mill workers in the interactions between the Tribune Company and the various unions representing its workers. The origins of North American paper workers' organizing date back to 1884, when skilled workers at a mill in Holyoke, Massachusetts, formed a "benevolent association" called Eagle Lodge. This helped to form the basis of the United Brotherhood of Paper Makers, a union chartered by the American Federation of Labor. As in many sectors, skilled workers in the paper-making industry chafed at organizing alongside unskilled workers, and in 1906 the different groups of laborers split into the International Brotherhood of Paper Makers (IBPM) and the International Brotherhood of Pulp, Sulphite and Paper Mill Workers (IBPSPMW). In 1948, the United Paperworkers of America merged with the IBPM to form the United Papermakers and Paperworkers, and in 1972 this union merged with the unions representing unskilled workers in the IBP-SPMW to form the United Paperworkers International Union. For most of their histories, North American paper manufacturing unions were international, as unions based in the United States followed the growing newsprint manufacturing industry in Canada and tried to organize workers there.[33]

As the various unions representing workers in the pulp and paper industry developed, the Chicago Tribune Company was at the forefront of supporting a unionized workforce in Canada. In August 1913, just after the Thorold mill went into operation, thirteen workers in the mill requested and received a charter from the International Brotherhood of Paper Makers to form Local 101 under the leadership of Matthew Burns, a paper machine operator. The following year, the IBPM reached "its first written agreement in Canada," when it negotiated a contract with the Tribune Company to cover the labor of Local 101 workers. The IBPM later hailed this pact as the "cornerstone of the union on Canadian soil. Its terms called for three daily shifts, each working 48 hours a week—with no work on Sunday—achieving the eight-hour day for the first time in the industry." In 1916, the company reached its first written agreement with the IBPSPMW Local 84, which represented the unskilled workers at the Thorold mill.[34]

The labor peace defining the Tribune Company's Canadian operations was noteworthy to parties on both sides of the bargaining table. As Matthew Burns recalled at the company's twenty-fifth-anniversary celebration in 1938, when he initially approached company executives in 1913 about forming a union, they were supportive, and once the union was established, the parties found it easy to cooperate productively. Over time, Burns claimed, the Tribune Company and its workers' unions had created a "record of achievement in relationships between employer and employee that a troubled world might well take note of at this time. . . . I am grateful to the Ontario Paper Company for the contribution that it has made to my life as an individual and I am grateful to the officials because of the contributions they have made to the cause of union labor generally." Company executive Arthur Schmon echoed these sentiments in remarking that the company had "followed the policy of collective bargaining with its workers. During all these years of friendly and happy relationship there has never been a strike or cessation of work within the mill itself." He attributed these positive labor relations to the basic fact that company policy "has always been to consider our employees as men, not as workers."[35]

These kinds of positive relations extended into the immediate postwar period, as the company's support of its workers' unionization continued alongside the expansion of production to Baie Comeau. In 1947, Matthew Burns, by that point the president of the IBPM, remarked that employers like the Tribune Company "create great employment opportunities for the people in a community and surrounding area. These employment opportunities enlarge the communities, bring forth new businesses of every description, increase the value of real estate. . . . Here all these people brought into the community because of the work opportunity, workers of various skills, business people of all kinds, doctors and lawyers, children, schools, and churches." In 1962, at the twenty-fifth-anniversary dinner for the Baie Comeau mill, Arthur Schmon noted that there was "nothing so fundamental in a community's welfare and the prosperity of a province as the stability of a worker's life. It is fundamental because it affects his family's happiness. The way we carried out our business assured good livelihood and security for those who came here—very few who came here had any capital. Those who planned this industry made Baie Comeau a place where citizens could invest their savings in homes, businesses, professions; where they could raise children and educate them without fear and uncertainty." As Schmon described his company's operations, he took care to note that its workers and their unions had been just as important in maintaining harmonious industrial relations and overall productivity. "One of the precious jewels of which we are very proud," Schmon noted, "is the long and friendly relationship between our Company and its employees. During the 25 years that our mill has been operating and running full from the day it started, there has never been a strike or work stoppage in the mill or in the woods operations. . . . This record is due in good measure to the wise and sound leadership of unions in co-operating with the Company in developing employment conditions beneficial to the employees, at the same time consistent and competitive with good business operations of this Company."[36]

In some ways, the quality of relations between workers and management of which Schmon boasted reflected the degree to which Robert McCormick's paternalistic practices worked. Numerous observers noted McCormick's role in creating and sustaining a well-functioning industrial community. For example, the Rimouski newspaper *L'Echo du Bas St-Laurent* wrote a laudatory editorial in 1942 stating that "today Baie Comeau is synonymous with McCormick." Another Quebec journalist described Baie Comeau as a "beloved child" over which McCormick exerted a "paternal dictatorship," stating that, for the town's residents, he was its "Père Noël." McCormick received personal correspondence from workers expressing similar sentiments. In 1947, Kathleen Beebe, a self-described "one-time resident of Baie Comeau," wrote McCormick to express her gratitude for what his company had done for her and the province. Beebe thanked McCormick for his "far sightedness," which she believed was "wholly responsible for the millions of dollars brought into Quebec. If it were not for Americans like you, where would Canada be today?"[37]

In the 1930s and 1940s, McCormick's standing in the town meant that he often received personal letters from residents asking for financial assistance. In October 1939, for example, Father Alphonse Anctil, a North Shore clergyman, wrote to McCormick that "I am a poor priest who is sick since two years," and he claimed that "my health does not permit me to take again charge of a parish." Anctil pleaded to McCormick that "if among your very numerous subscriptions to all kinds of good works, you desire to reserve a little share for the poor priest . . . your gift would be very much appreciated." In 1941, McCormick received a similarly poignant letter from a seventeen-year-old Baie Comeau resident named Gabriel Lepage asking for help paying for college. Lepage had finished high school but found himself working in the Baie Comeau post office ("I am not strong enough to do heavy work," he confessed), and he found the job unsatisfying. His parents could not afford to send him to a university, so he turned to McCormick after what he claimed were nine days of prayer. "Such a paternal heart I believe I have found in you," Lepage told McCormick, and he was "asking if a little part of your charities—which I know are numerous—may not be allotted to me, enough to pay for a college course in English. If you could grant me this favor, my gratitude would be without bounds, for gratitude lives forever in a heart which has been consoled." McCormick demurred and referred the matter to Arthur Schmon, who replied, "As a general proposition, I do not recommend any such contribution. The Town is too young for us to be taking on the responsibilities of giving superior education to the boys." There is no evidence of a reply to either Father Alphonse Anctil or Gabriel Lepage.[38]

In many ways, even though he instituted a number of paternalistic benefits for his Canadian workers, Robert McCormick ultimately had the same sort of distant relationship with them as would the head of any major corporation. McCormick kept a well-apportioned house in Baie Comeau, and even while in Chicago he wanted to keep some connection to Baie Comeau, as he instructed Arthur Schmon at one point to "please have one loaf of our bush bread, cooked in a lumber camp, wrapped up airtight and sent to me every week"; he would also request occasionally that "good salmon" be sent down to him in Chicago. On isolated occasions, however, McCormick shifted from being aloof to getting angry when he believed that residents seemed to be asking too much of him. At one point in early 1955, for example, he snapped upon learning that some North Shore residents were asking for municipal improvements that he believed were unnecessary and excessive. "When we were not there, it was just a little forest. We built them a big civilization," he ranted to Arthur Schmon, and yet some people were "bitterly ungrateful."[39]

Those who were fortunate enough to secure employment and housing in Baie Comeau found that the planning of everything around them was done down to the most minute details, including the exact colors for all of the various interior and exterior paints used for company and private buildings. For example, company officials decided that the local Roman Catholic Church needed to be built of some material that should clearly "make contrast with the plant building." Town planners devoted significant attention to the aesthetics of the housing stock. "Obviously," they noted, there was a significant financial incentive to be found in a "policy of standardization" in home design. "If all windows, doors, trim and other millwork are reduced to a minimum of size and appearance to permit quantity production and purchasing, it is obvious that the greatest value can be secured from the supplier." However, planners acknowledged, that same cost-saving standardization might have negative effects on employee morale. Too much homogeneity in housing stock might have a "serious effect upon the attitude of the employee and the promotion of satisfactory conditions as they are affected by living conditions." More housing variety would create a "more desirable appearance with resulting more satisfactory labor conditions." Robert McCormick maintained active and close supervision of town planning and was constantly instructing his Canadian employees on all aspects of the process. At one point, he went so far as to send Arthur Schmon a one-sentence memo to suggest that a "minor economy in the houses would be to install showers instead of tubs."[40]

McCormick's thrift was also accompanied by an understanding of the ways in which the design of the built environment afforded him a means of social control. In the overall design of Baie Comeau, he later remarked, "we wished a comfortable and sightly town," but he admitted that "one strange act of the burocracy was to require dining rooms in all of the houses, because most of the workers eat in the kitchen." This social control continued in other areas of town planning besides homes. Indeed, all of the proper names defining Baie Comeau were chosen with the specific and strategic intent of making the town seem sufficiently québécoise. The town's name was chosen to honor Napoléon-Alexandre Comeau, a naturalist and polymath who had spent most of his life on the North Shore working in various capacities as the "postmaster, telegrapher, deputy coroner, Dominion Government fishery overseer, and guardian of the salmon fishing" for the region in which he became one of the most prominent residents. If its name connected Baie Comeau to a recent provincial past, the street names invented a longer tradition for the new town. As company officials noted, the street names had been chosen explicitly "from the viewpoint of good relations" with residents and provincial officials to have "emphasis on the commemoration of national and provincial personalities and events with special regard for the French." To that end, street names commemorated such luminaries as Frontenac, Montcalm, Laurier, Champlain, and Cabot, among others.[41]

The newsprint supply chain that the Tribune Company had created in the four decades after 1912 undergirded an exceptionally successful newspaper corporation, and it created a continually moving circuit of raw materials from forests and newsprint mills in Canada to printing plants and then readers in the United States. As it pursued vertical integration, the Tribune Company sought to control and rationalize not only its industrial operations but also its labor force. Baie Comeau in 1950 in many respects represented the apex of this process, as the company created what was essentially a publicly authorized corporate colony that employed the principles of welfare capitalism to great effect in developing positive relations with its workers. Taken together, the various efforts made by the Tribune Company to shape workers' lives on and off the job aimed at creating a well-functioning community in a remote part of Quebec. But these efforts also were experiments in social control aimed at maintaining stable corporate operations that were vital to the printing of industrial newspapers in Chicago and New York. While these workers and their families remained mostly hidden from the view of many Americans, their labor was essential to the production of the *Chicago Tribune* and *New York Daily News*.

Conclusion

In recent years, perhaps the most influential scholarly description of the newspaper has come from anthropologist Benedict Anderson, who argued that newspapers create "imagined communities" of individual readers understanding themselves to be part of a larger whole through the act of reading. The newspaper reader is ever unaware of what his or her fellows are doing on any given day but still retains "complete confidence in their steady, anonymous, simultaneous activity" through the daily reading of news and information in a common language about events important to that community. "The date at the top of the newspaper," Anderson argues, the "single most important emblem on it, provides the essential connection—the steady onward clocking of homogenous, empty time." The "calendrical coincidence" of the newspaper creates the sense that the geographically and topically distinct reports represented therein are all part of the same orderly world, and it gives readers a common set of facts about current developments in culture, society, and politics.[42]

In Anderson's analysis, the "community" in question is a collection of people using the newspaper as a way to understand themselves as part of a group situated in place and time using a common language. There is much to be gained by following this line of thinking, as it offers a way to think about newspapers that is sensitive to the varieties of individual relationships with them. But it is also important to remember that at the same time that the newspaper was the basis for a community understanding itself in time and space, it was also an extraordinary feat of industrial mass production, and in many cases one requiring raw materials gathered in the global economy. In creating a corporation to operate under the conditions of industrial capitalism in the twentieth century, the Chicago Tribune Company built both a community of readers and a community of workers. In Baie Comeau, the Tribune built not only a mill to make newsprint but also schools, churches, hospitals, and houses. This was not in any sense an "imagined community," but rather a physically tangible planned community, the construction of which reshaped the natural environment of a previously remote region of Quebec. Manufacturing an industrial newspaper involved attracting workers and their families to this new community, and it involved creating not only industrial infrastructure to manufacture newsprint but also social and cultural infrastructure to manage and control the labor force involved in the production of that paper.

PART THREE

THE NEWSPAPER BEYOND
THE PRINTED PAGE

CHAPTER SEVEN

The Diversified Newspaper Corporation

On Tuesday, April 23, 1946, the *Chicago Tribune* hosted its annual spring luncheon in Le Perroquet Suite of the Waldorf Astoria Hotel in New York City. Guests included not only top *Tribune* executives but also leading editors and publishers from around the country, and the assembled elite of American newspaper publishing enjoyed cocktails and dined on a first course of stuffed tomato Gabriel with crabmeat, followed by hearts of celery, green olives, and salted nuts. For the main course, attendees enjoyed a new Waldorf salad, lamb chops, fondante potatoes, and a string bean sauté. Those wishing dessert were offered vanilla ice cream, strawberries melba, and petit fours, and the meal concluded with a demitasse. In most respects, this was an unremarkable gathering, and there is little that merits substantive discussion about a lunchtime meeting of corporate executives in a luxury hotel suite. A transcript of the events has not survived, but one can reasonably surmise that there was a general sense of decorum if not good cheer and that social networks were maintained, deals struck, and trade information shared as publishers navigated their way into peacetime prosperity.[1]

What have survived from the luncheon, however, are some curious physical artifacts. Despite the fact that the lunch was at the Waldorf Astoria, guests dined on paper plates that had a printed note on the corner indicating that "the stock used in this plate was produced by the *Chicago Tribune*'s paper mill. Ontario Paper Co., Limited, Thorold, Ontario, Canada." The event's short printed program was bound in an artisanal-looking cover, which the text noted was made "from the first trial run of a secret process to convert waste bark into paper." The process was "now in the laboratory stage," the program claimed, and "further refinement will make possible wide use of this new paper stock." The program's inner pages further celebrated the *Tribune*'s advanced and efficient production techniques, noting that the "body of this booklet consists of white waste newsprint from the *Chicago Tribune* press room and is 'core-strippings' or 'transit-damaged' stock." As luncheon attendees considered these production advancements, some lit cigarettes with the lighters at their tables, which the program informed them were filled with isopropyl acetate, a chemical that the Tribune Company had also recycled from the inking stage of its printing production process. The liquid was

At its 1946 spring luncheon, the Tribune Company showed off its newsprint manufacturing advancements in a variety of ways, including making the paper plates that guests ate from during the meal. Vol. 25179, Scrapbook 1940–50, QOPC, LAC. Used with permission.

a "highly volatile roto ink thinner which permits quicker absorption in roto printing and consequently higher speeds. By means of hoods placed over the printing cylinders the fumes of the isopropyl acetate are collected and condensation is achieved by charcoal distillation. Thus 75 to 80 percent of the original amount of thinner is recovered." The *Tribune* was at the forefront of recycling and using the waste products in the production of newsprint, and part of the aim of the luncheon was to tout the activities to peers.[2]

One might assume that the newspaper executive pausing to light his cigarette with a lighter filled with recycled isopropyl acetate would find these achievements noteworthy, at least briefly. But one also wonders how impressive or even palatable it was for this same executive to consider that he was not only reading from and eating off of the products of industrial waste but also actually eating and drinking food and beverage products derived from that same waste. The bark paper–covered program noted that the "special cocktails" people were drinking contained alcohol that had been "made from the sugars in waste sulphite liquor from the Ontario Paper Company's mill at Thorold," and that the rolls that people were eating were "leavened by yeast made from the sugars" also derived from that same appetizing-sounding liquid. At its annual spring luncheon, the *Chicago Tribune* unintentionally gave new meaning to the terms "news diet" and "newspaper consumer."[3]

The Tribune Company's 1946 celebration of experimentation and technical advancement was firmly in keeping with the company's tradition of industrial innovation in newsprint manufacturing. The food products at the luncheon were a part of this history, and continued developments on this front would become increasingly important for the company's overall performance in the coming decades. Perhaps most bizarrely, in the 1950s the Tribune Company's development of processes to make synthetic vanilla flavoring in the form of the edible chemical vanillin that it extracted from its newsprint waste would make the company into the world's leading producer of one of the world's most widely used commercial food flavorings.

The developments that led to these production innovations were reflective of the company's long history of industrial experimentation and innovation, but they were also the more immediate results of the particular geopolitical challenges of World War II. North American newsprint manufacturers faced shortages in many important materials, and this forced them to take conservationist and improvisational approaches to their industrial production. But the Tribune Company also had political reasons for embracing production innovations in support of the war effort. As a firm producing an aggressively nationalist and often anti-British newspaper through a supply chain stretching across the US-Canada border, the Tribune Company found itself in a rather controversial position after Canada entered World War II on the side of Britain in September 1939. For more than two years after that, the *Tribune* would face charges of impeding the Allied war effort, and even after the United States entered the war in December 1941, many Canadians remained critical of the paper's political views. Technological innovation in support of the war effort thus became not only a way of keeping the Tribune's manufacturing enterprise running smoothly but also a way of demonstrating to policymakers its patriotic commitment to the war effort. Partly to appease those who found its news to be ideologically distasteful, the Tribune Company during World War II undertook projects that would eventually lead to innovations that would shape the taste of food products around the world.

The Tribune Company's wartime manufacturing innovations initiated developments across the breadth of its supply chain which would continue into the Cold War in significant ways, as they began a process of diversification in manufacturing which would extend the company's industrial operations into a host of other sectors, some of which had little obvious connection to the newspaper business. Besides vanillin, the company also adapted the infrastructure of its core newsprint business to become a manufacturer of industrial alcohol and synthetic rubber in Ontario, and similar adaptations in Quebec would lead the company to partner in creating a multinational aluminum smelting operation, drawing its raw materials from West Africa.

Understanding the Tribune Company's activities in these spheres suggests new understandings of what it could mean to be in the newspaper business in the twentieth century, and it suggests a new way to look at the business of corporations that played a vital role in the daily construction of public understandings of reality. These corporations were industrial manufacturing firms, as this book has shown throughout. As a newsprint manufacturer, the Tribune Company was particularly innovative in creating a vertically integrated operation that involved building not only two state-of-the-art newsprint mills but also a townsite employing some of the leading midcentury principles of urban design and welfare capitalism. It is important to underscore that all of these efforts to innovate were structured by policy. The Tribune Company had begun operations in Canada motivated by tariff changes prior to World War I, and its Canadian operations were always structured through a relationship with the state. In the World War II and postwar periods, as the Tribune Company sought enhancements and expansions of its industrial operations in the interest of corporate profitability, the ways in which it could pursue these projects remained structured firmly and inescapably within its relationships to federal and provincial policymakers. Implementing corporate strategy, in other words, involved decisions made not only by managers but also by public officials.

World War II and the Continental Politics of Newsprint Manufacturing

Under Robert McCormick's leadership, the *Chicago Tribune* was one of the most conservative and isolationist of the major US newspapers. During the 1920s and 1930s, McCormick remained a vocal and resolute American nationalist and critic of the European powers, and he reserved a particular antipathy for Great Britain. Disdainful of monarchy and believing that the British class system created a sense of unearned merit among its upper classes, McCormick never tired of criticizing the country and its elites. As tensions in Europe increased in the late 1930s, McCormick's anti-British views manifested themselves in strongly articulated positions in favor of appeasement with Germany and avoidance of American military involvement. This was simply not a matter with which the United States should concern itself, McCormick believed.[4]

The isolationism and influence of the *Tribune* in the late 1930s were both considerable. At the time of the outbreak of the war in Europe, the *Tribune* had the largest nontabloid circulation of any paper in the country, and its wholly owned subsidiary the *New York Daily News* had the largest circulation of any paper in the United States. Together, as one journalist noted, the two papers "make up what is perhaps the most effective isolationist bloc in the daily press." This generated significant domestic criticism of the newspaper and even some public demonstrations against it. In Chicago in July 1941, the *Tribune's* perceived support of Hitler outraged so many that a group called the Fight for Freedom Committee held an anti-*Tribune* rally at Orchestra Hall in downtown Chicago, after which some took to Michigan Avenue to burn copies of the printed newspaper.[5]

The effects of the *Tribune's* isolationist position resulted not only in the burning of individual paper copies but also in some efforts to disrupt the industrial supply chain that it used to supply its newsprint. As tensions mounted in Europe in early 1939, the Tribune Company had a clash with US customs officials which the paper claimed had been motivated by US officials wanting to punish it for its politics. In one instance in May, the *Tribune* reported on what it described as an egregious instance of the Roosevelt administration's explicit " 'Get the Tribune' offensive." American customs officials had boarded a Tribune Company ship laden with newsprint coming into the United States from Canada. After examining the newsprint rolls, the customs officials determined that the paper "contained more than the 2 per cent ash content limit" required by the specifics of tariff regulations to be considered newsprint, which, unlike other kinds of paper, entered the country free of a duty. Since the rolls of paper at hand did not meet the legal definition of newsprint, the customs officials determined, the Tribune Company was therefore subject to duties amounting to $173,398. McCormick believed this to be outrageous and politically motivated harassment, and the *Tribune* noted that the duty was indicative of the "administration's chronic hostility toward one newspaper—this one." The Tribune Company appealed the customs assessment and eventually won, which it proudly reported in the newspaper. At the same that the *Tribune* was fighting the federal government over this tariff issue, President Franklin Roosevelt encouraged his friend Marshall Field III to start the morning *Chicago Sun* to compete with the *Tribune* and hopefully lessen its influence, and McCormick clearly and correctly understood the politics of this development.[6]

While the Tribune Company was defending itself against alleged grievances and perceived harassment in the United States, its isolationist and anti-British positions became increasingly obnoxious to many Canadians and Brits, especially between September 1939 and December 1941, a period in which Canada was at war with Germany and the United States officially was not. To some Canadians, the exploitation of their forests was disgraceful not only on environmental and economic grounds but also because of the particular corporation that was exploiting those forests. Many believed that the *Tribune's* editorial policies were repugnant and detested the fact that their natural resources enabled the paper's views to circulate in print. To many Canadians, it seemed appalling that one of the leading exponents of isolationism should be using their forests to promote that cause. Tribune Company officials were well aware of these criticisms and sought ways to deflect them. In December 1939, for example, Arthur Schmon told Robert McCormick that he was "somewhat concerned about the possibility that some editorial, at one time or another, may be objectionable to Canada." Though the *Tribune*'s political views were well established by the time World War II broke out, Schmon worried that some Canadians "might reasonably object" to its circulation in Canada, "on the grounds of hurting the war morale or efforts of its people."[7]

McCormick declined to take Schmon's advice that he moderate his tone, and before long, some Canadian newspapers began editorializing against the *Tribune* and its political views, for example, the *Ottawa Journal*, which in May 1940 claimed that, for "sheer malice against the British Empire and the Allied cause . . . [and] for absolute misrepresentation of Allied aims," the *Tribune* was "surpassed only by the propaganda of Herr Goebbels." The *Journal*'s editors pointed out that the Tribune Company's "manufacturing department" relied completely on Canadian materials and labor, and it was outraged that "Canadian forests and Canadian water power" were being used by a "bitter sinister enemy" that it claimed should at least be "barred from this country—if not dealt with more drastically." These sorts of editorials against the *Tribune* appeared across Canada. In Ontario, the *Peterborough Examiner* chastised the *Tribune* for its "hatred for and . . . scorn of the Allies and their war efforts" and suggested of its Canadian newsprint that it might be a good idea to "cut off the supply which keeps the *Tribune* operating." The *Lindsay Post* editorialized that it "goes against the grain to have Canadian forest products manufactured into paper which is destined to circulate anti-British journals in the republic to the south." The *Tribune* was giving Americans a "daily dose of . . . trash" in its pages, and the *Post* claimed that its coverage could "do more harm than all the Lindberghs and the noisy isolationists because it can influence more people directly." In Manitoba, the *Winnipeg Free Press* called McCormick a "stupid and unscrupulous man" and claimed that he was an "outright Anglophobe, a hater of Britain," with opinions that were "false, detestable and poisonous." Even in Quebec, where support for the war was not as strong, *L'Illustration Nouvelle* criticized the *Tribune*'s isolationist news and editorials and claimed that it was "remarkable that these articles are printed in newspapers using paper that is the product of our natural resources!"[8]

Over time, these press and private criticisms also filtered into the Canadian Parliament, especially as some policymakers began worrying about the porousness of their southern border in allowing pro-German publications to circulate. Member of Parliament T. L. Church, for instance, called for a bill to restrict what he called the "Nazi gutter press of the United States" from circulating in Canada. Church asked that the government ban newspapers and magazines publishing "articles detrimental to the cause of Britain and Canada and the prosecution of the war." Church named the *Chicago Tribune* and *Saturday Evening Post* as the worst offenders, and he subsequently began making official inquiries to Secretary of State for Canada Pierre-François Casgrain to find out whether anything had been done to restrict their circulation in Canada. Casgrain's office noted that all publications coming into Canada were being monitored by the Press Censors of Canada but replied that nothing in either of Church's feared publications had as yet warranted exclusion, though they were being closely scrutinized for "subversive material" and would be banned if necessary.[9]

In the Canadian Senate, P. E. Blondin and Arthur Meighen took up the attack on the allegedly subversive *Chicago Tribune* and *Saturday Evening Post*. Blondin claimed in June 1940 that the *Tribune* was the source of "vicious and virulent anti-Allied propaganda," and Meighen called it a "nauseating mixture of ignorance and malignancy from the first line to the last," claiming that it "will do more harm among ignorant people than anything else I know of published on the continent." Though unclear on what specifically ought to be done about the *Tribune*, Meighen did suggest that legislators might well consider restrictive policies against the paper. "I do not believe times are such that ordinary rules can be applied," Meighen concluded, "and I think we might well inject a little virility into our policy with respect to this kind of mendacity." Privately, Meighen supported a ban on the *Tribune* circulating in Canada, writing that the *Tribune* "has been permitted to contribute very heavily" to a "weakening and poisoning of a will to war in Canada" and believing that this constituted appropriate grounds for censorship. "No country at war can permit such a practice within its boundaries, if that country is true to itself." Moreover, Meighen argued, Canadian law allowed for the "heaviest of penalties" against seditious speech by Canadians. Why Canada "should allow the same thing to be done on a wholesale scale by outsiders" was, he claimed, "hard to understand."[10]

Arthur Schmon was deeply concerned by this outbreak of anti-*Tribune* sentiment. Though he noted that the initial *Ottawa Journal* editorial "did not carry through" into any anti-*Tribune* policies, the subsequent articles disturbed him greatly, as did the "discussion in the Canadian Senate and especially the remarks of Senator Meighen." Schmon reported that public comment was spreading and there were now "all kinds of suggestions of punitive measures—banning of the *Tribune*, stoppage of the exportation of newsprint, banning of goods of advertisers who advertise in the *Tribune*." Schmon urged Robert McCormick to consider finding ways of "building up better public relations in Canada if it can be done without affecting any policy of the *Tribune*. If we do not have a positive policy of this kind I am fearful that an animosity will be built up against the *Tribune*," which might result in policies constraining its industrial operations. Schmon reported that he had in Canada engaged a public relations firm to work behind the scenes in "constructive ways where our identity will not be known," but he also urged McCormick to consider ways of being more publicly conciliatory toward Canadians. "From a long pull point of view," Schmon remarked, "and even after war stops, it is necessary for us to have on record some favorable items so that we may conduct our business under normal conditions such as we did before this present war."[11]

Official assessments by Canada's Censorship Coordination Committee were more neutral toward the *Tribune* than public comments by politicians. One official remarked that it was possible that the editorial attacks coming from the *Ottawa Journal* were motivated less by politics than by the "ulterior motives" at play within a "private war which is said to be going on between certain newsprint interests." The *Journal*, the official speculated, might have been acting in support of newsprint manufacturers in Ontario and Quebec who were angered by the Tribune Company's "refusal . . . to join with them in agreements as to price and output." And, after reading the *Tribune* itself, the official remarked in February 1941 that he found little that directly "constituted a menace to Canadian war morale" or material of a "particularly offensive nature." Though there were within the *Tribune* "at least some unfriendly elements to Britain," the censors reported that they were "satisfied that they are not directly influenced by Nazi sympathies," and articles critical of the war effort were "offset by occasional very friendly references to the British and Allied cause." Ultimately, the paper was "one of the world's great newspapers," and banning it was a poor and unproductive idea.[12]

Though the Canadian government declined to formally censor or restrict the circulation of the *Chicago Tribune*, the fact that the *Tribune* was being scrutinized in this way created a general anxiety among other publishers throughout the United States. Given how important Canadian newsprint was to newspapers across the United States, some began to wonder whether their own news and editorials might create obstacles in their access to newsprint. In 1941, this motivated Cranston Williams, the president of the American Newspaper Publishers Association, to lobby Canadian officials to ensure that US newspapers in general would not be deprived of the newsprint that they needed. Williams reported that the "high officials" he spoke to in Canada had reassured him that no policies would be enacted that would "interrupt a steady flow of newsprint to United States publishers, regardless of what may be the attitude of any newspaper about any sub-ject on its editorial page or in its news columns."[13]

Despite these Canadian assurances that nothing would be done to limit US newsprint supplies, other wartime policies did have some significant effects on newsprint manufacturing and exports, most importantly in the form of policies prorating newsprint production. In 1942, the province of Ontario responded to war production needs by announcing plans to curtail the power available to the Tribune Company's Thorold mill in order to provide enough energy for the metal and chemical plants vital to the war effort and also operating on the Niagara Pen-insula. The Ontario government also announced plans that would force the com-pany to pool some of its newsprint production with other mills in order to ensure consistent supplies to newspapers across Canada. The Tribune Company's re-sponse to these Canadian policies mirrored the way it had approached its fight against US customs officials in 1939, as executives saw power rationing as the use and abuse of administrative authority in the service of punishing it for its politi-cal views. The Tribune was particularly suspicious of the politics of the power curtailment plan, noting that its Thorold mill was the only significant newsprint mill on the Niagara Peninsula and that the New York Times mill at Kapuskasing, Ontario, was not to be affected by the policy. In Thorold, where workers would be affected by any curtailment of production, Mayor W. A. Hutt agreed with the Tri-bune's speculation about the politics of the plan, remarking that the "interests who have been 'gunning' for the Ontario Paper company for years may now be able to achieve their aims under the cloak of necessity arising from the war."[14]

These concerns proved to be overwrought, and though the *Tribune* claimed that the Canadian government had plans involving the "complete stoppage of news print making at the Thorold mill," the Canadian Wartime Prices and Trade Board ultimately gave the company enough power to keep two of its five paper machines operating. To the Tribune Company, being forced to operate at 40 percent capac-ity was still an onerous restriction, and the Ontario policy persisted until late 1944, when provincial officials began easing power restrictions. The *Tribune* con-sistently blamed partisan politics for the situation, stating that it was subject to "violent attacks" because of its editorial policies, and it speculated that the restric-tions placed on its production in Canada were "not unpleasing to the Roosevelt administration in Washington." This was, the paper claimed, perfectly in keep-ing with the New Deal's attitude toward the press in general and the *Tribune* in particular. "A reduced visible supply of paper resulting from this situation," the *Tribune* editorialized, "produced in turn an excuse for burocratic distribution in the United States of newsprint paper, the one commodity which is absolutely necessary to the publication of a newspaper." The paper's efforts to fight against the "communistic New Deal" and in support of "American constitutional government and American private enterprise" had been "construed by the Canadians as being anti-British and anti-Canadian. A campaign of vilification against The Tribune has raged in various newspapers throughout the length and breadth of Canada," with the full support of the US federal government.[15]

Starting in 1939, the Tribune Company as a manufacturing firm faced particular challenges operating across the US-Canada border. With Canada, the home of its raw materials and newsprint production facilities, involved in the war on the side of the British for more than two years before the United States was, the *Tribune's* anti-British views came to be seen as increasingly objectionable to Canadians inside and outside of the government. In the United States, the *Tribune* maintained a continually and aggressively oppositional editorial stance against Roosevelt's domestic and foreign policies, and this placed it increasingly at odds with both official and public opinion. Even after the United States entered the war in December 1941 and the *Tribune* came to support the war effort on its editorial page, many in both the United States and Canada maintained suspicious and critical attitudes toward the paper. As the purveyor of an industrial newspaper, the Tribune Company in the early 1940s struggled to navigate the continental politics of newspaper publishing in wartime. The great and perpetual challenge that the company faced was the fact that its role as a source of news and opinion about the war was based on it being an international manufacturing firm operating under wartime conditions. As had been the case during peacetime, the company officials thus found themselves struggling to balance Robert McCormick's politically adversarial tendencies with the company's operational necessities.

The Practice of Wartime Innovation

As the Tribune Company navigated the shifting continental politics of World War II, one course of action that would prove to have both political and commercial benefits involved technological innovations in newsprint manufacturing. These developments had to do with chemical manufacturing operations built as adjuncts to newsprint operations and employing parts of the forest resources previously considered waste, and they were part of a broader pattern of innovation in the pulp and paper industry involving what historian James Hull describes as a "moving away from knowledge of the forest, of trees, and even of wood." As companies involved in cutting down trees to make paper rethought what it was that they were cutting down, this opened avenues of research and wood utilization in previously undeveloped and in many cases unexpected areas. What the pulp and paper industry evolved into was what Hull describes as "not a forest, not even a wood using industry. It was a cellulose processing industry with a knowledge base in chemistry and chemical engineering and much in common with other continuous flow chemical process industries." This was the case at the Tribune Company. One chemical engineer that worked for the company in Thorold for over three decades told me, "I never even have visited our forest operations. . . . There was nothing for me there."[16]

One can see this turn in thinking about trees in contemporary trade circles, where the tree came to be understood less as a singular living thing than as a collection of molecules that could be utilized in various kinds of industrial production, a shift in thinking that opened up a host of new possibilities for using forest resources. As the Canadian Pulp and Paper Association (CPPA) wrote in 1952, the tree "in its entirety is almost completely composed of carbohydrates, that great chemical family which includes virtually all the components of plant life together with products ranging from coal to alcohol, from sugar to shredded wheat, from shirts to plastics, and from honey to silk." Rather than just seeing the tree as a source of wood, many in the pulp and paper industry began understanding it as a source of cellulose, hemi-cellulose, and a substance called lignin binding these molecules together. "Because cellulose belongs to the great family of carbohydrates," the CPPA noted, "including sugar and flour, it can be converted into other carbohydrate products. Thus, under heat and pressure, together with acids, it can be turned into sugar." From there, it could be used to produce "rayon, nitro-cellulose, photographic film, cellophane, plastics, and other chemical products." The hemi-celluloses, on the other hand, "contain important quantities of fermentable sugars" that could be used to make ethyl alcohol and other chemical products that were used in the manufacture of synthetic rubber. And finally, the lignin could be used to make "resins, fertilizers, plastics, vanillin, alcohol, and other chemicals." As of 1952, lignin still remained a bit of a mystery to the pulp and paper industry. While the CPPA noted that lignin was "closely related to the carbohydrates, its chemical composition and molecular structure are unknown. It is the binder which gives strength to the tree and to other forms of plant life. It is thought that coal deposits largely owe their origin to the lignin in prehistoric plant and tree growth." Despite not yet being fully aware of lignin's composition, the CPPA was enthusiastic about its promise. "When scientists finally succeed in determining the exact chemical composition, then its uses will be multiplied." This would be the case in subsequent years, both in the general pulp and paper industry and within the Tribune Company's Canadian mills.[17]

During World War II, the Tribune began taking just the sort of broader approach to its newsprint production which the CPPA described. "Modern biochemistry," a *Tribune* science writer noted in 1943, "sees in a tree much more than just so much lumber." A tree was not just wood but a natural material composed of "tiny hollow cellulose tubes bound together by a cement called lignin." Through experiments, Tribune Company chemists found that sugar could be extracted from cellulose by "mixing it with acids under heat and pressure." Further processing through distillation yielded alcohol, "a base for synthetic rubber." Lignin, the other essential component of trees, was a substance consisting of "carbon, hydrogen, and oxygen," though in 1943 the "structure of its molecule is still a mystery to science." Potential applications, however, were numerous, including "making plastics, phenolic resins, fertilizers, vanillin, and a number of other substances." The transformation of a tree into paper was no longer the sole aim of the newsprint manufacturing process, but merely the middle phase of a process that was elongated to encompass the use of other parts of the raw materials previously considered simply waste.[18]

The Tribune Company's first forays into chemical processing involved the liquid waste generated in the papermaking process, of which there were substantial amounts. As the *Tribune* noted of its Thorold manufacturing, the "liquid drawn off sulphite pulp has always been discarded. It contained about half of the wood, everything except certain fibers. This always has been a complete waste and frequently a source of stream pollution." With the technical advancements in chemistry during World War II, however, the paper noted that this waste could be used to manufacture other sorts of products if the company could figure out ways of doing that. "With present knowledge," the *Tribune* stated, a ton of wood waste could yield roughly 1,100 pounds of sugar and nearly 600 pounds of lignin. "From the sugar can be made feeding yeast, baking yeast, pure glucose, ethyl and butyl alcohol, glycerin, and lactic, citric, butyric, and acetic acids. Other lignin products include tanning materials, pharmaceuticals, cosmetics, fireproof cast materials, insulating materials, and ash-free briquettes." During World War II, the Tribune Company created one of the pulp and paper industry's first chemical research departments to begin trying to develop these processes at its mill, and it would spend significant sums in this area in the postwar period. According to one estimate, the expenditures on research into uses for this waste sulfite liquor between 1944 and 1951 were $350,000. "Strong emphasis is placed on Research and Development at Thorold," the company noted in 1951. "Some day the newsprint industry, like the meat-packing industry, must recover and sell the maximum material from its raw material supply in order to be competitive."[19]

One of the first areas of research that bore significant results during the war was the process of refining the waste sulfite liquor into ethyl alcohol, and in June 1943 the company built a plant for this purpose, which it claimed was the "first of its kind in the western hemisphere." The company's consulting engineer remarked that the developments were the "beginning of a war against waste in agriculture and forestry. . . . It's the first practical step taken on the North American continent toward this end." Furthermore, the *Tribune* claimed that the "potential yield of alcohol from the sulphite liquor that is now poured into lakes and streams by woodpulp mills of Canada and the United States is estimated at 86 million gallons a year." In addition to being conservationist, the production of industrial alcohol from waste sulfite liquor rather than from grain would allow the estimated 120 million bushels of grain currently being so used to be utilized instead "to relieve a world food shortage."[20]

More important, given the context of World War II, was the possibility of using that alcohol to make synthetic rubber. Company experiments had made this a possibility in the summer of 1942, and in 1943 these projects picked up in earnest. In June 1943, Robert McCormick gave a radio address explaining the geopolitical importance of the process, and he stressed how useful the alcohol converted from waste sulfite liquor could be in the production of "rubber, explosives, and motor fuel." Rubber was particularly important, he noted, given the disruptions in Southeast Asian sources as a result of the Japanese capture of important islands in the Pacific. The Tribune Company quickly adopted its innovations within its mills, McCormick claimed, "so that they might produce alcohol for the war and for domestic consumption after the war, freeing the people of the American continent from the rubber monopoly of Java and Malaya." The potentials were seemingly limitless and the implications clear for both wartime and peacetime applications. "All waste wood can be converted into alcohol," McCormick told his listeners, and at present "only 37.6 per cent of a tree turns into lumber." If other companies were to follow his lead, they could help alleviate the "serious pollution of streams" happening as a result of paper manufacturing, and they could also reap new profits from parts of the production process which previously had been sunk costs. Ultimately, he claimed, "lumber and paper will be greatly cheapened and four billion gallons of alcohol a year will be produced and the rubber problem and a large part of the fuel problem of North America will have been taken care of." The aggressively antistatist McCormick closed his description of his firm's development of new ways of distilling alcohol from waste sulfite liquor with a celebration of the advantages of private enterprise over public works, noting that the "lesson of government interference, which has left us without tires to support cars or gas to drive them, and even food to eat, will have been learned, and men of genius will once more be freed to bring back to mankind the advantages of civilization which government interference has taken away from us."[21]

Later that year, the *Tribune* conducted a demonstration to promote its development of rubber manufacturing. In August, the company equipped an automobile with tires made from a synthetic rubber it was calling "Trib-buna" and planned for it to go on a three-month tour of the United States to demonstrate the rubber's performance. On the car's side, the company painted *"Chicago Tribune* Tire Test" and "Rubber from Raw Wood Pulp Waste Materials." In October, the car arrived in Hollywood, where it did a tour of the city. The *Tribune* reported that it had been "parked under the lights of the Cocoanut Grove entrance, at the Trocadero, and at the Mocambo. It took Hedda Hopper, former star and ace reporter, to the Brown Derby for lunch." The car then went to the Paramount lot, where it was photographed with the cast of the film *Rainbow Island.* In an article claiming that the process it used for making rubber involved a different and better method than the government's, the Tribune Company made clear that it was sharing all the data from its tests with the state.[22]

As it conducted these public relations campaigns in the United States, the Tribune Company also asked the Ontario government to lift the power restrictions on the company's mill so that it could help with the Canadian war effort. Noting that its alcohol plant could aid in the production of not only rubber but also a "great variety of war chemicals and explosives," the company claimed that "consideration should also be given to the public interest in a comprehensive way" when it came to including the Thorold mill in the power curtailment plans. There might be some "reasonable modification," the company noted, so that "the necessary war demand may be answered without completely closing down sections of a business which hitherto has contributed, and if preserved will in future contribute, in an important degree to the public interest of the country." In promoting its alcohol manufacturing in this way, the Tribune Company used its industrial operations to send a political message. Officials at the Kimberly-Clark Company, the New York Times's majority partner in the Spruce Falls Power and Paper Company, certainly believed this, as Kimberly-Clark executive Ernest Mahler remarked that the *Tribune* was making such a strong promotion of its activities simply because it wanted to be able to keep its mill operating at full capacity. "We fully realize," Mahler claimed, that "the publicity started by the Chicago Tribune has only a political and not a scientific background. If they would be permitted to make alcohol, naturally it would mean they would have to keep the Thorold mill in operation." Whatever its motivation for innovation, the Tribune Company expended significant energy and resources on it, receiving positive feedback from the US government. According to one account, the "War Production Board in particular showed interest" in the project, and it "encouraged" the company to continue its experiments in alcohol manufacturing. It did so with vigor, and in 1943 the company produced some 270,000 gallons of alcohol, a figure that had increased to 741,000 gallons in 1945.[23]

As the end of the war approached, company officials began discussing ways of extending and monetizing these operations in peacetime. In December 1944, Arthur Schmon remarked that the "subsidiary manufacturing" of alcohol and other chemicals was "desirable and fundamentally sound for many reasons." It would "stimulate plans for post-war expansion of existing sound industries in order to contribute to employment and the preservation of private enterprise," he claimed, and it would help create peacetime jobs for "employees returning to us after the war from the Armed Forces." Once the war ended, the company committed over $1 million to expansion of its alcohol production operations. "When completed," the *Tribune* proudly noted, "the new facilities will give permanent employment to more than 100 additional workers. Construction and installation will provide a total of 7,270 man months of employment, equal to 600 men working for one year." With the enhancements, production expanded to 972,000 gallons in 1948. To company executives, it was a smart strategic move, as Arthur Schmon remarked in December 1945, to "diversify the operations so that . . . earnings will not depend entirely on newsprint. Our experience during the war showed the soundness of this policy." In the postwar period, the company began selling significant quantities of its alcohol production to private clients that used it to make various consumer products. As Arthur Schmon told one audience in 1946, the company's alcohol "may now be one of the ingredients in your wife's perfume, or might be in your anti-freeze, or in the paint that you use on your house."[24]

This alcohol was one way of creatively reusing the waste sulfite liquor, but it by no means exhausted the potential or volume of the waste products left over from newsprint manufacturing, and Tribune Company chemists continued experimenting with ways of utilizing them. After the alcohol was extracted from the waste materials, the newsprint mill was still left with "lignin, a vegetable compound the potentialities of which have never been touched." As one company official later remarked, the company was always interested in "finding commercial uses for the huge quantities of sulphite liquor which were literally 'going down the drain.'" By 1947, chemists in the Tribune Company's mills were experimenting with further uses of the lignin, and they thought they just might be able to make synthetic vanillin, or commercial grade vanilla flavoring, from the by-products of its newsprint manufacturing. These experiments would be stunningly successful. Some years later, the company noted, its researchers "undertook a twofold task of developing a marketable product and at the same time to further 'clean up' the effluent being added to the local natural water courses." What they developed was a process of and facility for making vanillin, which eventually made the company accountable for "approximately forty per cent of the world's supply," making it "the largest single producer in the world."[25]

The Refined Taste of Vanillin

Vanilla is one of the most popular flavors in the world, ubiquitous to the point that describing something as "vanilla" is to call it commonplace or boring. In the natural world, however, vanilla is anything but that. Vanilla beans come from the only species of orchids that are edible. They are also notoriously difficult to cultivate, and one scholar of the bean calls it "the most labour-intensive agricultural product in the world." Vanilla can only be grown in tropical climates near the equator, mostly in Madagascar, Tahiti, and parts of Mexico. Producing it involves pollinating by hand the flowers on the single day in their lives on which they open. If the grower misses the window, the flower falls off the vine and dies. After ripening for roughly nine months, the beans are then cured for up to six months under close observation. The combination of the labor-intensiveness of production and the limited locations where it could be done have long made the product prohibitively expensive for industrial food producers. But it has become, as one food writer describes it, "by far the single most popular flavor in the world," and companies needing the flavor have sought more consistent and affordable alternatives.[26]

Just as newsprint manufacturers found alternatives to cotton rags as their primary raw material starting in the second half of the nineteenth century, so too did food producers find alternatives to natural vanilla flavoring. In 1875, the German chemist Ferdinand Tiemann made synthetic vanillin using coniferin, a substance found in crystal form on pine cones. In 1891, a French scientist discovered a way to make vanillin from eugenol, a chemical found in cloves. By the 1930s, chemists had developed a way to make vanillin from lignin, a substance that had been part of the waste material generated by paper production. Vanillin manufactured from lignin proved to be an ideal way for paper manufactures to develop new lines of business serving clients in the food industry, as the amount of paper being produced was at an all-time high and supplies of the basic chemicals readily available. Over time, synthetic vanillin came to be a much more plentiful and affordable source of the flavoring for industrial food producers. According to one 1962 estimate, a "pound of commercial vanillin sells for a fourth the price but has 13 times the flavoring strength of a pound of vanilla beans."[27]

As one of the largest North American producers of newsprint, the Tribune Company had significant internal supplies of lignin to utilize and thus found itself advantageously positioned to move into the commercial flavoring market. They were not the only paper manufacturer to be doing this, as the Wisconsin-based Marathon Paper Company was by 1936 partnering with the Salvo Chemical Company to manufacture vanillin from Marathon's waste sulfite liquor, but they would become the most innovative and successful at it. By 1947, the Tribune Company had developed its vanillin process to the point that it was starting to consider how to begin commercial production. "Laboratory and pilot plant work has proceeded to the point where we are able to design a commercial plant for making vanillin," as an internal research report noted. In early 1948, Arthur Schmon claimed that the company was "prepared to build a manufacturing plant" at a cost of $700,000.[28]

As a substance derived from the lignin waste left over in newsprint manufacturing, vanillin was produced in two primary grades through "successive extractions with organic solvents, acids, and alkalis." The first of these was USP (United States Pharmacopeia) grade vanillin, an aromatic substance sold as white crystals and fit for human consumption. The second form of vanillin was known as "technical vanillin," which had a purity of between 97 and 99 percent and was sold in "yellowish chunks." This was the form often produced for the export market, as USP vanillin was prone to spoilage and contamination and the technical vanillin could be converted into USP grade vanillin for "only a few cents a pound." Technical vanillin could also be used in manufacturing a host of other chemical products, such as pharmaceuticals, and industrial purchasers not involved in food production purchased this grade.[29]

The Tribune Company's vanillin manufacturing involved a patented method employing a series of chemical processes. At the start of production, workers took the liquid waste from newsprint manufacturing, added lime, and heated the solution to 345 degrees Fahrenheit, all the while forcing air through the mixture and vigorously agitating it. This produced a slurry that was then transferred to a centrifuge and spun to remove solids. With the vanillin remaining in the liquid "in the chemical form of its calcium salt," carbon dioxide and sulfuric acid were introduced to the solution, and this released the vanillin from the calcium salt. Next, a chemical called toluene was added, which extracted nearly all of the vanillin from the solution, and the toluene was then extracted to be reused in later batches of vanillin. The remaining liquid was then treated with sulfur dioxide gas, caustic soda solution, and sulfuric acid, at which point "the vanillin separates along with other materials as a crude dark brown oil containing approximately 80% vanillin." This solution was "distilled under vacuum" to produce a 97 percent pure technical vanillin that the Tribune Company called Lioxin. For the vanillin to be manufactured into edible USP vanillin, this technical vanillin was "purified to flavouring grade vanillin by crystallization." The Lioxin was "dissolved in warm distilled water and activated carbon added," which removed "colouring matter and when the carbon is filtered off, there is left a clear solution which is cooled and partly evaporated under vacuum so that flavouring grade vanillin comes out as crystals.

The latter are separated by centrifuging and the wet crystals dried in a vacuum oven after which the crystals are screened and packaged."[30]

In 1952, the Tribune Company had honed this chemical process and announced that it was investing $1.3 million in a new vanillin manufacturing plant that would produce some 400,000 pounds annually. The company presented the development as a strong contribution to the Canadian economy by an American corporation. Having built "North America's first successful plant for the production of industrial alcohol from waste sulphite liquors," the company had already provided "great value to the war time synthetic rubber program." Now, they were promising a "completely new source of national income, including United States dollars, to assist Canada's economy," and their aim was to continue diversifying their operations going forward. The hope, the company claimed, was to "see the company's paper mills surrounded by associate plants processing materials now discarded as useless."[31]

By 1959, the Tribune Company had become the global leader in vanillin production, and a significant portion of the company's output went to the United States. In 1960, the company exported some 421,000 pounds of vanillin from Thorold to the United States, and in 1961 it accounted for about 94 percent of all US vanillin imports. The vanillin was used in a variety of ways, and the US Tariff Commission estimated that roughly one-third to one-half of it was purchased by "countless small producers of bakery goods, candy, and flavoring materials throughout the country. The remaining sales are made direct to manufacturers of chocolate, candy, flavoring extracts, flavors, cookies, frozen desserts, and numerous other products." The Tribune Company's vanillin business expanded through the end of the decade, and it began making significant sales beyond the North American market. By 1961, Tribune Company vanillin sales had expanded to distributors and firms in Australia, Great Britain, the Netherlands, Japan, Mexico, and Switzerland.[32]

In the 1960s, the company's vanillin business continued to boom. By mid-1963, it was producing 1,750,000 pounds of vanillin every year. Though it was sold to food manufacturers for a variety of uses, one estimate noted that "if the company's production went into ice cream exclusively it would flavor nearly 10 billion gallons of ice cream." In 1964, the Tribune Company announced that, because of technical improvements in the production process, it had increased its vanillin production capacity to 3 million pounds a year. By the end of the 1960s, the company claimed that its Thorold vanillin plant was "now the biggest of its kind in the world, with an output of 4 million pounds year." Beyond vanillin's applications in flavoring, there was also "growing demand for it for pharmaceutical purposes," and some manufacturers were using the Tribune Company's technical vanillin to

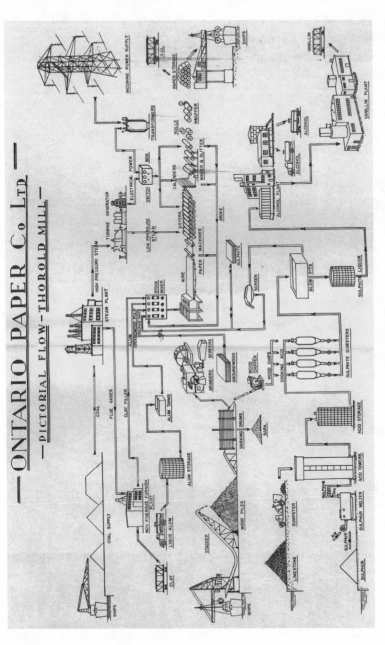

This diagram of the Tribune Company's newsprint production in Thorold shows not only the process of logs being converted into paper but also the various chemicals that were used throughout the process to manufacture both paper and ancillary products such as alcohol and vanillin. Vol. 25793 (acquisition no. 2006-00392-5, box 504), folder Ontario Paper Company Historical and Background Material, pt. 2, QOPC, LAC. Used with permission.

develop a variety of products, including tranquilizers and L-Dopa, a drug used to treat Parkinson's disease. In response to this increasing demand from the pharmaceutical sector, the Tribune Company developed direct relationships with drug manufacturers. For example, in 1962 Merck purchased 100,000 pounds of technical vanillin to use in the development of a new drug called Aldomet, which it hoped would be useful in the treatment of high blood pressure. In 1974, the Tribune Company reached a $100 million agreement with Merck to supply vanillin for the next decade, with Merck wanting some 4.8 million pounds annually by 1976. The rising global demand for vanillin pushed the company to keep expanding production, and the company estimated that it could sell as many as 9 million pounds if it could produce that volume.[33]

For the Chicago Tribune Company, World War II innovations paved the way for significant ancillary businesses for decades after the end of the war. In producing alcohol and vanillin, the company streamlined industrial operations, reduced waste, and developed successful and profitable lines of business in sectors with little obvious connection to the newspaper business. The Tribune Company did not set out to get into the business of manufacturing alcohol or food flavoring, and it is quite possible that had the war not intervened the company would have kept these innovations in-house or perhaps simply have never pursued them. In doing so, however, the company demonstrated its adaptability, both as an industrial manufacturer and as a political actor. The Canadian state did not directly compel the Tribune Company to become a vanillin manufacturer, but it did indirectly compel the company to demonstrate its commitment to a war effort that the isolationist Robert McCormick did not support. Innovation became for the Tribune Company a way to demonstrate at least temporary patriotism.

Though many political elites in the United States and Canada took a dim view of the *Tribune*, it remained a hugely successful newspaper financially. As had been the case since McCormick had taken over the paper in 1911, this success was built on advances in the paper's industrial manufacturing operations, and this would continue after the war. In general in the post–World War II period, the company remained strongly committed to what chemists on the research side of the business described as a top-down culture of innovation in all aspects of the enterprise. In the decade after World War II, the *Chicago Tribune* expanded its industrial operations at an impressive rate. Less than a month after the war ended, the company announced that it would invest $1.7 million to expand newsprint operations at Thorold and $2.8 million at Baie Comeau. In the breadth and range of its industrial operations, the Tribune Company was one of the most innovative newspapers in the world in the ways that it expanded and diversified its operations. The company diversified in Ontario into the manufacture of industrial alcohol and food flavoring beginning in the 1940s. In Quebec starting in the 1950s, the company used its position as an industrial newsprint producer to expand into aluminum production.[34]

From Paper to Power

The Tribune Company's path to aluminum manufacturing began with expansion of its hydroelectric generating capacity. Starting in the 1910s and 1920s, the Tribune Company's initial impetus for North Shore development had been the allure of forest resources, and the company developed hydroelectric power on the Outardes River in order to exploit them. In the post–World War II period, the company's diversification into aluminum production was part of its expansion of hydroelectric generating operations onto the Manicouagan River, the mouth of which was roughly 10 miles northeast of the Outardes. The Tribune's project of reorienting operations toward the Manicouagan began during World War II, when the company in 1944 completed the purchase of timber limits along the river from Anglo-Canadian Pulp & Paper Mills (see chap. 5). The newly acquired limits offered significant additional supplies of pulpwood, and damming the Manicouagan would allow the company to create the additional power generating capacity on the North Shore which it needed for its mill and the growing city of Baie Comeau. By October 1945, the company was beginning surveys and estimating costs for construction of the dam on the first falls of the river.[35]

The Tribune Company's power development on the Manicouagan began at a particularly important moment in Quebec's history, as provincial leadership was in the process of developing a new approach toward hydroelectric power generation as part of a rising mood of postwar nationalism. Across Canada, the development of hydroelectric power was, like all natural resources, subject to provincial rather than federal jurisdiction. And, as in many areas of resource and industrial policy, these laws were very different in Ontario and Quebec, the major industrial provinces in Canada. In Ontario, provincial officials built a strong public power movement in the early twentieth century, starting with Niagara Falls, the source of the power that the Tribune Company's Thorold mill used. Under the leadership of Adam Beck, Ontarians succeeded in the drive to "forge a public power movement" through the creation of the Hydro-Electric Power Commission of Ontario, the publicly owned utility that was producing three-quarters of the power in the province by 1932.[36]

In Quebec, in contrast, hydroelectric power remained under only nominal state control. Prior to 1907, the province simply sold waterpower rights, either directly to buyers or through public auctions. Starting in 1907, the province began using the "technique of alienation by long lease," often terms of fifty to one hundred years, through which it was "able to exercise a slightly greater control over water-power development," as the province in theory could cancel leases if lessees did not fulfil their conditions. Even with this new policy approach to hydroelectricity, however, regulation remained slight. The situation changed dramatically in 1944 under the leadership of the Liberal premier Adélard Godbout, whose government from 1939 to 1944 represented the only years between 1936 and 1960 when Maurice Duplessis's Union Nationale party was out of power. In April 1944, responding to public outcry over its high electrical rates and poor service, Godbout nationalized the Montreal Light, Heat and Power Company, "the only power company in Quebec that has been built up on the basis of an exclusively urban market." In the process, the government also set up the Commission hydroélectrique du Québec, also known as Hydro-Québec.[37]

When Duplessis returned to power in late 1944, he initially declined to pursue Godbout's more aggressive approach to the regulation of hydroelectricity in the province, but the continuation and evolution of Hydro-Québec's activities did structure the immediate post–World War II context of the Tribune's Manicouagan River development. As Arthur Schmon noted in December 1945, "Government ownership of power is a political football" in Quebec. In many ways, these new circumstances forced the Tribune Company to stress publicly that it was not trying to enter the public utility business, but rather that it was aiming to use the power generated by its new dam solely as a private party.[38]

This was in many respects in keeping with the company's long attempt to function as a publicly subsidized corporation operating as independently as possible across an international border. Mindful of the new politics of hydroelectricity in Quebec, Tribune Company officials began planning a new North Shore hydroelectric dam that would dwarf and replace the one they had built on the Outardes River in 1937. The Tribune Company began developing more specific plans for this Manicouagan dam throughout 1946 and 1947, and the company found itself balancing its own designs on the river with those of the province, as some Quebec officials saw North Shore power development as a crucial engine for postwar economic growth. As the parties negotiated terms for the dam's construction, Arthur Schmon later remarked, provincial officials "insisted that the power dam be built in such a manner that it would not interfere with the full utilization eventually of the power resources of the First Falls, therefore it became an expensive dam." In effect, the company was told that if it wanted to build any kind of dam on the Manicouagan River, that dam would have to take full advantage of its hydroelectric generating capacity at the first falls, the planned location near Baie Comeau.[39]

The Tribune Company's negotiations with the province over permission to build the dam were aided significantly by its relationship with Bishop Napoléon-Alexandre Labrie, one of the most powerful figures on the North Shore. As a missionary and priest, Labrie had been a fixture in the region for most of the time that the Tribune Company had operated there. When the Vatican created a new North Shore diocese in 1946, Labrie was made bishop and began designing a townsite to be the seat of his diocese. He set his sights on a large parcel of land near Baie Comeau which was owned by the Tribune Company, and he offered the company the assurance that he would lobby provincial officials in support of their proposed dam if they would arrange to give him the land that he wanted for his townsite. Corporate correspondence reveals a stark exchange of promises between the interested parties. As Arthur Schmon told Robert McCormick, he had "settled with Bishop Labrie the proposals we would present to the Quebec government in connection with land for the new townsite." The company would give to the provincial government a roughly 400-acre parcel of land that it owned, and the government in turn would publicly "grant the land free to the Bishop." In return for the land to build a townsite, Schmon claimed, the bishop "agreed to speak to the Prime Minister that he favors the development of power at Manicouagan."[40]

The bishop did his part in writing to Maurice Duplessis that the Tribune Company "wishes to obtain, as soon as possible, the authorization to build a dam on the Manicouagan River. I am convinced that the time has come to grant this permission." The current Outardes dam, he remarked, would soon be insufficient to satisfy the power needs of the mill and Baie Comeau. A new dam on the Manicouagan, on the other hand, not only would help expand the mill's production capacity but also could provide power for the bishop's new town and potentially even aid in the North Shore's industrial diversification. As he told Duplessis, a "greater development of power would probably draw other large industries. The outlook is not of minor importance for a county which can only foresee an industrial development. It is always with regret that we see other regions benefit from raw materials, while our workers must satisfy themselves with the less profitable tasks."[41]

Duplessis responded positively to Bishop Labrie's lobbying efforts, and with church, state, and private interests in harmony, Duplessis in September 1948 agreed to draft a bill allowing the Tribune Company to develop power at the Manicouagan's first falls. With the bill headed to the legislature with Duplessis's support, Arthur Schmon continued his private efforts to encourage public lobbying by others in favor of the project. Schmon wanted to use the "Bishop and the merchants" to push for the power facilities that the company wanted. "We will be in the public eye and any agitation that starts or is stimulated by us or if we become closely identified with it will be recognized as our method for accomplishing something for our own selfish purposes. Therefore, we will be arousing opposition to our project. For the time being, until we get out of the public eye, we must keep in the background."[42]

To provide additional support for the company's interests, the bishop wrote a public letter published in the Quebec City newspaper *Le Soleil* in December 1948 in which he argued that the province should support the Tribune Company's plans for the North Shore, as the new hydroelectric capacity would inaugurate a new era of progress and development. It was "divine Providence," Labrie proclaimed, that had given the North Shore tremendous natural resources, and it was up to Quebec's leaders to help the Tribune Company continue to exploit them. It was a "foreigner," Labrie declared, who had "recognized all that the North Shore could offer, and has had sufficient vision to invest his capital in such a significant enterprise." The benefits of the Tribune's investments were "too numerous to recount," Labrie claimed, and overall they had encouraged exactly the "kind of progress" that the province needed. Ultimately, Labrie argued, a new Manicouagan dam would "serve not only the region, but also will be a contribution to the prosperity of the province as a whole."[43]

The bishop's letter was greeted warmly by Tribune Company executives. "It is most extraordinary how sometimes a seed bears fruit," Arthur Schmon wrote to Robert McCormick in forwarding him the bishop's published letter. "This seed of thought had been planted quietly," he noted, but it had blossomed wonderfully for the company. "This is exactly the political factor which I have been counting on," Schmon claimed, and he forecast that the bishop's letter "will obtain for us the right to expand." Schmon did caution that the company would have to take care in the coming months to perpetuate the notion that the bishop's lobbying campaign was being done of his own volition rather than with company encouragement. "The danger," Schmon remarked, was that Duplessis "and other interests may think at this time that we are behind the Bishop's thinking for our own selfish purposes. It may be embarrassing to him."[44]

Ultimately, the company's private deal making and public lobbying campaigns proved successful. Duplessis, with a wide majority in the provincial legislature, shepherded the law through in March 1949, remarking that the Tribune Company was "rendering a great service to the province by harnessing the Manicouagan." The dam would help the company expand, Duplessis promised, and it would aid in rural electrification in the region and in creating power for Hauterive, the new town that the bishop was planning. Duplessis added that since this development was entirely funded by the private sector, "we do not pay anything and everyone profits." The legislation was passed on March 10, 1949, granting the Tribune Company a concession to build a dam on the first falls of the Manicouagan River. Legislation in hand, the Tribune implemented plans immediately to begin construction. Given the complexity of the task, the survey and planning were time-consuming, lasting some sixteen months.[45]

Construction began in 1951, and in a front-page story the *Tribune* called the dam a "New Niagara." According to company estimates, the Manicouagan's maximum flow was greater than that over Niagara Falls, the former estimated at 300,000 cubic feet per second and the latter at 267,000. The company's new dam, which Robert McCormick boastfully decided to call the McCormick Dam, would be a massive structure, with the company claiming that it would be 4,630 feet long and have a maximum height of 96 feet. Building it would necessitate using some 280,000 tons of concrete. In April 1951 CBC correspondent Ralph Marven visited the construction project and reported to his listening audience that "it'll be a big dam" requiring some fifteen hundred men and $15 million to build. The following month, McCormick traveled to Baie Comeau to participate in a ceremony marking the start of his eponymous dam's construction, which was broadcast over the radio back in Chicago. After Bishop Labrie blessed the construction site, Maurice Duplessis was given the honor of triggering the first dynamite blast to begin the work.[46]

Dam construction was completed in June 1952, and the *Tribune* described the achievement as a "private enterprise project which will supply electricity to this part of the country." At the ceremonies inaugurating the dam in July 1953, after Bishop Labrie blessed it, Maurice Duplessis pressed a ceremonial button to commence operations, and Arthur Schmon remarked that the dam "represents the turning of another key to unlock the natural resources of Quebec. This project is an asset not only to the company, but to the entire district, the province, and the country." Maurice Duplessis added a laudatory address and used the occasion to celebrate the public-private partnerships involved in exploiting Quebec's resources. As the *Montreal Star* noted, the remarks were "sort of a capsule sermon on the possibilities Quebec holds out to enterprise and the benefit the province itself derives from development of its still largely dormant resources." According to Duplessis, the "building of this project focuses public attention on the advantages of private enterprise working for the good of the people. . . . Anyone who says that we do not want American capital does not know what he is talking about." To the Americans, Duplessis pitched that the "possibilities of the province of Quebec are unlimited. . . . We have the resources. You have the money, Quebec has the best government. Let us work together." At a dinner that evening, Duplessis

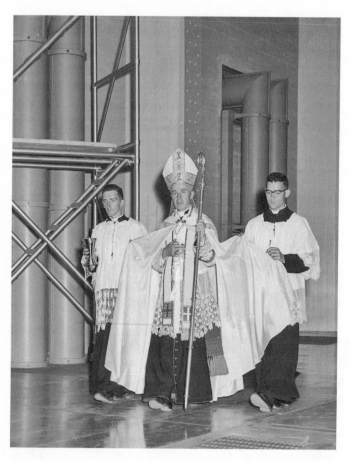

In July 1953, Napoléon-Alexandre Labrie, the bishop of the Gulf of St. Lawrence, blessed the McCormick Dam as it commenced operations. Vol. 25180, Scrapbook 1951–54, QOPC, LAC. Used with permission.

told the 215 assembled guests, "We're somewhat alike, the Colonel and I. We're both criticized, but we both do some good work." In the United States, *Time* magazine reported on the dam's opening, remarking that "along with his many other distinctions, Chicago *Tribune* publisher Robert R. McCormick is probably Canada's largest single foreign investor. His Canadian holdings, worth some $50 million by his own estimate, are scattered from western Ontario to the St. Lawrence River." All of this constituted, the magazine quipped, "Canada's McCormickland."[47]

The success of the McCormick Dam construction soon motivated Hydro-Québec to begin its own work in the region, which otherwise still "remained undeveloped in spite of its enormous water power potential." By the end of 1952, Hydro-Québec had begun plans for power generation on the nearby Bersimis

When it was completed in June 1952, the so-called McCormick Dam provided a tremendous new source of hydroelectricity to Quebec's North Shore. This photo shows an aerial
view of the dam and surrounding site on June 11, 1952. Vol. 25187, Scrapbook, QOPC,
LAC. Used with permission.

River, and the province would extend its dam construction projects in subsequent
years upriver on the Manicouagan.[48] As a private sector project accomplished
with strong support from the province of Quebec, the McCormick Dam was very
much in keeping with the kind of business-state relations that the Tribune Company had cultivated in the province for decades. In soliciting the support of
Bishop Labrie, the Tribune Company demonstrated how deft and adaptable it was
in navigating the political conditions in Quebec and using influential individuals
to its advantage. In the case of the McCormick Dam, the construction was also in
keeping with a midcentury attitude in both the United States and Canada that
massive dams were instruments of economic and regional development, and in
this way it was in line with other midcentury "megaprojects" such as the St. Lawrence Seaway. The dam gave the Tribune Company the hydroelectric power that
it needed for its present and future newsprint operations, and it also provided
excess capacity that would stimulate the further development and diversification
of its Baie Comeau operations into new sectors in the ensuing years.[49]

Brokering Hydroelectricity, Smelting Aluminum

As part of its arrangement with the Quebec government, the Tribune Company could not sell power generated by the McCormick Dam on the open market, and this created some practical concerns about its ordinary operation. Perhaps the primary challenge facing any power producer in the 1950s was that power had to be used when generated because it could not be stored cheaply and substantially. To have a facility generating power at a capacity above immediate needs was to have a facility wasting operating costs. In order to make use of its excess generating capacity, Tribune Company officials began seeking other companies to move to the region to use it. This need to sell excess power harmonized with some of the company's broader aims of diversifying the area's economy, as the company believed that "unless it is possible to attract new industry to Baie Comeau, not only is the future limited for the young men and women now reaching manhood and womanhood, but there can be little if any attraction for additional growth to the municipality. The answer to possible stagnation in this case is additional production, providing extra work for those already growing up in Baie Comeau, as well as an attraction for others to become part of the community." To find corporate partners to utilize the newly available power from the McCormick Dam, the company entered into negotiations with roughly forty firms that it sought to attract to the North Shore. Virtually all were in the mining and mineral processing sectors, for example, steel production and copper and zinc refining. As the company noted, the "only products which can be manufactured with advantage at Baie Comeau are those consuming large quantities of power per ton of product, or utilizing wood as a raw material."[50]

Ultimately, the best partnership opportunity for the Tribune's Quebec operations came from across the Atlantic with a British multinational mining firm wanting to build an aluminum smelter in a place with cheap power and political stability. Globally, aluminum was a booming business after World War II. Lighter and more flexible than steel, aluminum enabled advances in a host of industries and sectors, and it proved to be a perfect fit for the political and infrastructural conditions on the North Shore since, like pulp and paper production, one of the key elements in aluminum smelting was cheap power. Aluminum is readily available in the earth's surface, and indeed estimates are that it composes roughly 8 percent of the earth's crust, trailing only oxygen and silicon as constitutive elements. The problem for producers, as the sociologist Mimi Sheller notes, is that "aluminum is extremely difficult to get in pure form." Aluminum can be found in some form in shale, slate, and granite, but the primary source of raw material that became practical for mass production is bauxite, an ore that is named after the French village of Les Baux, where geologists first discovered it. In the late nineteenth century, chemists in France and the United States developed processes of smelting aluminum from bauxite which were scalable to mass production, and the industry boomed in the following decades. During and after World War II, the increased amount of global aluminum production reconfigured global bauxite supply networks, as industrial producers sought to exploit the significant sources that had been found in Africa and in the tropics. Bauxite mining is done in open pits, and the process leaves behind a highly toxic "red mud" that can be devastating to local ecosystems. After mining, bauxite is processed at a nearby location into aluminum oxide, also called alumina, after which it is shipped away to smelting sites located near cheap power sources and made into aluminum. Throughout the smelting process, aluminum producers depend on massive supplies of raw materials and electric power. According to one midcentury estimate, producing one ton of aluminum in ingot form required "about four tons of bauxite, a little under two tons of coal, about one-fifth of a ton of caustic soda, half-a-ton of carbon and from 18,000 to 20,000 units of electricity." This power component was, after the raw material itself, the essential part of the process, as the British Aluminium Company noted that the "most important factor in the production of aluminium is cheap electricity, and the cheapest source of electricity in most parts of the world is water-power." In 1960, a *Newsweek* report estimated that the "prodigious amounts of cheap hydroelectric power" required in aluminum production were so great that the "power needed to produce a single ton of aluminum would light an average home for a generation."[51]

When the Tribune Company began planning to partner with a British aluminum firm, it continued a history of American participation in North Shore power development and aluminum smelting which had begun with the American industrialist James Duke in the early 1910s. Starting in the previous decade, Duke had embarked on aggressive attempts to expand his operations in North and South Carolina, forming the Southern Power Company on the Catawba River in 1905 as part of his aim of diversifying into chemical manufacturing. After 1910, this plan to use hydroelectric power as a corporate strategy of diversification drew Duke to Quebec's Saguenay River, which emptied into the St. Lawrence roughly 120 miles upriver from Baie Comeau. In 1925, Duke merged his Canadian interests with the Aluminum Company of America (Alcoa), the main corporation involved in making the Saguenay region into an aluminum production center. Following the merger, Alcoa created the Canadian subsidiary Aluminum Company of Canada, and when James Duke died in 1926, the Aluminum Company of Canada took over greater control of these Saguenay operations and power rights.[52]

North Shore hydroelectric power became increasingly important for global aluminum production prior to World War II, accelerating quickly after 1939 as the war disrupted the international supply chains of raw materials in similar ways to those experienced by newsprint manufacturers. The major British producer, British Aluminium, lost control of the bauxite sources it had in southern France, and the Nazi occupation of Norway ended British Aluminium's production there. Aluminum was essential to the war effort, as it went into airplanes and a host of other military hardware. With the demand for aluminum rising during the war, the global locus of production shifted to Canada and employed materials that could be moved around the Western Hemisphere. The major Canadian producer, the Aluminum Company of Canada, got much of its bauxite from British Guiana and Dutch Surinam, and with the support of the state, it also greatly expanded its power generating and smelting capacity along the Saguenay River during the war. Prior to the war, the company's usual bauxite supply route to Quebec involved boats going north from these South American sites and around Nova Scotia prior to coming south up the St. Lawrence.[53]

German U-boat attacks in the North Atlantic disrupted this route, and the Tribune Company had its first indirect experience with aluminum smelting when its ships were used to move bauxite along the St. Lawrence in order to assist the war effort. Company ships on the route between New York City and Baie Comeau were required by wartime regulations to transport war-related cargoes on some legs of their voyages, and like the Aluminum Company of Canada's boats on the Atlantic, Tribune boats also were subject to the threat of German U-boat attacks on the route from Baie Comeau to New York City. To meet both challenges, Tribune Company boats laden with newsprint from Baie Comeau began avoiding the Atlantic and instead traveled up the St. Lawrence to New York ports, where the newsprint was then loaded onto train cars bound for New York City. The boats then went to Oswego and were loaded with bauxite that had been brought there by boat and rail from British Guiana. Tribune Company boats finally took that bauxite back downriver to the Aluminum Company of Canada operations on the Saguenay on the return trip to Baie Comeau. As a result of these kinds of creative supply chains, Canadian production eventually accounted for 90 percent of British wartime aluminum supplies, and more than one-third of the total used by the Allies. By the time the war ended, Quebec's North Shore had taken a leading role in the global aluminum market, and the Tribune Company's Manicouagan project both benefited from and contributed to postwar development.[54]

Having gained some exposure to the aluminum smelting business during World War II, and having built a massive new hydroelectric dam after the war, the Tribune Company would find in British Aluminium the ideal partner for North Shore industrial expansion. British Aluminium was formed in 1894, and its operations had been international in scope for much of its existence. From its base in the United Kingdom, the company in 1907 began moving some operations to Norway. Its bauxite supplies initially came from southern France, but the company expanded around the world to include sites in British Guiana in the 1920s and the African Gold Coast (present-day Ghana) in the early 1930s, the latter of which was the company's most important raw material location after World War II. The primary challenge beyond maintaining sufficient bauxite supplies, the company noted, was finding affordable power. "In view of the primary importance of cheap power in the production of aluminium," the company claimed, "the industry must always be on the look-out for power developments remote from civilization, unlikely to be required for normal electrification purposes."[55]

In 1950, British Aluminium and the Aluminum Company of Canada entered into discussions about building a dam and an aluminum smelter on the Volta River in the Gold Coast. Both companies had bauxite concessions there, and the Volta River had since World War I been looked at as a promising source of hydroelectric power for aluminum smelting. The main problem in the post–World War II period, as one account later described it, was that the lack of "stability of local conditions" failed to provide foreign corporations with the confidence to commit the necessary capital for building a dam and smelter. During the 1950s, the prevailing "pre-independence passions" helped bring Kwame Nkrumah to power as the first prime minister of the Gold Coast, and Nkrumah completed the transition away from British colonialism in 1957 by declaring independence for Ghana. This political turmoil delayed the construction of the facility on the Volta River which would become the Akosombo Dam, ultimately not beginning until after 1960. During the early 1950s, these delays frustrated British Aluminium's efforts to quickly increase production, and they began to seek alternative sites for a new smelter.[56]

By late 1954, this had led them to the North Shore of the St. Lawrence River and a partnership with the Tribune Company, and the two parties began discussions to form a joint venture. British Aluminium had, as one Tribune analyst noted, "extensive bauxite deposits on the African Gold Coast and in the West Indies" and was looking for a place to smelt this raw material into aluminum. The company was stymied in developing at home by the "lack of cheap power." It had been considering building its new smelter in Norway or the Gold Coast, both of which had abundant potential for hydroelectric development, but economic and political factors stood in the way. Tribune Company officials met with British

Aluminium officials and found that they were "reluctant to invest in low cost power sites in Africa and Norway because of the political factors and the length of time to develop the power. They want to start production within two years. They fear Africa because of the black man's ascendancy and the insecurity of any investment made there. Norway is socialistic and the interference of the government is terrific." Conditions in Quebec seemed more amenable to the company, and the combination of cheap power and political stability convinced British Aluminium to build operations on the North Shore.[57]

British Aluminium eventually partnered with the Tribune Company to form a company called Canadian British Aluminium (CBA), thus bringing aluminum production into the newspaper company's corporate structure. The company was incorporated in 1955; over the following four years CBA planned to spend an estimated $110 million on industrial development around Baie Comeau, and company officials estimated that this created approximately one thousand jobs. In the partnership, British Aluminium owned 60 percent of CBA, with the Tribune Company owning the balance. Smelter construction continued throughout 1957 and extended the process of the industrial reshaping of the North Shore. The CBA smelter went into operation in December 1957, and the first ingot was produced on December 23. One journalist noted that the "Baie Comeau spirit has carved power projects, pulp and paper industries and now one of the world's great aluminum producers out of the wilderness." The Francophone press was similarly laudatory, as *L'Aquilon* editorialized that the new aluminum smelter was "marvelous" and an indication that the "contribution of foreign capital" was the motivating force behind the North Shore's industrial development and economic growth.[58]

On June 14, 1958, Quebec premier Maurice Duplessis spoke at a public ceremony marking the official opening of the aluminum smelter, and he touted the benefits of industrial development on the North Shore. "Here we believe in freedom of enterprise; we believe in private enterprise," Duplessis claimed. "We don't believe in the government interfering all the time, and we don't believe in state ownership as a general rule. . . . Here you can bank on the future, you can think in terms of the future. . . . During election, or after election, the policy is always the same. . . . Free enterprise is the basis of incentive, the basis of good working conditions, the basis of labor unions, and the basis of real progress." As he had in 1953 at the ceremonial opening of the McCormick Dam, Duplessis used the occasion of the smelter's opening to tout his promotion of economic development and to attack critics of his policies: "American and British interests have come together to provide the province with a Canadian industry. Quebec provides the power and light. We are pleased with this triple alliance. Foreign

By the early 1960s, the McCormick Dam had enabled the creation of Canadian British Aluminium and the construction of its smelter. As this aerial photo from October 1962 shows, the overall scope of industrial operations in Baie Comeau had expanded considerably in the quarter century after the town's initial development. Vol. 25197, folder Baie Comeau, Townsite (3 of 3), QOPC, LAC. Used with permission.

capital should be welcome in the province of Quebec, and it can count on the cooperation of the provincial government, whatever the critics say." Moreover, to potential foreign investors Duplessis made it clear that some "extremists speak against foreign capital, but you should not worry about that. The province of Quebec believes that American capital must be and is welcome in the province of Quebec."[59]

By 1962, with the smelter in operation, the overall industrial footprint of the Baie Comeau operations had greatly expanded. One report even noted that there were some "traffic problems" in the area because of all the people and activity. Newsprint remained the basis of the local economy, though that was changing with industrial diversification. This expansion and diversification came through the power generated at the McCormick Dam, and exploitation of the North Shore's hydroelectric resources would become one of the most important issues in Quebec during the 1960s. As the Tribune's private dam proved immensely successful as a way to create regional and economic development, provincial officials began planning several publicly owned dams upriver on the Manicouagan, including one that would be one of the largest in the world. These, some Quebec officials believed, might do for the entire province what had been done in Baie Comeau.[60]

Navigating the Quiet Revolution

As the Tribune Company expanded its power generating capacity and newsprint manufacturing operations on the North Shore, it proved adept at navigating the politics of the Quiet Revolution, the term used to describe Quebec in the decade or so after Jean Lesage became premier in June 1960. Lesage, a Liberal, represented a Quebec electorate seeking major political reforms and social changes after the long period of conservative rule under Maurice Duplessis. Under Lesage's leadership, the government embarked on significant civil service reforms, aimed at rooting out political corruption, and in general worked to modernize the political and economic structures of Quebec society in a broader effort to promote the interests of its Francophone population. Hydroelectric power was central to this process, as many politicians and citizens saw it as the means through which the province could achieve greater political independence from the rest of Canada, by way of promoting internal industrial and economic growth.[61]

The Tribune Company's expansion and development plans continued throughout the 1950s with Maurice Duplessis in office, but Duplessis's sudden death in 1959 forced the company to adapt to swift political changes in the province, particularly those affecting hydroelectric power. One Tribune Company executive remarked in April 1960 that he was concerned about "recent events in this political trend toward nationalization. . . . If Mr. Duplessis were still alive, this talk of nationalization would be only talk—but his place has been taken by men who do not have the same conviction about Quebec maintaining a sound balance between privately owned and publicly owned power facilities." In July, another executive speculated that it was "pretty obvious" that Hydro-Québec had designs on nationalizing the McCormick Dam, and he urged the company to "waste no time in developing every argument opposing nationalization."[62]

Despite the Tribune Company's internal concerns, the new Quebec government ultimately showed little interest in nationalizing the McCormick Dam. However, Hydro-Québec's actions on the Manicouagan did have some significant effects on the Tribune Company's operations, as the province announced in mid-1960 that it would spend some $3 billion on a fifteen-year project to build five dams on the Manicouagan and Outardes Rivers, where the Tribune Company had long been operating. Construction on a massive Hydro-Québec project upriver from the new McCormick Dam on the Manicouagan River would create a reservoir initially called the Duplessis Reservoir and later named the Manicouagan Reservoir, and one of the Tribune Company's concerns was that this would limit the river's flow downstream to the degree that it would not be able to run the McCormick Dam to its necessary capacity. The company immediately began a lobbying campaign, claiming that the so-called Fifth Falls project, later known alternately as Manic-5 and the Daniel-Johnson Dam, would decrease the river's flow and "jeopardize" the McCormick Dam's operations, thus having "a drastic effect upon the industries dependent upon it for power." The company claimed that its "truly pioneering" developments on the North Shore had earned it the right to an equitable solution with the province to its problems. "It is obvious that the development of Canadian British Aluminium came about through great effort and expense on the part of" the Tribune Company, one official argued to the province. The development, the company claimed, "never could have been accomplished" by the public sector because it "involved large expenditures, financial participation, and the providing of a variety of 'know-how' not available" to anyone working in it. According to internal engineering estimates, the proposed Hydro-Québec development would cut the Manicouagan's flow and thus the power available to 37 percent of prevailing capacity for six or seven years, and the Tribune Company essentially wanted Hydro-Québec to guarantee that it could keep its power supply at the same price, even if Hydro-Québec had to provide that power somehow as the Manicouagan flow decreased. Eventually, the Tribune Company was able to get the province to provide it with discounted power via Hydro-Québec to offset the loss of internal generating capacity from the altered and limited flow of the Manicouagan.[63]

As it adapted to the province's plans for the river, the other major problem that the Tribune Company faced was that the dams along the Manicouagan affected the company's timber limits, which straddled the river from Baie Comeau up to the point where the fifth falls dam was to be built. As the company told provincial officials in May 1960, the Manicouagan was the "life-line of our present operations, and the key to future developments, at our paper mill in Baie Comeau. Our main pulpwood resources lie within the Manicouagan watershed, and the Manicouagan itself is the means by which the pulpwood is floated to the mill. Any change, therefore, in the character of the river, particularly on the large scale planned in the proposed power development on the Manicouagan River, presents our Company with serious problems not only for the present but also for the future with respect to our operations." Hydro-Québec's plans to dam the Manicouagan would, the company claimed, "flood a total area of 854.7 square miles containing about 3,700,000 cords of merchantable pulpwood. This will extend for some 200

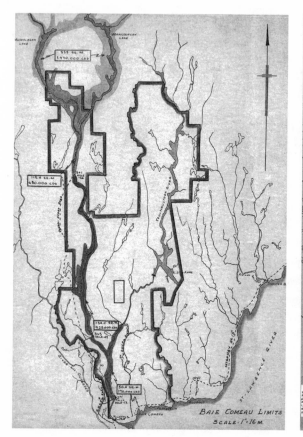

These two maps show the evolution of the Tribune Company's North Shore timber limits between 1960 (*left*) and 1975 (*right*), a period in which Hydro-Québec undertook major hydroelectric development plans upriver from the McCormick Dam. *Left,* "Memorandum Presented by the Quebec North Shore Paper Company Concerning Problems to Immediate and Future Operations Created by the Flooding of Timber Limits and Crown Land along the Manicouagan River," May 4, 1960, vol. 25325 (acquisition no. 2006-00392-5, box 36), folder First Falls Expansion, pt. 1, QOPC, LAC. Used with permission. *Right,* Ontario Paper Company Limited and Quebec North Shore Paper Company, "Forest Properties in Ontario and Quebec," Nov. 1975, vol. 25397 (acquisition no. 2006-00392-5, box 108), folder Forest Properties in Ontario & Quebec—Allowable Annual Cuts, QOPC, LAC. Used with permission.

miles along the river. Of this area, about 300 square miles, containing some 1,700,000 cords, are within limits held under lease by our Company for many years." To exploit this valuable pulpwood, the company asked to be "permitted to carry out the necessary salvage operations in the areas affected and that this approval be given as soon as possible to enable maximum pulpwood to be salvaged."[64]

Public observers of the problem noted that there was as much as $275 million worth of timber that had to be harvested prior to the area being flooded. One of the problems with simply cutting all the wood before flooding was that it would create storage problems and inundate the pulpwood market. "The large pulp and paper companies have their own forests where annual cutting is carried out on carefully established plans," one Quebec journalist noted, and these annual cutting rates were instituted not only for conservationist reasons but also to control the supply of pulpwood on the market. This dual challenge was the "crucial point: Our forest is a rich resource that must be protected." But, at the same time, "when the market suffers from an over-production, it is the beginning of a chain reaction" that would have significant effects on one of the province's most important industries. After negotiations with provincial officials, the Tribune Company eventually was able to arrange to purchase some 300,000 cords of wood that it would cut above the fifth falls at a bargain price, and it was able to have the stumpage duties removed on wood that it would cut downriver prior to the flooding. Over time, the company was also able to acquire timber limits from the province replacing the land that the province needed for Manic-5 construction.[65]

All of these negotiations and adaptations were in keeping with the generally cordial relationship that the company had with provincial officials during the Quiet Revolution. Even after Duplessis's death, company officials were successful at navigating political changes in Quebec. Besides a genial relationship with Jean Lesage, company officials also remained cordial with René Lévesque, Quebec's minister of natural resources and a strong proponent of nationalizing the province's hydroelectric resources. Lévesque believed that, as historian Caroline Desbiens notes, "too many private utility companies remained under the control of the Anglophone 'colonial barons,'" and he was a strong advocate for a "profound 'decolonization' of the energy sector." In 1962, Lévesque "proposed the acquisition and nationalization of the eleven private power companies still under English control and the development of standard rates for every energy user across the province," and he continued to push for even greater government control of the province's energy sector. That fall, Lévesque's lobbying convinced Jean Lesage to call an election that became "essentially a referendum on the continued expansion of hydroelectric development." Public support for Lévesque's ideas was strong, with one of the most prominent examples being the October 1, 1962, headline in *La Presse* reading "L'eléctricité: clef qui nous rendra MAÎTRES chez nous," or "Electricity: The Key to Making Us MASTERS in Our Own House." The election was a resounding success for Lesage and Lévesque, as the Liberals won sixty-three of ninety-five seats in the Legislative Assembly. Almost immediately afterward, "Lévesque's acquisition plan for eleven private energy companies went ahead, and Hydro-Québec was mandated to take over any remaining sites with hydroelectric potential that had not yet been conceded."[66]

One Tribune executive reported in April 1962 that he had met with Lévesque eight times over the previous year, and he remarked on "one of the best and friendliest conversations I have had with him," in which he lobbied Lévesque to exclude the McCormick Dam from his nationalization plans. In response, Lévesque "reiterated his absolute convictions about public ownership, on which he does not deviate one iota—but he clarified that by saying he did not necessarily feel the same way about our plant which is not a distributer in the full sense of the word, and which is basically established to supply the manufacturing needs of its own industry." Ultimately, the Tribune Company succeeded in convincing Lévesque that they should be kept out of nationalization plans, and they found that his stern nationalist and anti-corporate remarks in public masked a much more friendly private attitude toward private enterprise and, more importantly, the Tribune Company.[67]

Through its well-developed political connections and skills, the Tribune Company was able to navigate the new circumstances created by this rising tide of hydro-nationalism. Even after Lesage won an election in which hydroelectricity was essentially the sole issue, the company was able to protect its interests and operations. As Arthur Schmon noted in December 1962, the "power rights of investor-owned companies were the claimed single issue of the recent provincial election." Even though Lesage "did state at the beginning, and throughout the election campaign, that this issue did not touch companies generating electric power for their own industry requirements," the Tribune Company was going to face unique challenges because it was "distinct from other companies within this integrated group" since it was operating "on the same river with Hydro-Québec." Schmon, however, was confident that any and all problems would be resolved in a reasonable manner. "For the Premier, as head of a government, the retention of a sound climate for capital investment, both past and future, should and would mean a great deal more than the technical problems of operating several power developments on the same river."[68]

Schmon's confidence was the well-earned result of decades of close partnership with the Quebec government, and the groundwork that the company had laid in working with the Duplessis government from the 1930s to the 1950s helped the company continue to evolve and prosper in the era of the Quiet Revolution in the early 1960s. This adroit maneuvering in Quebec followed a similarly deft series of political moves in Ontario during World War II, and Tribune Company officials displayed a high degree of diplomatic capability in navigating the changing political contexts of its increasingly diverse Canadian manufacturing operations. Building on its wartime projects and by skillfully partnering with provincial developmental imperatives, the Tribune Company was able to both diversify its operations into new and profitable sectors and remain in the good graces of the public officials whose largesse enabled the overall industrial enterprise to flourish.

Conclusion

Across the breadth of its Canadian operations, Tribune Company newsprint production increased throughout the postwar period, and the company also extended the industrial diversification it had begun in the 1940s. Some of this was the result of an ongoing commitment to technological experimentation, but evolving commercial imperatives also motivated the company to continue developing new lines of business. In the first decades of operation, as one Tribune Company executive noted, the Canadian subsidiary had "only one major objective," and that was supplying newsprint to the parent company's US newspapers. The expansions into alcohol and vanillin production after World War II were the "natural developments of the principal objective—supplying newsprint as cheaply as possible."[69]

By the mid-1960s, this push toward diversification became stronger, as an internal assessment noted that the company as a newsprint manufacturer was tied to a sector that was already mature and perhaps even soon to be in decline. Diversification of the Canadian operations was advisable, the report noted, because of "the fact that the whole of our income with the exception of that derived from vanillin or power is still dependent on the success of the North American desire to advertise." If the company wanted to continue to prosper, it was essential to figure out "how wood still available at Thorold or Baie Comeau can be turned into products which are not dependent upon advertising." In other words, the company needed to develop strategies to diversify in case newsprint usage was to decline as a result of a loss in advertising spending, something that had happened in recent decades during the Depression and which some feared could happen again.[70]

From the 1940s to the 1960s, in acting on both these sorts of internal calculations and in response to the significantly changing political contexts of Ontario and Quebec, the Tribune Company showed itself to be adept and adaptive as a manufacturing firm. As the business remained grounded in using trees and hydroelectric power to produce newsprint, the company found innovative ways of expanding and diversifying its operations into other sectors. One manifestation of this strategy led the company in its Ontario mill to adopt a project that would make it into the world's largest producer of vanillin. In Quebec, the company leveraged the expansion of its hydroelectric generating capacity into aluminum smelting. In both provinces, the process of manufacturing an industrial newspaper resulted in an industrial diversification that was always rooted in and structured by politics and policymaking.

CHAPTER EIGHT

The Industrial Newspaper and Its Legacies

In July 1954, Robert McCormick decided to celebrate his birthday with a lavish party in Baie Comeau. McCormick invited most significant company officials, as well as a number of personal friends. Arriving prior to the day of the festivities, McCormick toured the area and his company's operations. The party itself had been planned around a picnic at nearby Lac St. Anne, but weather intervened and the gathering was forced indoors in Baie Comeau. However, surrounded by family, friends, and business associates, McCormick celebrated his seventy-fourth birthday and achievements on the North Shore. "An accordionist and a string quartet—woodsmen in colorful lumberjack shirts and neckerchiefs—played French Canadian ballads," the *Chicago Tribune*'s report noted. "The birthday cake was 3 feet square, topped by a miniature model of Col. McCormick paddling a canoe, as he did when he surveyed the forests and streams of the Quebec north shore years ago in search for pulpwood limits." McCormick received numerous gifts and tributes, and one Toronto newspaper executive called him a "bulwark of America and builder of Canada." This would be the Colonel's last birthday. Robert McCormick died on April 1, 1955.[1]

In the United States, prominent Republicans hailed McCormick's achievements. President Dwight Eisenhower called him a defender of the "free press so essential to our freedom," and Vice President Richard Nixon celebrated him as a "vigorous and uncompromising fighter for the program that he believed was best for this country." Former president Herbert Hoover added that McCormick was a "brilliant and generous personality," and General Douglas MacArthur praised him as a "comrade in arms of world war whom I held in the highest esteem."[2]

There were numerous assessments of Robert McCormick's life and achievements across the United States and Canada, and the overwhelming majority commented on the vigor and stridency with which he articulated his political views. In the United States, McCormick's death elicited comment in the major national weekly magazines. *Time* called his paper the "No. 1 fortress of personal, daily journalism in the U.S." and noted that McCormick "put the mark of his eccentric, sometimes pugnacious personality into every column of the *Tribune*." *Life* noted in its obituary that McCormick's death "closed . . . a career that was the last great bulwark in the U.S. of personal journalism." *Newsweek* noted that he "dominated The Tribune utterly, from broad policy down to minor news stories," and remarked that one of the things that was particularly galling to his critics was that the paper "fought the colonel's battles in every one of its columns instead of just on the editorial page."[3]

In Canada, newspapers located in regions where the Tribune Company operated lauded McCormick's contributions to these areas of the country. *L'Aquilon*, a Quebec paper, called his death "a great loss to French-Canadians on the North Shore" and noted that McCormick had a "great affection for the French workers who helped him to establish his industrial empire in our region." In St. Catharines, Ontario, where the nearby Thorold mill's operations had brought decades of investment in jobs and manufacturing to the region, the *Standard* remarked that McCormick "played a leading role in Canada's business and economic development through his newsprint mills, transportation, power development and forestry operations. . . . In Canada we accord him tribute as a benefactor."[4]

Canadian newspapers in cities without direct economic ties to the Tribune Company were a bit more grudging in offering praise. The *Globe and Mail* in Toronto, for example, noted, "Wrong and prejudiced, he might often be; weak and compromising, never. . . . Colonel McCormick was an individualist in a time of collectivism, an isolationist in a time of internationalism, a non-conformist in a time of conformity. . . . In his death, the United States has lost a patriotic American of the old school, with the very real virtues of the old school. Whatever may be said of his opinions, there can be no question of his fortitude and integrity." The *Ottawa Journal* similarly noted that McCormick was a "salient, forceful, colorful figure, a part of the American scene. For his own peculiar traits he will be missed." However, having praised McCormick's vigor in articulating his opinions, the *Journal* was rather less charitable toward their content. "Politically it was impossible to take the Colonel seriously; in his rantings, his feverish and warlike vocabulary, his almost grotesque assumptions of military lore, he was more comic than convincing."[5]

Some other assessments of McCormick's career went beyond emphasizing his editorial stridency and noted that his strategy of vertical integration in newspaper production had afforded him the high degree of self-sufficiency that enabled him to articulate views that many people across North America considered eccentric and even dangerous. As the trade journal *Editor & Publisher* noted, McCormick was a "man of honest convictions and integrity, a courageous and patriotic fighter for the principles he believed best for America, a jealous guardian of the free press, and a journalist with genius and vision." These aspects of McCormick's public persona were reflected and enabled, the magazine went on, by the fact that he was a brilliant and innovative manufacturer. "His vision is best exemplified in the extensive newsprint manufacturing enterprise which he developed to insure a source of supply for his publishing properties." McCormick, the magazine argued, "built more than a newspaper. He created an institution."[6]

Robert McCormick built a company that was at the time of his death one of the most important newspaper corporations in North America, and this institution, both as a producer of public information and as an industrial manufacturer relying on forest resources, had significant effects in both the United States and Canada. Over more than four decades, McCormick had built an industrial newspaper firm relying on vertically integrated operation to manufacture the two highest-circulating newspapers in the United States, and he died at a moment when this business model was exceptionally successful. The *Chicago Tribune* had the highest circulation of any standard-sized daily newspaper in the country, with roughly 900,000 copies sold daily and 1.4 million on Sundays. The only paper with a higher circulation remained the *New York Daily News*, which sold some two million papers daily and 3.6 million on Sundays. The two papers also led the industry in advertising revenues and had combined to sell more than $100,000,000 of advertising in 1954.[7]

This moment would not last. As this and the following chapter will show, the Tribune Company's industrial newspaper operations would change dramatically over the course of the following few decades. When McCormick died in 1955, it was at a time when mass-circulation printed newspapers formed the core of a small but powerful group of gatekeepers in the dissemination of public information. As was the case during McCormick's lifetime, the *Chicago Tribune* remained one of the most important corporations in this history, and its evolution and eventual dissolution as an international manufacturing firm offer a way to view some of the most important developments in the recent history of the American news business. If the Tribune Company was representative of the success of the industrial newspaper in the first half of the twentieth century, it was also emblematic of its travails in the latter half of the century. Its struggles were not without consequence in both the United States and Canada, and the company's operations would create significant legacies and have lasting effects on the people, politics, and landscapes of North America.[8]

Expanding a National Newspaper Chain

Though Robert McCormick died in 1955, he exerted a durable influence on the *Tribune*'s politics, which remained consistent in the following decades. In 1971, for example, editor Clarence Peterson commented that the paper "used to be a good, strong, unfair voice of conservatism. Some of the old guard didn't even know what fairness was. The new guard does." But, as the *Wall Street Journal* noted, "the new guard doesn't ask that the paper change its politics, merely that it be responsible." As Peterson confessed, he wanted the *Tribune* to be a "good, strong, fair voice of conservatism." As the *Tribune* retained McCormick's conservative editorial sensibility, the parent company began expanding into new areas of the country through acquisitions of other existing newspapers. As Americans were on the move to the suburbs and away from northern cities like Chicago and New York, the Tribune Company tried to follow them.[9]

The company's first foray into owning a paper beyond the *Chicago Tribune / New York Daily News* operation actually began in 1949, before McCormick's death, when the Tribune Company acquired the *Washington Times-Herald* after a strange and contested estate settlement following the death of Robert McCormick's cousin, Eleanor "Cissy" Patterson, who had owned the paper for a decade.[10] McCormick's ownership of the *Times-Herald* was brief and marked by clashes with his niece Ruth McCormick "Bazy" Miller, whom he had installed as the paper's editor. Miller resigned in 1951 because of disagreements with McCormick over politics. As Miller stated, there had been "things I wanted to do that I wasn't permitted to do. . . . I won't be anything but an independent publisher. . . . I know his view so thoroughly that I knew what the bounds were within which I could operate. . . . There is some difference in our political beliefs. I am for the same people the Colonel is for, but I am for more people. I have broader Republican views than he has." After Miller's resignation, McCormick took "personal charge" of the *Times-Herald*, but this ultimately lasted only three years. In 1954, McCormick sold the paper to the *Washington Post* for a reported $8.5 million, a tidy profit over the $4.5 million purchase price. McCormick claimed that the sale was prompted by the fact that he was now seventy-four years old and "his doctors had advised him to do less work because of severe illnesses in the last year."[11]

When McCormick died the following year, the terms of his will established the conditions for an initially smooth transition at the Tribune Company. McCormick had no biological children, but he was survived by a group of loyal executives operating with a sound financial and institutional footing. Chesser Campbell took over as Tribune Company president, and the majority ownership of the company remained with the McCormick-Patterson Trust, which had been established in the early 1930s, with the aim of ensuring that a controlling interest in the company remained with the family. The trust stipulated that it was to dissolve twenty years after the deaths of both Joseph Patterson and Robert McCormick. Patterson had died in 1946, and McCormick's 1955 death meant that the trust was set to exist until April 1975, as it would. With the arrangement in place and the business operating profitably at the time of McCormick's death, the transition into a new management structure went smoothly. As one executive remarked at the 1956 meeting of shareholders, "We are going on as usual."[12]

In New York, with Joseph Patterson having died a decade earlier than Robert McCormick, the *Daily News* already had executives in place to navigate the transition to new management after its founder's death. To satisfy a demand for its printed product which remained strong and growing in the mid-1950s, the *Daily News* began a $12,000,000 expansion of its printing plant in 1956. The paper could well afford the expense. As president and publisher F. M. Flynn pointed out, the *Daily News* "had its largest advertising revenue last year" and continued to be the country's leading generator of advertising. Despite the paper's continuing success, Flynn did note that some significant changes were afoot in postwar New York City, the most important of which was suburbanization, and the *Daily News* management was struggling to address this development. "One of the worst things for New York newspapers is the moving of people to the suburbs," Flynn claimed. "Here in town, they could always find three newsstands around the corner but when they move out they do not find them and the shopping centers do not have newsstands. Local papers are delivered to homes in many suburban areas. Where such families formerly bought two New York City papers, they now often get only one and add it to the local suburban paper." Flynn and the *Daily News* management took steps to address readers' residential migrations, and by the end of the decade they believed that the paper had responded successfully to the new market conditions. As one 1959 advertisement touted its popularity with readers outside of the city, "1,900,000 News readers are suburbanites," and "when New Yorkers move to the suburbs, The News is one habit that goes along. It is preferred breakfast fare in Stamford and Saddle River as well as Stuyvesant Town. It gets first reading from commuting husband and stay-at-home housekeeper."[13]

As Tribune Company management adapted to changing market conditions in New York, its first post-McCormick acquisition came in 1956 with the purchase of the *Chicago American* from the Hearst Corporation for a figure estimated to be "between $11 and $13 million." The *American* acquisition left Chicago with two pairs of competing newspapers: the *Tribune* and *Sun-Times* in the morning and the *Daily News* and *American* in the afternoon. Prior to the deal, *Daily News* publisher John S. Knight had aimed to buy the *American* and combine it with the *Daily News* to make what he hoped would be the "largest evening newspaper in America," but he later noted that he was unsuccessful because he was "thwarted in that by my friends" at the *Tribune*. Unable to expand the *Daily News* in the manner he had hoped, Knight sold the paper three years later to *Sun-Times* publisher Marshall Field.[14]

To produce the newly acquired *American* in a more cost-effective way, the Tribune Company began printing the afternoon paper at the *Tribune* plant in 1959. In 1961, the company moved the *American's* editorial staff into the Tribune Tower as well. Publishing these two papers was a major organizational and industrial feat, as it meant that the *Tribune's* presses were working constantly. "From a production standpoint," one observer noted, "the integrated operation of a seven-day morning paper and the American's six-day evening publication, plus its Sunday edition, represents one of the most complex publishing cycles in the newspaper field." To take advantage of this same kind of production efficiency, the competing *Sun-Times* also began printing the newly purchased *Daily News* in its printing plant. In effect, by the early 1960s, Chicago had four major daily papers published by two companies in two buildings, and this concentration prompted one observer to remark that "Chicago's four newspapers are controlled by one man and a ghost."[15]

Despite its efforts to economize on the *American's* production, the Tribune Company found itself "steadily losing money and circulation" with the paper throughout the 1960s. In order to stem the losses, the Tribune Company in early 1969 remade the *American* into a tabloid that it renamed *Chicago Today*. This rebranding did little to turn around the redesigned paper's fortunes, nor did other strategies such as raising the newsstand price or deploying "heavy promotion for circulation and subscription which led to publication of pop posters, a popular gimmick which served only as a temporary stopgap." In 1974, the Tribune Company stopped publishing *Chicago Today*. In 1978, the *Sun-Times* stopped publishing the *Daily News*, creating the newspaper market that continues to define Chicago today, with the *Tribune* and *Sun-Times* competing as morning papers.[16]

The newspaper market in postwar Chicago evolved in roughly the same way that it did throughout the United States. Though overall newspaper advertising revenues remained robust, the total number of papers went down through mergers and consolidations aiming to create greater production efficiencies and eliminate local competition over advertising. Chicago's six daily newspapers in 1946 had contracted to four by 1960, and the ensuing years saw ever-increasing corporate consolidation and ever-decreasing local competition. These developments were replicated throughout the country. Between 1940 and 1960, the percentage of cities with two or more daily newspapers declined from 23.4 percent to 16.4 percent, and this shrank by half again to 8.2 percent in 1986. As publishers in many northern cities were struggling to retain readers, many Americans were moving to the South and West as part of postwar migrations to newly expanding Sunbelt cities. For some newspaper firms, the response to this was to create or expand newspaper chains, a strategy that would allow publishers to operate more papers in more cities through the centralization of some corporate functions and the standard-ization and serialization of some newspaper content.[17] In the Tribune Company's case, the move to follow readers into newly burgeoning cities and regions began in earnest in the 1960s as the company expanded outside of Chicago and New York. In 1963, the company bought the *Fort Lauderdale News* and the *Pompano Beach Sun-Sentinel*, and in 1965 it bought the *Orlando Sentinel* and the *Orlando Star*.[18]

As the Tribune Company was expanding its US newspaper holdings after Robert McCormick's death, the company also continued expanding the scale and scope of its industrial operations in Canada. By mid-1958, the company employed almost eight thousand workers in Canada and had an annual payroll of over $25 million. In 1968, the company began a $54 million expansion of the Baie Comeau mill. The project employed some seven hundred workers and added a fourth papermaking machine that increased overall newsprint manufacturing capacity from 350,000 to 530,000 annual tons. In order to fund this expansion, the com-pany sold its stake in its aluminum smelting business, though it retained its ownership stake in the McCormick Dam, which remained vital to North Shore industrial newsprint operations. When the new Baie Comeau paper machine was installed in 1970, it became one of the three largest papermaking machines in the world. During this period, the company also improved and expanded its mill in Thorold after considering the possibility of closing it down in 1977. The company eventually decided to stay in operation in Thorold and built a new $260 million mill with the help of $21 million in subsidies from the federal and provincial governments. Some 330 jobs were lost in the mill's renovation, though it would have been 1,600 had the mill closed.[19]

These mill expansions resulted in dramatic increases in newsprint manufac-turing in the postwar period. In Thorold, production went from 173,698 tons in 1950 to 312,927 tons in 1991. In Baie Comeau, production went from 116,967 tons in 1940 to 153,014 tons in 1950, and by the 1980s the mill was manufacturing over 400,000 tons of newsprint per year. The Tribune Company's overall newsprint production in Canada more than doubled from the 1950s to the 1980s, as table 7 shows. Because of industrial expansion of the newsprint facilities and the fact that newsprint consumption in the United States was decreasing, the company found by the 1960s that it was able to become entirely self-sufficient in its manu-facturing and publishing. As one assessment of operations in 1964 noted, "The newsprint business has reached a turning point in two ways—the consumption of our three main newspapers has ceased to grow rapidly . . . [and] we can manu-facture the total requirements of our papers in New York and Chicago, some-thing which has never been possible before."[20]

TABLE 7

Annual newsprint production tonnage at Thorold and
Baie Comeau mills, 1913–91 (selected years)

| Year | Annual production | | Total |
	Thorold	Baie Comeau	
1913	5,713		5,713
1914	31,707		31,707
1915	34,189		34,189
1920	69,970		69,970
1925	95,283		95,283
1930	112,580		112,580
1935	146,168		146,168
1938	154,927	92,247	247,174
1939	156,340	110,204	266,544
1940	157,890	116,967	274,857
1945	136,791	128,529	265,320
1950	173,689	153,014	326,703
1955	190,668	163,952	354,620
1960	206,916	178,086	385,002
1965	190,200	260,700	450,900
1970	210,700	316,800	527,500
1974	232,400	448,100	680,500
1984	264,000	404,000	668,000
1991	312,927	430,083	743,010

Sources: Carl Wiegman, *Trees to News: A Chronicle of the Ontario Paper
Company's Origin and Development* (Toronto: McClelland & Stewart, 1953),
352; memorandum, Newsprint Production, Quebec North Shore Paper
Company, vol. 25316 (acquisition no. 2006-00392-5, box 27), folder Baie
Comeau 25th Anniversary, pt. 2, QOPC, LAC; OPC—Thorold Mill
Newsprint Production, vol. 25793 (acquisition no. 2006-00392-5, box 504),
folder Ontario Paper Company Historical and Background Material, pt. 1,
ibid.; Background Information for the Ontario Paper Company, June 24,
1975, vol. 25779 (acquisition no. 2006-00392-5, box 490), folder
Background Information for the Ontario Paper Company Organization,
ibid.; *The President's Report*, various years, vol. 25796 (acquisition
no. 2006-00392-5, box 507), folder President's Reports 1983–92, ibid.
   Note: The Thorold mill started operation in September 1913, and the Baie
Comeau mill began production in 1938.

On the one hand, this was the realization of the aims that Robert McCormick
had set forth more than a half century earlier. On the other hand, one 1974 estimate showed newsprint sales accounting for 90 percent of the Canadian subsidiary's profits, a worrisome statistic given that, as one internal assessment noted, the
"pulp and paper industry's financial performance over the past ten years is generally regarded as poor." Another manager lamented in 1976 that newsprint manufacturing was a "low-growth business, but it is still and is likely to remain our
major business." One strategy that the company employed was to begin selling
its newsprint to other publishers on the international market; by the mid-1970s
it was selling roughly 28 percent of its production to other newspapers. The

company sold much of its surplus to US newspapers in the Midwest and Northeast, but it also began exporting significant quantities to the United Kingdom and India.[21]

Ultimately, the Tribune Company in the two decades after Robert McCormick's death employed two distinct strategies to develop its operations in North America. On one front, in the United States, the company purchased established newspapers in newly burgeoning regions of the country in order to follow readers out of northern cities. On the other hand, company officials also began to sense that the newspaper business might be headed for stagnation and even contraction, and some began to look warily at aging and expensive industrial plants in Canada. Having used industrial newsprint manufacturing and vertical integration to great success from the 1910s to the 1960s, the company began to realize in the 1970s that this strategy may have outlived its utility.

The End of Industrial Labor Peace

As officials within the Tribune Company began to reassess the strategy of vertical integration which it had used for decades in Canada, they found that relations with the company's employees began to fray in the 1960s and 1970s. In some ways, this had to do not just with ferment among unions in the papermaking industry but also with broader political changes in Ontario and Quebec. Whatever their motivations, in the 1960s and 1970s many in the Tribune's Canadian industrial workforce began to express dissatisfaction with the terms of their employment, and the effects of their attempts at raising workers' consciousness about the politics of their employment by a US corporation extended beyond the context of the shop floor. Overall, workers' activities in the 1960s and 1970s contributed to a significant change in the generally positive tenor of labor-management relations which had defined the company's operations in Canada for the previous half century. To some workers, the Tribune Company no longer seemed like a benevolent American employer, but rather an example of the US domination of North America.

This line of thinking was especially evident in Quebec, where critics of the company and the kind of investment that it represented became increasingly aggressive in the 1960s. After Maurice Duplessis's death in 1959, anti-colonialism became a much greater part of Quebec's political culture. To some degree, the Liberal party benefited from this sensibility, and René Lévesque's push for nationalizing hydroelectricity in 1962 both was based on and helped to feed this mood. As one *Le Devoir* editorial noted, Quebecers increasingly saw themselves as "subject to the economic domination of foreign capital." This belief would become increasingly prominent throughout the decade, and many Quebecers began to see the province as part of a global anti-colonial movement in step with those in such countries as India, Ghana, Morocco, Cuba, and Algeria.[22]

As anti-colonialism flourished in Quebec, corporations that were operating along the North Shore, including the Tribune Company, began coming under increasingly vigorous and pointed criticism. In 1955, a priest named Paul Duclos remarked in the North Shore newspaper *L'Aquilon* that the corporations in the region had "reigned as masters over a little colonial empire." These companies operated at such distance from public scrutiny, another Quebec newspaper argued in 1957, that workers were "controlled exclusively by companies from which they must get everything—shelter, food, transportation, and a livelihood. . . . ALL IS QUIET on the operations, meaning that the employees are hired, fired, exploited, transferred at the employer's pleasure, and nobody hears about it." Within these general attacks on North Shore corporations, the Tribune Company increasingly became the subject of specific and pointed criticism in Quebec. In 1960, the newspaper *La Côte-Nord* published a letter assailing the company's power in the region. "Everything is monopolized there," the letter claimed, "and Quebec North Shore Paper Company is master and God, and the company is the dictator." In 1962, a correspondent for *La Presse* remarked that "for some time" North Shore residents had "blamed the Quebec North Shore Paper Company for all their problems."[23]

As these public sentiments circulated in critical newspaper columns, anti–Tribune Company sentiment also mounted in the workplace. As the company had been generally supportive of unions over the previous fifty years, it found much of its workforce unionized by the 1960s. In March 1961, the company experienced the "first union demonstration of any importance to take place in Baie Comeau since the founding of this town, 25 years ago," when more than 250 union members gathered publicly to support the contract demands of Office Workers' Union Local 361, which had been founded in 1958. The workers who showed up in solidarity at the demonstration included leadership from the major mill and woods workers unions, and their members came out in the interest of helping the office workers secure a better contract.[24]

This demonstration marked the beginning of a significant shift in the company's relations with its workers and their unions across Canada. In June 1965 at its Thorold mill, the company experienced the first strike in its history, when workers from United Papermakers and Paperworkers Local 101 began an illegal "wildcat strike" while contract negotiations were ongoing. Just before 6:00 p.m. on June 15, one of the mill's lead machine operators called the mill's superintendent to inform him that they were "pulling the plug at 6 P.M." Mill managers rushed to the site, met the workers "on their way out," and told them that this was an illegal strike, but the workers ignored the threat and walked off the job. By 8:00 p.m., "Papermaker pickets were patrolling at all gates." Despite calls from Local 101's president for the workers to return to work because the "strike was illegal," they refused, and soon a number of members "of the other eight unions represented at the plant were not crossing the picket lines." Within two days of the initial wildcat strike at Thorold, more than one thousand company workers were honoring the picket lines and refusing to show up for work. The ongoing labor dispute widened several days later when one thousand workers at the Baie Comeau mill went on strike seeking wage increases of their own. The strike ultimately lasted twenty-four days before workers and management agreed on a contract providing wage increases.[25]

To some company executives, the 1965 strike represented a significant turning point in labor-management relations. According to a 1966 internal report, the company's relations with its Canadian workers had generally been positive, "until recently." At Thorold, however, there had been over the past five years a "growing estrangement between the unions . . . and the company." Some unions were seeing the rise of "militant" leaders, and thus the management was "confronted with an environment in which traditional company philosophy of union relations is no longer effective. Quite naturally most management representatives are unable to adjust completely to this new climate. Company representatives genuinely believe, and with considerable justification, that they have treated employees fairly in the past, and that the 'stab in the back' atmosphere created by the Unions, allegedly representative of employee reaction, is quite unjustified." As company officials became increasingly wary about and adversarial toward its unions in Ontario, they faced in Quebec the additional challenge of addressing the rise of French Canadian nationalism in the 1960s. In Quebec the interjection of language and nationalist politics into labor relations created increasing conflicts on and off the job in Baie Comeau. As one observer noted, the "fact of English-speaking management of predominantly French-speaking employees" created a "potential risk, because of the current social and political evolution in the province of Quebec, in maintaining 'outside' management of the Baie Comeau operations."[26]

In the face of changing business circumstances and labor-management relations, the company experienced another work stoppage in Thorold in late September 1973, when some two hundred members of trade unions representing plumbers, electricians, and machinists at the mill went on strike for higher wages. The strike shut down the mill for twenty-seven days before the dispute could be settled. In September 1975, the company was hit with additional strikes in Thorold and Baie Comeau and was forced to halt all newsprint production in Canada.

With the unions pushing hard for wage increases, negotiations dragged on for months, and it was only in late January 1976 that workers ratified a new contract and newsprint production began again. The strikes ultimately had a significant effect on the Tribune Company's overall profitability, as the parent company actually lost money in the first quarter of 1976. As an April *Chicago Tribune* report noted, its robust American advertising revenues had been "offset by strike-related losses suffered by its newsprint/forest products group" in Canada.[27]

The labor peace that had defined company operations from their beginnings in the 1910s had dissipated by the 1960s and 1970s, and the increasing tension between labor and management and the resulting strikes created a great deal of concern and bewilderment among company officials. One Ontario executive was puzzled by the changing tenor of industrial relations at the bargaining table. "Our employees get along well with each other and have appeared to get along well with our supervisors and management," the executive noted in 1976. "Employees' recreational and social functions continue and to some extent have expanded in this period." The executive reported that he had "spoken personally with a great many of our employees from all areas of the plant" and was "continually assured" that the company had provided a "good place to work." How could it be, he wondered, that "when negotiations come along, everyone is . . . mad at the Company?" Ultimately, the corporate culture that Robert McCormick had cultivated proved difficult to sustain after his 1955 death and in the changing context of Canadian political culture, and relations between labor and management began to take on an increasingly adversarial character in subsequent decades. As was the case in manufacturing industries across North America, the Tribune Company's industrial newspaper operations seemed to be entering a transitional and perhaps even declining phase in the 1960s and 1970s.[28]

Brian Mulroney, Baie Comeau, and North American Free Trade

Though Tribune Company labor-management relations suffered in the 1970s, Robert McCormick's lingering presence and influence in Baie Comeau still filtered into policies that affected all Canadians later in the twentieth century. Ultimately, his company had substantial effects on the entire nation of Canada in the ways that its operations significantly influenced the life of Brian Mulroney, who became prime minister in 1984, signed the Canada–United States Free Trade Agreement (FTA) in 1988, and helped pave the way for the subsequent and expanded North American Free Trade Agreement (NAFTA). Mulroney's father, Benedict, was one of the first Tribune Company employees in Baie Comeau, having arrived in the city after an itinerant career as an electrician along the North Shore. Settling in the newly developing community, Ben Mulroney and his wife began a family, and

Brian was born in Baie Comeau in 1939. Ben Mulroney would spend his adult life working in the Tribune's newsprint mill, but the family wanted their son to have more opportunities than the town provided. Thus, at age fourteen Brian was sent to boarding school in New Brunswick, after which he attended college at St. Francis Xavier University in Nova Scotia and then law school at Laval University in Quebec City. Brian Mulroney spent summers in Baie Comeau, and he remained a regular follower of the town's rapid growth. Indeed, his own life history closely follows Baie Comeau's expansionary phase from the newly built town in 1939 that was organized around newsprint production to a place that was by the 1960s a diversified center of resource exploitation. After leaving the region for schooling and the start of his professional career, Brian Mulroney reconnected to it in 1977 when he became the president of the Iron Ore Company of Canada, a company that had developed operations along the North Shore in the decades following the Tribune Company's arrival.[29]

The arc of Brian Mulroney's life and career is one that intersects in a number of important ways with the history of US-Canada trade in the twentieth century. As chapter 2 showed, in 1911 the United States and Canada nearly entered into a reciprocity agreement that would have enabled duty-free trade in a host of commodities. The US Congress approved the pact, but Canadian voters recoiled from it, as many believed that it was a significant step toward political supplication to and perhaps even territorial annexation by the United States. In the election called in 1911 as a referendum on the reciprocity agreement, Wilfrid Laurier's Liberals lost to Robert Borden's Conservatives, thus bringing to an end Laurier's fifteen-year term as prime minister and pushing US-Canada free trade off into some indeterminate future. Despite Canada's rejection of the broader reciprocity agreement, US policymakers did allow some Canadian newsprint to enter the United States free of a duty, and the Underwood Tariff in 1913 extended this duty-free status to all newsprint manufactured in Canada. This change in the tariff on newsprint provided a tremendous boost to the Canadian economy in the coming decades as the pulp and paper industry grew, mostly to supply the US market. Over the next thirty years, the Tribune Company's operations would expand as well, from the construction of the Thorold mill to the creation of Brian Mulroney's hometown.

As Brian Mulroney began what would be a successful management career in the mineral extraction business, Canada in general was developing closer economic ties to the United States. Outside of the pulp and paper industry, perhaps the most important midcentury manifestation of this was the Canada-US Automotive Products Trade Agreement signed in 1965 by US president Lyndon Johnson and Canadian prime minister Lester Pearson. The agreement, also known as the "auto pact," created a "borderless North American auto industry" by allowing many newly manufactured vehicles to cross the border free of duties. Most of the pact's terms were agreeable to US-owned auto firms, and the US-Canada trade in automobiles and auto products jumped from $10 billion in 1973 to $60 billion in 1986.[30]

As the auto pact expanded on a robust midcentury trade in pulp and paper, some Canadian economists began to worry that Canada's economic dependence on trade with and foreign investment from the United States had reached troubling new levels. In 1968, the Task Force on the Structure of Canadian Industry, a group set up by the federal government and headed by the liberal Canadian economist Mel Watkins, noted that the "extent of foreign control of Canadian industry is unique among the industrialized nations of the world." For the Watkins group, this development was important not only in its economics but also because of its politics. The "most serious cost to Canada" of US foreign investment, the Watkins report claimed, was the fact that the US government tended to "regard American-owned subsidiaries as subject to American law and policy with respect to American laws on freedom to export, United States anti-trust law and policy, and United States balance of payments policy." This, the report claimed, represented the "intrusion of American law and policy into Canada by the medium of the Canadian subsidiary," and this development "erodes Canadian sovereignty and diminishes Canadian independence. It implies that the American-based multi-national corporation is not multi-national but American. It creates political costs for Canada from American direct investment that seriously undermine the economic benefits."[31]

Others analyzing this trend at the same time took an even more critical view, perhaps most prominently the radical Canadian political economist Kari Levitt (the daughter of economic historian Karl Polanyi), who in 1970 criticized US corporations operating in Canada as "manifestations of a new mercantilism of corporate empires which cut across boundaries of national economies and undermine the national sovereignty of the hinterland countries in which their subsidiaries and branch plants are located." There were significant political and economic consequences to US foreign investment, Levitt argued, and she attacked Canadian economists who advocated lower tariff policies enabling American corporations to gain easier access to Canadian natural resources. While this strategy might promote short-term economic growth, Levitt argued, it did so at the expense of Canada's long-term economic and political development. The Canadian continentalist argument "totally misses the point," Levitt claimed, as it "fails to provide an effective answer to the concern that economic integration with the United States, in the context of an economy dominated by branch plants and subsidiaries, will weaken internal integration within Canada, will perpetuate the 'technology gap,' and deprive this country of the 'dynamic comparative advantage' accruing to indigenous technological advance and innovation." Perhaps more troubling to Levitt was the possibility that the US-Canada trade relationship might evolve into a loss of Canadian national sovereignty. "The most bitter harvest of increasing dependence and diminishing control," Levitt concluded, "may yet be reaped in the form of the internal political balkanization of Canada and its piecemeal absorption into the American imperial system."[32]

Though these public criticisms resonated with many Canadians, they did little to slow the development of US-Canada trade, which grew dramatically over the twentieth century as a vibrant trade in pulp and paper products, including newsprint, was joined by cross-border automobile manufacturing. After Brian Mulroney became prime minister in 1984, this trade relationship would draw even closer. Mulroney had first tried entering politics in 1975 and 1976 when he campaigned for the leadership position of the Progressive Conservative Party, and he routinely used his Baie Comeau background as a way of fashioning his political identity. As a trio of journalists noted after covering Mulroney's ultimately unsuccessful initial run for party leadership, "You can take the boy out of Baie-Comeau, but you can't take Baie-Comeau out of the boy." After working in the private sector for several years, Mulroney returned to politics in 1983 with a second and successful run for leadership of the Progressive Conservative Party. As a politician, the *New York Times* noted in 1983, Mulroney was "well-coiffed, well-tailored, with a resonant voice . . . [and] the television presence of an Academy Awards show host," and he often used his Baie Comeau background to establish both his appeal and his legitimacy. Mulroney, according to the *Times*, "sought to project himself as a model Tory and a new kind of Canadian politician. For one thing, in a party that has its share of high-born patricians, Mr. Mulroney has stressed his working-class roots as the son of an electrician in the pulp and paper town of Baie Comeau." Throughout his campaign in Quebec in 1983, Mulroney's "stress on his Quebec origins and ability to speak French" aided him greatly, and his perfect fluency in English allowed him to address audiences outside the province with authority as well.[33]

In 1984, the Liberal prime minister Pierre Trudeau retired, and Mulroney's Progressive Conservatives defeated a Liberal Party led by John Turner in the federal election held later that year. Mulroney's North Shore roots remained prominently on display during the campaign, as Canadian observers Rae Murphy, Robert Chodos, and Nick Auf der Maur noted that year that "much of the basis of Brian Mulroney's political philosophy was formed in Baie-Comeau," and they expected these ideas to be applied throughout Canada under his leadership. What Mulroney had learned on the North Shore, they claimed, was the belief that there were "huge pools of private capital" from the United States ready to be "tapped and put to work." The state's job, as he had learned in Baie Comeau, was to facilitate the movement of that capital to Canada. "Deals had to be struck and concessions given, but the work was done and, for people of Brian Mulroney's generation, for the most part the good times did roll. Baie-Comeau was, in some ways, a perfect little microcosm of Brian Mulroney's best of all possible worlds."[34]

When he was campaigning in 1984, Mulroney had not made free trade a central part of his party's platform, and his economic message was oriented toward generally promoting economic growth rather than pushing for greater trade with the United States. However, almost immediately after assuming the office of prime minister in 1984, Mulroney began seeking closer economic relations with the United States. In his first official visit to New York City, Mulroney remarked at a speech at the Economics Club of New York that "Canada is open for business again," and he added that his government was placing "the highest priority" on improving trade relations with the United States.[35]

One of the more infamous episodes in Mulroney's early cultivation of closer relations with the United States had some connections to his youth in Baie Comeau. In his autobiography, Mulroney related that, as a child, he often listened to his mother "sing soft and sorrowful Irish ballads" in the house. One day in Mulroney's early teens, Robert McCormick arrived in Baie Comeau with a party of revelers, including the boxer Jack Dempsey. McCormick's wife's "new favourite song" was "Dearie," and Mulroney was tapped to sing it for him at a gala event. "I recall standing on a table in the commissary," Mulroney wrote, "as the Colonel and his few hundred guests cheered the official debut of my professional singing career. Colonel McCormick congratulated me warmly and placed a crisp American fifty-dollar bill in my hand, which I promptly ran home to give to my mother."[36] Decades later, Mulroney performed in the presence of another prominent US Republican, President Ronald Reagan, with whom he sang "When Irish Eyes Are Smiling." In 1985, Mulroney and Reagan, both of Irish ancestry, met on St. Patrick's day in Quebec City at the so-called "Shamrock Summit," the televised conclusion of which featured the two men and their wives singing the song onstage. As the Canadian historian Robert Bothwell acidly noted, the performance was "a piece of unalloyed sentimentality that would for years afterward be considered a landmark of bad taste among Canadian intellectuals. (It seems not to have been noticed by American intellectuals.)"[37] Despite the scorn from some in Canada, the overall event, as one account describes it, "recaptured an amity missing since the 1950s" between leaders of the United States and Canada.[38]

In ways more directly relevant to policymaking, Mulroney was also motivated to seek free trade with the United States by the work of the Macdonald Royal

Commission, which had been constituted by Prime Minister Pierre Trudeau in 1982 as part of a major review of Canadian trade policy. The creation of the committee was a sharp departure from the policies of the previous decade within the Trudeau government, which showed scant interest in promoting free trade with the United States. In the 1980s, however, the Macdonald Commission's work put free trade on the Canadian policy agenda in a major way. After extensive study by a group that ultimately encompassed almost three hundred scholars and researchers, the Macdonald Commission delivered in August 1985 a nearly two-thousand-page report that offered strong recommendations for the Canadian government to pursue a free trade agreement with the United States. The Macdonald Commission's report began a new public debate in Canada about a highly divisive issue. According to Gallup polls taken in 1953, 1963, 1973, and 1983, the percentage of Canadians thinking that the country would be better off with free trade with the United States was 54, 50, 56, and 54, respectively. As Mulroney began promoting free trade with great vigor in 1985, this divided public opinion soon made it the most important issue in Canadian politics.[39]

In the United States, the *Chicago Tribune* was a staunch supporter of a free trade agreement, a stance that was consistent with both its support of the Republican Party and its standing as a multinational industrial newspaper. In October 1985, the paper editorialized that a potential free trade agreement was a "good idea, important for the economic health of both nations." The *Tribune* saw the agreement as a significant step toward a newly emerging globalized economy defined by the free movement of capital and commodities. The US-Canada agreement, the *Tribune* argued, might set a powerful example to the world: "Mr. Mulroney and Mr. Reagan have shown the way in committing to the world's largest free-trade zone. It is not a perfect arrangement in that both countries refused to give up some subsidies and other barriers. But it is a major step toward ending creeping protectionism in the world, and it demonstrates to others that nations can bilaterally negotiate free trade pacts that cover sensitive areas like investment and services. Legislators in the U.S. and Canada should not erase this example and should not prevent the benefits of free trade from touching their citizens." In an active but less public manner, executives at the Tribune's Canadian subsidiaries were also staunch supporters of free trade. As a 1988 company newsletter noted, "Where do we stand on free trade? The mill started up at Thorold as a result of the lifting of the tariff on newsprint in 1911. Our Company and the community have prospered as a result, and we hope it will continue."[40]

As negotiations toward a free trade agreement continued, the fall 1988 Canadian federal election became a referendum on North American trade in very similar ways to those in the 1911 election. During a Canadian election campaign that the

New York Times described as "emotional and bitter," Mulroney vigorously championed the benefits of free trade. According to one journalist who followed him on the campaign trail, a regular trope during his appearances involved him locating "at least one opposition heckler in the crowd." Mulroney would then single the person out with a remark such as, "You are the fearful and the timorous. You want to hide Canada behind the poverty of protectionism. Well, by God, we're going to have so many free-trade jobs here . . . that we're even going to have one for you." In contrast, Mulroney's opponent, the Liberal John Turner, railed relentlessly against the free trade agreement. In a televised debate, as the *New York Times* described it, Turner claimed that Mulroney was "selling Canada out to the United States by negotiating a free-trade agreement that opened this country of 25 million people to economic and cultural domination by those feared and resented foreigners to the south."[41]

Ultimately, Brian Mulroney triumphed in an election that was as much a referendum on free trade as the one seventy-seven years earlier. Immediately after winning the election, Mulroney announced that he would "move 'at the first opportunity' to use the parliamentary majority he won in Monday's election to formally approve Canada's free-trade agreement with the United States." He did so, and the law went into effect on January 1, 1989. US-Canada trade increased even further after the passage of the FTA, as the value of merchandise exports from Canada to the United States rose from $101.9 billion in 1989 to $271.5 billion in 1998 and imports into Canada from the United States rose from $88.1 billion to $203.3 billion over the same period. In the wake of the passage of the FTA, some officials from Mexico began to seek their own free trade agreement with the United States, and in June 1990, the United States and Mexico announced that representatives would begin negotiations on a free trade agreement. After some initial hesitation, Canadian officials joined in what became trilateral negotiations. These discussions continued over the following two years, and the *Chicago Tribune* remained committed to promoting the emerging neoliberal order during the NAFTA debates, editorializing that "free trade—within North America, throughout the world—is the future. The U.S. must face this future, not flee from it."[42]

Though US-Canada trade increased dramatically after the FTA, both countries soon felt the effects of a global recession. Mulroney's popularity dropped significantly, and in early 1993 he had what the *New York Times* reported was the "lowest support in opinion polls for any Prime Minister in this century." Rather than face a federal election with such scant support, Mulroney resigned from office in February 1993. The Progressive Conservative Kim Campbell served a brief term as Mulroney's replacement, but Jean Chrétien's Liberals scored a resounding victory in the October 1993 federal election. In the United States, Democrat Bill Clinton won the US presidency in late 1992, but these leadership changes in Canada and the United States did little to slow the momentum toward NAFTA's passage and implementation, and the pact went into effect on January 1, 1994. Starting in the Mulroney years, free trade became an unavoidable part of Canadian life. As one academic noted in 1998, "Though most Canadians probably now are resigned to the FTA and even to NAFTA, they are just that: resigned. . . . Free trade has formed a grim triumvirate with death and taxes, two other phenomena regarded as unpleasant, unwelcome, and unavoidable."[43]

Given the increasingly strong economic relations between the United States and Canada over the course of the twentieth century, and given the broader context of late and post–Cold War neoliberalism, it seems relatively unsurprising that the United States and Canada agreed to a free trade pact in 1988. But it was by no means inevitable, as the 1911 election demonstrated. Canadians had been given a path to North American free trade then, and they rejected it. That they chose that path in the 1980s had a great deal to do with the political skills of Brian Mulroney, and these skills and his appeal had at least something to do with his Baie Comeau background. On the campaign trail and in public office, Mulroney routinely used his Baie Comeau upbringing as a way of promoting the alleged benefits of free trade. In 1987, for example, Mulroney praised Baie Comeau as a place that taught residents to have "respect for those who were raised in a different language, or a different faith. The only test was whether you could pull your weight and make a contribution to the life of the community." Besides providing a model for a linguistically and culturally harmonious Canada, Baie Comeau also provided a vision for Canada's place in the global economy. In Baie Comeau, Mulroney claimed, "we have always known the importance of international markets: we lived by them. We've always known the importance of foreign investment: we saw it in our jobs and in our very lives. We saw the importance of securing those markets and securing new ones, and we have never feared for our sovereignty or our identity." And in Baie Comeau, as he wrote in his 2007 memoir, Robert McCormick "influenced every part of Baie Comeau life. . . . Though derided today by many as some sort of twentieth-century Republican robber baron, Colonel McCormick was in fact a larger-than-life figure who dreamed the great dreams necessary to carve a thriving community out of bedrock and forest on the North Shore, while building a newspaper empire in the United States."[44]

In applying the lessons learned from one of America's foremost newspaper publishers, Mulroney would help reshape the histories of North American trade and of Canada's role in the global economy. This is not to practice unlicensed psychoanalysis, nor is it to suggest that we can or should draw a straight line from Brian Mulroney's childhood to NAFTA. But there are clear and important connections between Mulroney's early life in a town built by the Chicago Tribune Company and the actions that he would take as an elected official. Ultimately, one of the Tribune Company's most important legacies in Canada involved the personal links between Robert McCormick, a paper mill employee named Ben Mulroney, and then his son Brian Mulroney. Enabled by free trade policy, the Tribune Company had built a well-functioning industrial plant and community in Baie Comeau. To Brian Mulroney, his own background was evidence of the benefits of free trade, and while in government he succeeded in institutionalizing those beliefs in Canada.

Newspapers, Monuments, and Shadows

As Brian Mulroney applied the beliefs that he developed through growing up in a town built by the Chicago Tribune Company in shaping policies affecting the lives of all Canadians, those who lived and worked on the North Shore and in Thorold experienced the company's legacies in more immediate ways. In its Quebec and Ontario industrial activities, the Tribune Company shaped not only the social life of surrounding communities but also many aspects of their built, natural, and even olfactory environments, and these effects were not always positive. In 2015, one longtime Thorold resident remarked to me, "I remember somebody telling me that Thorold had a very strong bond with Baie Comeau because they both smelled the same," owing to the occasional "awful stench" and "rotten egg smell" emanating from the Tribune Company's paper mills. Indeed, the company became one of the worst polluters in the Thorold region as an industrial newsprint producer. In 1954, the company estimated that it was dumping some 50,000 pounds of "mill effluent" into the Welland Canal every day. One company chemist warned that the company was dumping so much waste in the canal that there was a danger "that the sludge would settle out in the canal before it had proceeded too far. I suspect that a shoal, or reef, would very rapidly develop where our sewer enters the canal." With the plant's vanillin waste, it was also "possible that a considerable stench of hydrogen sulphide would result" and continue to plague the area, making it smell like rotten eggs.[45]

The company's deleterious effects on the Niagara Peninsula's environment continued for years after this assessment. As a 1970 *Tribune* article noted, its refining of waste sulfite liquor and vanillin manufacturing was "responsible for turning the waters of the old Welland Canal into a cola color and forming masses of foam that cascade over the crumbling locks." The canal had become a "favorite setting for Canadian television camera crews doing documentaries on Canadian water pollution and environmental issues," as it was topped by "thick foam" that "sometimes covers the waterway from bank-to-bank and extends for a hundred feet or more." The "tan foam and reddish-brown waters of the canal form a striking contrast with the green lawns and beds of red roses in the park." One member of the Ontario Water Resources Commission called it "probably the worst-looking watercourse we have" in the province. "It is the problem that constantly plagues Ontario government. It is an embarrassment." In an attempt to control this pollution, the Tribune Company in 1969 added a "clarifier" to its Thorold operations which eliminated wood fiber from paper waste, and in 1970 it was preparing an "ion exchange plant which will remove 40 tons a day of dissolved solids from the mill effluent and the mill is installing an evaporation and burning process for removal of foam and color from the effluent." Ontario officials took note of this, as "George A. Kerr, the Ontario government cabinet minister in charge of pollution control, praised Ontario Paper Co. two weeks ago as one of the most cooperative companies in the province."[46]

In the Thorold / St. Catharines area, the Tribune Company was active in shaping the cultural and physical surroundings of the community in other ways that were more positive. Indeed, the *Thorold News* editorialized in 1950 that the Tribune Company had been instrumental in the area's industrial growth. "We in Thorold know Colonel McCormick as a man much better than we do as a publisher. We know that he turned a cow pasture into a site for one of the most thriving industries in the Niagara Peninsula." After McCormick died, Arthur Schmon continued the company's commitment to developing the region. Schmon was the chairman of the Founders' Committee of Brock University, which opened in 1964. To this day, the tallest building on campus, Schmon Tower, carries his name. Schmon's presence in philanthropy and community life went beyond his activities at Brock, so much so that, when he died in 1964, the *St. Catharines Standard* remarked that the city "lost its leading citizen." The *Standard* went on to say that Schmon's death "brings to an abrupt close one of the most illustrious careers in the North American papermaking industry. It also ends a life of dedication to community work, philanthropy, education and church affairs unrivalled in St. Catharines history."[47]

The Tribune Tower in Chicago remains a monument to Robert McCormick on one end of his international newspaper empire, and there also remain two significant monuments to him on the other end of this empire, on Quebec's North Shore. The first of these is the massive McCormick Dam. The second is a more curious sort of symbol, in the form of a statue on Baie Comeau's waterfront of a life-size Robert McCormick paddling a canoe. When the statue was dedicated on

In Baie Comeau, Robert McCormick was and is memorialized by a statue of him paddling a canoe. Vol. 25181, Scrapbook 1955–60, QOPC, LAC. Used with permission.

July 30, 1956, a host of local and provincial dignitaries came out for the ceremony, during which Baie Comeau mayor J. A. Duchesneau remarked that McCormick "was to a great extent responsible for the prodigious developments that have been so advantageous to this region." Bishop Napoléon-Alexandre Labrie added, "We, the children of Saguenay County . . . have reason to rejoice together in raising a monument to him. This symbol of eternal gratitude we justly owe to great men, he has indeed signally deserved. We require this visible sign to inform the present generation and those to come after them that all there is here has grown from the seed he sowed." Quebec finance minister Onésime Gagnon remarked that McCormick made the area "into one of the province of Quebec's richest jewels. Colonel McCormick was looking for and discovered vast forest and hydro-electric resources and in so doing he opened to civilization a territory, the fantastic mineral resources of which have since attracted the attention of leading industrial-

In this aerial photograph taken in 1960, McCormick's memorial appears in a wider visual perspective also showing his company's newsprint mill. Vol. 25193, folder Baie Comeau Mill Exteriors 2/2, QOPC, LAC. Used with permission.

ists of Canada, the United States, Great Britain, Germany, and other countries." Gagnon touted the community of Baie Comeau as the lingering tangible evidence of what McCormick had done. "Perhaps the greatest living monument to Colonel McCormick is the town of Baie Comeau itself. . . . Baie Comeau today unveils a monument to this great man but is in itself the real monument."[48]

McCormick's statue is a monument to his arrival on the wild North Shore rather than the corporate reorganization of its space and the industrial exploitation of its resources. The newspaper executive who spent most of his adult life in boardrooms and mansions is memorialized in Baie Comeau by a bronze statue of him paddling a canoe. An aerial photo taken in January 1960 puts that statue— and McCormick's achievement—in a rather different perspective. In this image, the statue appears at the lower right of the frame, set off from the industrial activities that clashed harshly with the natural landscape McCormick's company had exploited and transformed. A columnist in the *Windsor Daily Star*, a paper published in a border city passed regularly by boats traversing the Detroit River laden with the *Tribune*'s newsprint, captured the statue's multiple and competing meanings in noting that, at its dedication ceremony, there was no discussion of the *Tribune*'s "outspoken scorn for everything British—except, of course, newsprint resources in a British Commonwealth country." As the author concluded, "statues are nice to have around, if one does not look too closely at the shadows they cast."[49]

Conclusion

If we do look closely into these shadows, however, we can see that the Tribune Company's effects on the politics and landscape of Canada were lasting, meaningful, and diffuse. Robert McCormick's dam and statue in Baie Comeau are perhaps the most prominent and overt physical monuments to the Tribune Company in Canada, and Brian Mulroney did much to keep McCormick's name a part of Canadian political discourse. Along Quebec's North Shore, after McCormick's death the Tribune Company also became an avatar of further development. Starting in the 1930s, its operations demonstrated the region's industrial potential, and the company cultivated relationships with provincial policymakers which other companies would build on and benefit from. Starting in the mid-1950s, the presence of hydroelectric power and infrastructure on the North Shore had begun attracting other development and industries to the Baie Comeau area. In 1958, the *Financial Post* noted that the "once-slumbering land on the St. Lawrence is poised today for a burst of growth," and it published a report entitled "Great New Sinews for the Canadian Economy" which called the North Shore the "Hub of a New Empire." Of particular note for the *Financial Post* was the "power potential" of the dams that would later be built by Hydro-Québec and the "vast deposits" of iron ore on the "Quebec-Labrador frontier."[50]

Though other companies were undertaking these new development projects, the Tribune Company remained known on the North Shore for decades as the initiator of the region's development. In 1969, for example, one journalist visiting the region noted that the "spirit of Col. Robert R. McCormick is still evident in the nearby woodlands as the north shore of the St. Lawrence grows fat on United States investment." Baie Comeau, he claimed, remained "the envy of Quebecers because of its success in attracting investment capital." In 1974, Quebec historian Robert Parisé added that Baie Comeau was "l'avant-garde de l'expansion" of the North Shore, which he also called one of the most promising regions in the country for industrial development. And this, Parisé remarked, was the Tribune Company's doing. He called Robert McCormick "le visionniare" and declared that he was the "giant who truly launched the North Shore's industrial development." In 1987, another Quebec historian, Pierre Cousineau, called the postwar North Shore mineral boom "un nouvel Eldorado québécois" and argued that

Baie Comeau was the basis for it. For decades after Robert McCormick's death, he remained a part of the region's history and a touchstone for discussing the ongoing development of the North Shore and its resources. McCormick left a legacy that can be found in the landscape, built environment, policies, and popular memory of Canada, if one understands where to look for it.[51]

CHAPTER NINE

The Problem of Paper in the Age of Electronic Media

As a communications medium and a commercial product, the printed newspaper proved strikingly durable in the twentieth century, even as new forms of media were made available to and adopted by consumers on a mass scale. Radio emerged as a consumer technology in the early 1920s, and in 1925 roughly one in ten homes in the United States had a radio set. This jumped quickly past two-thirds of US homes by 1935, and by 1950 more than 95 percent of US homes had a radio set. Television was adopted even more quickly, with 2.3 percent of US homes having a television set in 1949, 9 percent by 1950, and 87 percent by 1960. After that point, nearly every home in the United States had both a radio and a television set, and the advertising revenues in both media increased along with the size of the audience.[1]

All the while, daily circulation of newspapers more than doubled from 27.8 million to 58.9 million between 1920 and 1960 as radio and television were emerging. On average, these newspapers increased in size as publishers printed more pages of news and advertising, and the average daily newspaper expanded from twenty-seven pages in 1940 to sixty pages in 1978. As table 8 shows, this increase in the size and circulation of newspapers in the post–World War II period motivated massive increases in newsprint consumption. As had been the case in the first half of the twentieth century, newspapers used much of this newsprint after 1950 to print advertising, which continued to generate huge revenues even as television grew in popularity. As advertising revenues in all media skyrocketed in the post–World War II period, newspapers as late as 1990 led all other media in terms of total advertising revenue generated. The current financial crisis facing the newspaper industry is a strikingly recent development. Alongside a variety of new media, the printed newspaper remained a central part of both North American public culture and the media business well into the late twentieth century. Through the end of the Cold War, the printed newspaper remained the leading medium of advertising in the United States and the core of the overall media marketplace.[2]

Though printed newspapers remained central to the business of media throughout the twentieth century, they continued to present huge operational challenges

TABLE 8

Daily newspaper circulation, annual newsprint consumption, and advertising revenue in newspapers and broadcasting in the United States, 1950–95 (selected years)

Year	Daily newspaper circulation	Total newsprint consumption (tons)	Newspaper advertising revenue ($)	Radio advertising revenue ($)	Television advertising revenue ($)
1950	53,829,000	5,937,000	2,070,000,000	605,000,000	171,000,000
1955	56,147,000	6,638,000	3,077,000,000	545,000,000	1,035,000,000
1960	58,882,000	7,426,000	3,681,000,000	693,000,000	1,627,000,000
1965	60,358,000	8,550,000	4,426,000,000	917,000,000	2,515,000,000
1970	62,108,000	9,727,000	5,704,000,000	1,308,000,000	3,596,000,000
1975	60,655,000	8,395,000	8,234,000,000	1,980,000,000	5,263,000,000
1980	62,200,000	10,088,000	14,794,000,000	3,702,000,000	11,488,000,000
1985	62,800,000	11,507,000	25,170,000,000	6,490,000,000	21,287,000,000
1990	62,300,000	12,125,000	32,281,000,000	8,726,000,000	29,073,000,000
1995	58,200,000	11,261,000	36,317,000,000	11,338,000,000	37,828,000,000

Sources: *Historical Statistics of the United States, Earliest Times to the Present, Millennial Edition*, ed. Susan B. Carter, Scott Sigmund Gartner, Michael R. Haines, Alan L. Olmstead, Richard Sutch, and Gavin Wright (New York: Cambridge University Press, 2006), tables Dg267–74 and De482–515; Newsprint Association of Canada, *Newsprint Data: 1961* (Montreal: Newsprint Association of Canada, 1961), 11; Canadian Pulp and Paper Association, *Newsprint Data: 1975* (Montreal: Canadian Pulp and Paper Association, 1975), 7; Canadian Pulp and Paper Association, *Newsprint Data: 1997* (Montreal: Canadian Pulp and Paper Association, 1997), 15.

because of how important and costly newsprint was. During this time, US publishers faced the perpetual problem of obtaining paper in adequate supplies and at affordable costs. After 1911, Canada was the primary source of this newsprint, though US publishers and manufacturers engaged in consistent and persistent efforts to break their dependence on foreign paper. Various attempts at alternative material for pulping never succeeded on a mass scale, for example, bagasse, peat, straw, corn stalks, and kenaf.[3]

One source of new raw materials which did prove feasible was recycled newsprint, and the Tribune Company adapted its production to this new source in a significant way in the 1970s and 1980s. The basic technologies and practices to produce newsprint from recycled paper were in place by the early 1930s, though it would take decades to make the process commercially viable. As early as 1932, Tribune Company executives were discussing the concept of manufacturing newsprint from recycled paper, but two significant impediments remained: the cost of transporting the wastepaper to the mill and the cost of removing the black ink from the pulp to produce a sufficiently white sheet. As Arthur Schmon noted, "irrespective of the merits of the process itself," using freshly cut wood was simply cheaper than using recycled paper. "Without even attempting to go any further," Schmon noted, "you can readily realize that the freight and handling costs, plus the cost of power for beating the paper, irrespective of any cost of use of chemicals, would far exceed the price" of pulpwood.[4]

The company's experiments with recycled paper picked up momentum during World War II, when wartime restrictions on labor and raw materials curtailed pulpwood cutting and newsprint production, though de-inking remained perhaps the biggest practical challenge. In 1943, Arthur Schmon wrote to Robert McCormick that company experiments had "showed definitely that we could use up to 6 percent wastepaper without discolouring." In early 1944, the company produced 200 tons of paper using between 5 percent and 10 percent recycled paper, and it used that newsprint to print an entire day's edition of the *Tribune*. Officials within the company were satisfied with the quality of the paper, though they noted that the main problem was that the paper would not look as clean and white as competitors' products, and this might hurt sales. "If the whole industry manufactured newsprint with a certain percentage of waste paper," Arthur Schmon noted, "then discoloured paper would be used universally without adverse competitive effects among individual publishers."[5]

These wartime experiments did not lead to immediate practical changes to the *Tribune*'s Thorold mill, but the promise of using significant quantities of recycled paper as a raw material was realized when the Tribune Company in 1983 inaugurated what it described as a "virtually new mill" at Thorold, which it had spent more than $200 million updating. The new mill had a state-of-the-art de-inking plant that was able to process old newspapers into pulp at a cost comparable to using wood. The new mill took advantage of what the company called the "urban harvest" of waste newsprint, claiming that it would be the "largest recycler of old newsprint in Canada" once the mill opened. The company planned to use recycled newsprint for roughly one-quarter of its pulp, and the recycling program that it set up across southern Ontario was the "largest ever undertaken in Canada." By 1987, expansions in programs and the plant had increased annual recycling capacity to 180,000 tons. By that point, the mill made nearly half its pulp from this recycled paper, and its Ontario recycling network remained robust. "Programs supplying newspapers for our recycling operation have been set up in more than 50 communities across Ontario, many of them being curbside pick-up services," the company noted. "The recycling of paper eases the pressure on municipalities to dispose of this material and helps conserve our precious forest resources." In many ways, this collection of waste materials to make paper was a return to the eighteenth- and nineteenth-century practice of rag picking, but now on a much larger scale.[6]

By the late twentieth century, these scales of manufacturing and using newsprint more generally could be massive. As table 7 shows, by 1984 the Tribune

In the post–World War II period, newsprint remained the foundation of the US newspaper business, and publishers still required massive quantities of it. This 1958 photo shows rolls in the Baie Comeau mill awaiting shipment to New York to print the *Daily News*. Vol. 25194, folder Baie Comeau Newsprint Production (1 of 2), QOPC, LAC. Used with permission.

Company was producing 668,000 tons of newsprint annually, and this increased to 743,010 tons by 1991. Besides the Tribune Company, other major newspapers reported similarly significant newsprint usage during the 1980s. The *New York Times*, for example, in 1982 used some 339,000 tons of paper, which it estimated was the yield from 6,000,000 trees. Because of the need to maintain steady and substantial supplies of newsprint, newspaper publishing remained an immense industrial enterprise.[7]

One transaction that brings into relief the degree to which newspaper publishing retained its industrial core into the 1980s involved the Tribune Company's purchase of the Chicago Cubs in 1981 for $20.5 million. Though the purchase of a baseball team took the Tribune Company into a new line of business, the acquisition at the time remained within the context of shaping a media corporation with an industrial newspaper at its core. While the Tribune might have spent $20.5 million for a baseball team in 1981, it was simultaneously spending more than ten times that sum to update its newsprint mill in Thorold. And, at the same time, the company also began modernizing its Chicago manufacturing operations and relocating its printing to a new site just northwest of downtown. Planning for the new printing plant began in 1979, and on September 19, 1982, the company stopped printing at the Tribune Tower, an event that meant that, as one account put it, the "last manufacturing plant on Chicago's 'magnificent mile' closed its doors for good." Most of the old printing equipment was to be "sold for scrap," as the new $186 million facility featured $80 million worth of new presses. One glowing report described the new plant, which the company called the Freedom Center, as the "most modern newspaper facility in the country." In the early 1980s, as the Tribune Company was paying $20.5 million to buy into Major League Baseball, it was spending hundreds of millions of dollars updating its industrial plants in the United States and Canada so that it could continue to manufacture and sell large quantities of printed newspapers.[8]

For US newspaper publishers in the post–World War II period, the problem of paper remained a central element of their businesses, and their operations remained oriented around obtaining and using massive supplies of it in costly factory settings. Developing new ways of recycling paper was a somewhat successful strategy for industrial newspapers like the Tribune Company to control manufacturing costs and raw materials. This was, however, only a partial solution to the perpetual problem of paper, and publishers throughout the twentieth century sought ways of using emerging electronic media to provide better means of delivering their products. Publishers, including the Tribune Company, experimented with facsimile broadcasting in the 1930s and 1940s and with screen-based delivery systems in the 1980s, at each moment searching for alternatives to paper-based newspaper delivery.

In the case of facsimile, publishers gravitated to it because they believed that it might dramatically alter the economics of producing a newspaper. As two journalists noted in the *American Mercury* in 1945, a huge advantage of facsimile was that it "might cut in half the present cost of manufacturing and distributing a newspaper. There would be no need for rotary presses . . . for stereotyping equipment, delivery trucks, postal payments, newsboys—perhaps not even for newsprint." Throughout the newspaper industry, there were "economies facsimile would effect by applying electronics to a manufacturing process which has not changed in any essential respect since the invention of the rotary press by Hoe in 1846." For publishers, the benefit of facsimile was that it might provide the same content to readers while significantly lowering manufacturing and distribution costs. "The actual assembly of the contents of a newspaper—the gathering, assessment, writing and presentation of news, the typography and make-up of the paper and the solicitation of advertising—would remain largely unaffected by facsimile. The work of reporters, editors and compositors would be carried on substantially as in the past, save, perhaps, for adaptation to new deadline demands. Facsimile would not necessarily create a new kind of newspaper; it would merely modernize the method of delivering a newspaper to its readers."[9]

These attempts at creating new delivery mechanisms for printed newspapers would continue in later decades as publishers sought to adapt their print products to screen-based delivery systems while they still had, as one industry observer noted in 1980, "little boys dashing around on bicycles delivering their newspapers," a delivery network that connected back to the newsboy system of the early 1900s. Columnist Abe Chanin, however, predicted in *Editor & Publisher* that there might be significant changes afoot toward the end of the twentieth century because of electronic publishing. As of 1980, he noted, "as far as totally burying the printed newspaper, the electronic information systems are more as an adjunct than as a killer—at least for a long while." This problem of paper has been perhaps the most important challenge that publishers have faced in the twenty-first century, and it is important to understand, as this chapter will show, that this challenge is not in substance all that recent of a development. Rather, it is the latest episode in a longer history of newspapers struggling to find alternatives to paper for distributing their products.[10]

The Promise of Facsimile Broadcasting

Today, facsimile is considered by most people to be an obsolete technology presenting irritating obstacles in making what should be the simple transmission of a copy of a document from one machine to another. The rare and usually frustrating occasions that many have to use a fax machine do not leave them pining for the medium's return. It is little known, however, that the now-maligned technology of facsimile was in its developmental stage considered to be potentially a *mass* medium. Indeed, one of the striking things about reading assessments of 1940s media is the fact that facsimile was often talked about as part of a technological trinity with television and FM radio representing the wave of the future of media. As NBC executive Lewis Titterton noted in 1941, there were "three infant services: television, the amazing art of transmitting moving images through the air; facsimile; and FM, or frequency modulation, broadcasting," which promised to transform the media business. For many newspapers, facsimile offered the chance to deliver to readers a printed copy of the newspaper without all of the industrial apparatus that they needed to do so on newsprint. Facsimile might transfer the cost of printing to the reader, and this was to many publishing companies that had struggled for years to maintain an industrial printing operation a highly seductive possibility.[11]

Speculation about the possibilities offered by facsimile emerged even in the earliest days of radio broadcasting. In 1927, for example, Associated Press executive

Milton Garges predicted that soon a "home-printed newspaper" could be delivered via facsimile. "For some time," Garges claimed, "only minor difficulties have stood in the way of sending whole pages of newspapers from one continent to another." To journalist Silas Bent, the wireless distribution of news via facsimile could mean that the reader rather than the publisher printed the newspaper, and this change might disrupt the entire industry. This shift, Bent pointed out, "would mean scrapping the huge printing plants and abandoning the clumsy distributing systems now maintained by the daily press." This "does not seem a fantastic forecast," Bent claimed, and he envisioned the future newspaper to be one "printed, folded, and ready in the household of a morning or an evening" after wireless delivery to a home receiver. If the technology could be developed properly, Bent speculated, it might radically reorient the industrial economics of newspaper publishing. "There is no reason," Bent wrote in 1929, "to suppose that we may not in time have compact machines in our homes, driven by invisible waves, which will print and illustrate the tidings of the world in readiness for our breakfast-tables. Every man's newspaper could thus be manufactured in a corner of the dining room, and the frightfully expensive and clumsy system of distribution now utilized by the press would disappear."[12]

Like Silas Bent, other inventors and entrepreneurs involved in developing facsimile broadcasting often used the delivery of the printed morning newspaper as the point of reference for their distribution system. In 1935, for example, *Broadcasting* magazine described efforts to develop home facsimile as an "attachment for the ordinary receiver" that would receive the transmissions and print the daily newspaper overnight. "This receiver would work while its master slept," *Broadcasting* noted, "for the plan would be to have it 'print' the newspaper during the night, so that in the morning there would be laid down in a basket under the set a series of 'flimsies' of tissue-paper texture carrying all of the newspaper features 'photographed' by radio, with headlines, comics, fashions and display advertising."[13]

The Federal Communications Commission (FCC) authorized experimental facsimile broadcasting in 1936 in hopes of motivating additional experimentation, though progress proved halting. Slow printing speed remained a significant impediment to using facsimile to deliver full newspapers. In 1938, an RCA engineer considered a Sunday edition of a "prominent metropolitan paper" and found that its "total printed area" was 50,300 square inches. "At an arbitrarily assumed speed of 10 square inches per minute for facsimile transmission," he calculated, "it would take about 83 hours or more than three days to transmit continuously all this material, even assuming that some of the fine, 6-point, type could be satisfactorily transmitted." Daily papers had, he claimed, "about 30 per cent of the area of the Sunday editions, and would therefore require somewhat more than one day to transmit, by area, according to the above assumption." Gene Wallis, a *Dallas Morning News* staffer, spent six months as the paper's facsimile editor in late 1939 and 1940, after which he came away similarly pessimistic that facsimile could ever really take off as a mass medium. The time to print was too slow, Wallis believed, and people simply preferred printed newspapers as material objects from which to read the news. His paper averaged one hundred pages on Sunday, and Wallis estimated that, in order to send a complete edition via facsimile, "we would have to start running the machines continuously from Friday to supply Sunday morning's paper on Sunday morning." In addition to the challenge of speeding up the transmission of newspapers via facsimile, entrepreneurs also faced the challenge of developing receiving sets that were cheap enough that consumers would purchase them. The Crosley Corporation developed a receiving and printing device in 1939 that it claimed "would retail for less than $150," a figure that would be equivalent to roughly $2,500 in 2017.[14]

Still, despite these obstacles, facsimile broadcasting retained significant promise for many observers in the late 1930s. As journalist Ruth Brindze noted in 1938 in the *Nation*, facsimile was "potentially the most socially significant invention since the development of the printing-press." Many publishers thought the same, and they continued to experiment with facsimile papers and used public demonstrations to stimulate public interest. In St. Louis, the *Post-Dispatch* began transmitting a facsimile paper on December 7, 1938, after placing receiving sets around the city in public places such as department stores and hotels. In California in February 1939, the McClatchy newspaper chain began transmitting a paper that it called the *Radio Bee* to some one hundred homes in which the company had installed sets, and the company planned to move these receivers to a different set of homes three months later. In their grassroots promotion, the company sent eight-page newspapers in the middle of the night at the rate of twenty minutes per page. More prominent demonstrations took place at the New York World's Fair in May 1939. Often remembered as the site of the first major public demonstrations of television, the fair was also the site of public demonstrations of facsimile broadcasting, as the *New York Herald Tribune* partnered with RCA to print a three-page paper called the *Radio Press*. The year after the New York World's Fair, the FCC authorized commercial facsimile broadcasting, a decision that prompted facsimile entrepreneur John Finch to call it a "new deal for radio" and to predict that the medium would be commercially viable within five years. In the 1940s, a number of other newspapers attempted to incorporate facsimile into their organizations, including the *Milwaukee Journal*, the *Cincinnati Times-Star*, and the *Buffalo Evening News*.[15]

The *Chicago Tribune* invested significant sums in facsimile (as was the case with many new technologies) despite not being entirely convinced of its utility. The company's earliest experimental broadcasts took place in March 1938, and they continued in subsequent years even while showing only modest commercial potential. As Robert McCormick noted in July 1945,

> I can conceive how any news conscious person on a hunting or fishing trip, or the crew of a ship, possibly people in an automobile, would like a facsimile newspaper. I can conceive that a farmer not receiving a daily paper would like to find one at breakfast time and again at supper time. It does not seem probable that it can compete with the newspaper when the newspaper can be delivered, for, after all, the facsimile machine will be an individual printing press. It hardly seems likely that a million facsimile newspapers, printing one newspaper per day, can be produced in economic competition with printing presses printing 40 to 50 thousand copies an hour.

Despite this skepticism, McCormick believed that the Tribune Company needed to be involved to stay at the forefront of the news business. However facsimile developed, he surmised, "newspapers had better get there first."[16]

Following this strategic logic after the conclusion of World War II, the Tribune Company began investing significant resources in developing its facsimile broadcasting operations. In April 1946, the paper claimed to have spent some $250,000 over the previous two years to bring it to the "threshold of public service." The company began public demonstrations the following month, when it sent a copy of a facsimile edition from the Tribune Tower to Robert McCormick's home in Wheaton, some 29 miles away. This edition comprised four roughly letter-size pages with four columns of news on each page. The transmission took 28 minutes. Robert McCormick claimed to be doing the broadcasts in keeping with the company's long-standing practice of experimentation with new technologies. "I do not know what facsimile is anymore that I knew what radio was 20 years ago . . . but we are going to find out all about it. There is no doubt that radio is constantly developing. FM, television, facsimile are all new. We can't resist these advances. We've got to go with them. Facsimile may prove too costly. The recorders cost more than $400 now and the paper used for the printing is expensive. We don't know who will use it."[17]

The Tribune's public demonstrations continued throughout the spring of 1946. In late May, the company installed a facsimile receiver at the Museum of Science and Industry in Chicago, and it sent an edition hourly from 11:00 a.m. to 5:00 p.m. to demonstrate to museumgoers how the technology operated. In June, the company put receivers on display at its downtown offices on Dearborn Street, hoping to attract the attention of passersby. Ken Clayton, the *Tribune* staffer in charge of setting up these public facsimile demonstrations, found that many observers were unimpressed by them, mostly because the facsimile newspaper just reformatted news that they had already seen in a regular printed newspaper or heard on the radio. "What's new about that?" Clayton claimed that one observer asked of a facsimile *Tribune* around lunchtime. "I read that stuff this morning." Clayton began to rethink the presentation of facsimile news, and in June 1947 he performed new demonstration transmissions "using a revised format." Instead of a four-page paper with reformatted news, the new editions were one-page editions made on the hour "covering news from around the world as it developed. . . . At these demonstrations the editions were posted as they came from the recorders so that onlookers could see how, by means of facsimile journalism, they would be able to maintain constant contact with the entire world from their own living rooms."[18]

Despite all of these optimistic hopes for facsimile broadcasting among publishers, consumers were simply never sufficiently attracted to it, mostly because the paper product that they could receive electronically was seen by most to be radically inferior to the one they could buy on the newsstand. "The quality of the home facsimile print doesn't compare with the old-fashioned, or inked, variety," one journalist noted in 1946. "Although pictures reproduce quite well over home receivers, 'faxed' print looks somewhat quavery." Facsimile could provide news bulletins in between print editions of the newspaper, but so could radio, and consumers seemed satisfied with that. As one observer noted in 1948, facsimile had been "alternately hailed as the beginning of a revolution in news dissemination and damned as a ridiculous way to do bad printing," and he believed that the broader trend in public opinion was toward the latter. "Experts doubt that the novelty of facsimile newspapers alone will provide sufficient interest to induce purchase, or even rental, of many recorders. Newspapers are more complete, and radio news is just as fast." As a consumer good, facsimile never found a niche in the media market.[19]

Another major consumer criticism of facsimile was that the cost of paper never came down, and the idea of paying more for something inferior to the printed newspaper seemed absurd.[20] For publishers, one of the prime attractions of facsimile was that it transferred the printing costs to consumers, and most consumers ultimately rejected the idea of taking on these expenses given the low quality of the product that they received. As one estimate in 1948 noted, facsimile paper cost roughly one cent per page, meaning that a consumer was paying four cents for a four-page "newspaper" composed of short items that could be had over the radio for free. Given that a full printed copy of the *Chicago Tribune* cost four cents and the *New York Times* three cents in 1948, this seemed egregiously expensive. Industry observers continued predicting that costs would go down, but consumers seemed uninterested. Some promoting facsimile tried to set up sponsorship arrangements with corporations, the idea being that a logo could be printed on all sheets, or perhaps that the reverse side could have preprinted advertisements. These strategies never succeeded, and by the end of the 1940s facsimile was essentially done as a mass medium. As one observer noted, facsimile broadcasting had "returned to the cocoon, after fluttering mothlike around the spotlight of postwar invention." By the early 1950s, most facsimile broadcasters had ceased operations, and by 1952 only six stations had licenses from the FCC to do so. Two years, later, even those stopped.[21]

In the long run, facsimile failed as a mass medium, but it did have a significant global presence in the 1980s and 1990s as a point-to-point medium. Perhaps motivated by the widespread adoption of facsimile machines in homes and offices, several newspapers in the late 1980s again made short-lived efforts to use facsimile to distribute news, including the *Minneapolis Star-Tribune, St. Paul Pioneer Press*, and *Hartford Courant*. In 1990, the *Chicago Tribune* launched a facsimile paper called *Tribfax*, but it lasted only four months. By this time, other technologies had emerged that might do better what publishers had hoped facsimile would do and allow them to be "newspaper" publishers without costly industrial plants and tons of newsprint. Ultimately, one can see in facsimile the ways that publishers were pushing against the boundaries of the printed page using the technologies that were available to them. Though some of these efforts can seem quixotic or misguided today, they were for historical actors often entrepreneurial and forward thinking, and facsimile's failures had less to do with poor business strategy than they did with the fact that the technological limitations of the medium never allowed publishers to create a cost-effective alternative to the printed daily newspaper.[22]

Electronic Newspapers from Paper to Screen

Before, during, and after experiments with facsimile broadcasting, newspaper publishers also experimented with using video screens in order to deliver news electronically instead of on newsprint. In television's primordial phase, for example, some publishers looked at it not just as a way to send motion pictures but also as a way to send text and photographs, like a newspaper. In 1930s, William S. Hedges of the *Chicago Daily News*, the newspaper that also owned radio station WMAQ, remarked that television "may be regarded as a potential ally of the publishing and news distributing business. Its ability to flash headlines, to send facsimiles of newspapers or to transmit photographs through the air may some day be put to practical use in the newspaper field." For Hedges, early television promised to be something less like an offshoot of radio—broadcasting with pictures in addition to sound—and more like an offshoot of facsimile, as it might send the images of newspaper pages that could be read on the screen. Television's mainstream history obviously developed in a very different way, but Hedges's notion of the screen as a way to deliver something like a "newspaper" came closer to being realized in the 1970s and 1980s in technologies like Viewtron.[23]

A prominent assessment of these emerging technologies came in 1971, when journalist Ben Bagdikian and economists at the RAND Corporation completed a study of US media, which included consideration of the available "alternative methods of production and distribution" for newspapers. High manufacturing costs remained the biggest problems facing the newspaper business, Bagdikian found. In one particularly illustrative example, he claimed that it was "not unusual" in Los Angeles County that the total circulation of the city's Sunday newspapers "cost about $22 million to produce, cost the consumers $500,000 to buy, and, the next day, require municipal services to haul away five million pounds of discarded newspapers." For Bagdikian, perhaps the most practical way of generating lower manufacturing costs would be to transfer some of them to telecommunications companies and consumers. A newspaper that could be delivered via "paper-to-home direct transmission" would "begin to follow the economics of broadcasting in the sense that most of the cost of hardware would be borne by the customer. The laying of cable connections and the provision of home-receiving devices presumably would be part of a system supported by someone other than the newspaper." In other words, if a newspaper firm could find a way to distribute its textual news through a medium other than paper, it might revolutionize the business by removing the need to maintain costly printing plants and newsprint supplies. "Once all newspaper information is stored digitally in a computer," Bagdikian believed, "the economics of electronic transmission will begin to look tempting to counter the high costs and inconvenience of making documents in the home."[24]

In the early 1970s, other industry figures began expressing similar hopes that the screen might do what the facsimile printer had not in creating widespread demand for a newspaper delivered to homes electronically rather than physically. As one observer noted in 1971, "Central to almost any future development in newspaper . . . is of course the computer." It was not clear as of 1971 what this computer-driven future might look like, but these early ideas about screen delivery hinted at the paths along which the devices would develop in subsequent decades. "It is difficult to discuss the future of newspapers without introducing the prevalent concept that the American consumer in future years will have at his disposal some form of communications center in his home. Not a communications center as we think of a radio or tv station, but a transceiver unit in the home which will give the consumer access to a broad range of information, entertainment, and marketing sources."[25]

Seeking to stay at the forefront of these developments, entrepreneurs and media corporations continued developing new ways of distributing the news electronically. The Dow Jones Corporation began offering via television and computer screens an electronic version of its news and market data in 1974, believing that, as one of its editors, Peter Schuyten, noted, the newspaper was becoming "an increasingly costly product" to manufacture and distribute. Schuyten added, "I don't know whether everything in a newspaper is worth the paper it's printed on. I mean that literally." As Schuyten saw it, there was certain information printed in newspapers which was simply better and more cost-effective to distribute electronically. "Statistical data like stock quotes and sports' scores," for example, "might be better carried in electronic form because of increasing newsprint and publication costs." By 1981, Dow Jones had 25,000 subscribers to its electronic news service, and the company's experiments with electronic publishing continued that year as it began offering a version of the *Wall Street Journal* sent electronically to cable subscribers. The fledgling service also offered to readers a degree of interactivity in the form of an interface that was a "keyboard device connected to their tv set."[26]

The idea of circulating a "newspaper" via a box connected to a television set became the dominant form of experimentation with alternative distribution methods in the 1980s. Newspaper firms, as one report noted, began to see electronic news distribution not as a competitive threat but rather as a "means to beat the enemy at its own game," and perhaps the most important company to undertake an institutional response was Knight-Ridder. In 1980, the company created a wholly owned subsidiary called the Viewdata Corporation of America, which developed an interactive technology called Viewtron. In partnership with the telecommunications firm Southern Bell, Knight-Ridder spent $1.3 million developing the Viewtron system, which it planned to test-market to a small number of families in Coral Gables, Florida. Through the partnership, Viewdata would provide the necessary hardware to the pilot program homes, and Southern Bell built a "special set of lines for the experiment so as not to interfere with the families' regular telephone service." As Knight-Ridder executive James Batten described the system, it would have "12,000 frames of information, each frame being a screenful of text and graphics. . . . Stories will range from one to ten to 15 frames. . . . The system puts the viewer in the driver's seat. He asks for what he wants, when he wants it. When he's ready to move on, he moves on." As *Editor & Publisher* described this new interactive screen-based system, "Viewtron users will be able to do everything from buying a lawn mower to taking Spanish lessons while they keep up with the latest news from the *Miami Herald* and the Associated Press."[27]

The Coral Gables test started on July 14, 1980, in twenty-five local households, each of which was equipped with a "19-inch color television set modified to receive the Viewtron information, a typewriter-like keyboard, a remote-control keypad resembling a hand calculator, and an electronic control unit connected to a regular telephone line." Each individual person wanting to use the system was given an identification number and a "confidential password" affording access to content on the television screen in the form of "20 lines of information with 40 characters per line." These were drawn from an ASCII character set, and there were no graphics or sound. Knight-Ridder was encouraged by this initial test and subsequently planned significant expansion of the project. While estimating that it would spend some $18.5 million on it by the end of 1983, Knight-Ridder remained completely committed to selling printed newspapers as well. As CEO Alvah Chapman stated in 1982, print would remain the heart of the enterprise, even with these Viewtron experiments. "We'll be much more involved in electronic communications," he claimed, but "print will still be the major thing we do." As had been the case with facsimile, consumers did not take to Viewtron, and by 1986, in the face of high costs and low success, Knight-Ridder had abandoned its experiments with electronic news distribution. Reid Ashe, the head of Knight-Ridder's Viewdata subsidiary, commented that the "market is thin" for electronic news distribution as of the mid-1980s, and he claimed that Viewtron offered "no prospect for being a mass medium in the foreseeable future." As had been the case with facsimile, the printed newspaper in the 1980s still remained the most cost-effective way to deliver textual news.[28]

On the reader's side, one of the biggest problems with these new electronic forms of news distribution was that, like facsimile, they simply functioned poorly in replicating the experience of reading the printed newspaper. As Walter Mattson, the president of the New York Times Company, noted in 1982, the classified ads in a Sunday *New York Times* alone would take up some 1,200 screens of information if presented electronically. This would take "62 hours to browse through," he estimated, a practical hurdle that at the time was simply insurmountable with existing technology. As *Editor & Publisher* concluded in 1986 of these screen-based delivery systems, "even though modern electronics" could deliver the information contained in a printed newspaper "almost instantaneously" to the home and without paper, "no one has found a way to reduce the time and expense required" for readers to "absorb" all of the information in a newspaper that way. "Specialization, yes. The equivalent of a daily newspaper, no." Electronic distribution, the magazine ultimately believed in 1986, would find at best a niche rather than a mass market.[29]

Throughout the 1980s, newspapers spent a great deal of energy and money attempting to develop alternatives to their core printed product. Most of these experiments were abandoned by the end of the decade. Though mostly unsuccessful, these various projects did help pave the way for the subsequent practice of delivering news via the personal computer in the following decade. If we see this 1990s shift as emerging out of these failed 1980s experiments, we can see the development of digital news distribution not as a rupture in the newspaper's history but rather as a realization of previously unsuccessful projects. Ultimately, the internet was not so much the invention of a new vision as it was the successful implementation of failed visions from the previous decades, including facsimile and Viewtron, of putting the text and images of a newspaper onto surfaces other than newsprint.[30]

The Industrial Newspaper Becomes the "Information and Entertainment Company" and "Content Curation and Monetization" Firm

As had been the case with facsimile broadcasting, the Chicago Tribune Company remained actively experimental and expansionary with other forms of electronic news distribution throughout the twentieth century. The company had been a pioneer in radio broadcasting back to the early 1920s, when it purchased a broadcasting station and secured permission from federal regulators to change its call letters to WGN, an acronym for its immodest print slogan "World's Greatest Newspaper." Throughout the decade, WGN expanded its transmitting power and geographic reach. By 1928, the company was broadcasting at 25,000 watts, and in 1930 it received a license to broadcast at 50,000 watts, the highest power allowed by the federal government. This increase in power allowed WGN to be a station that "served the entire middle west," just as the company had wanted to do with its print circulation in blanketing the territory it called "Chicagoland."[31] Ultimately, what the company accomplished through radio was to demonstrate that publishers could readily and profitably adapt their businesses to new forms of media in addition to and alongside printed newspapers. Radio showed that the media business was not defined by zero-sum competition between media, and publishers learned and demonstrated that the introduction of a new medium could be used to create new revenue streams for firms that remained defined by industrial newspaper production.[32]

The Tribune Company continued its adoption of new media in the post–World War II period with facsimile and with television broadcasting on station WGN, which went on the air in April 1948 as Chicago's second commercial television station. This diversification into new media extended to the Tribune Company's New York newspaper property, as the *Daily News* obtained a television license, picked the call letters WPIX, and began broadcasting on June 15, 1948, as the fifth station in the city. On Christmas Eve 1966, in a strange and almost certainly unintended convergence with its history of forest product exploitation in newsprint production, WPIX became the first station to broadcast a burning Yule log on television. As the *New York Times* reported on what it called the "television industry's first experiment in nonprogramming," WPIX planned for a three-hour broadcast that would "revive one of the season's oldest customs—the burning Yule log. From 9:30 P.M. to 12:30 A.M., there will be nothing on the screen but a crackling fire to dream or dress the tree by. And, thanks to the wonders of electronics, ashes and smoke have been eliminated along with the commercials."[33]

All of these various broadcast outlets were part of the company that operated under the conditions of the trust that Robert McCormick had established prior to his death. When that trust expired as scheduled in 1975, company control became newly dispersed, as 153 individual shareholders, mostly relatives in the McCormick and Patterson families, could attempt to reshape the company's direction or even sell their shares. By the early 1980s, the company was still not publicly traded, but private stock sales of the roughly seventy-five hundred shares (the price in 1981 was estimated to be between $78,000 and $90,000 per share) had swelled the number of shareholders to four hundred. This diffusion of ownership was kept in check by what a 1982 *Fortune* profile of the company called a "self-perpetuating, tightly controlled management structure," and the Tribune Company remained controlled by owners of large blocks of shares.[34]

At the time, the company's overall print circulation remained among the highest in the country. Though the company's eight newspapers made it only the eleventh-largest chain in the country in terms of print properties in 1977, the chain's total circulation (3.1 million) trailed just Newhouse (3.2 million) and Knight-Ridder (3.5 million), and these chains owned twenty-nine and thirty-two papers, respectively. The Tribune Company's newspapers were also still generating significant revenues, with the *Daily News* generating roughly $350 million in 1981. However, some troubling trends began to develop around this time for the company. In 1981, the *Daily News*, while still having the "largest single-market circulation" in the United States, began losing money, to the tune of an estimated $12.5 million during that year. As a chain, the company's 1981 circulation had slipped from third to fourth when it fell behind Gannett. At the *Tribune*, advertising linage declined 10 percent in 1980 and another 5 percent in 1981, and circulation declined as well.[35]

So, while *Fortune* called the company in 1982 a "family-owned newspaper, TV, radio, wood-products, and baseball conglomerate," significant changes were soon afoot. Early that year, the Tribune Company publicly announced that it wanted to sell the *New York Daily News*. The paper was losing money, and its management blamed its financial troubles on the fact that the paper had eleven labor unions that "refused to budge on contractual arrangements they had won over the years." Rupert Murdoch, the publisher of the *New York Post*, offered to buy the *Daily News* but was rebuffed by the Tribune Company, which ultimately decided to hold onto the paper after reaching an agreement with the unions representing its workers. The return to profitability never came, and the Tribune Company eventually sold the paper in March 1991 to the British publisher Robert Maxwell after claiming losses of more than $100 million in the previous decade.[36]

In terms of the Tribune's corporate evolution, more important than the attempt to sell the *Daily News* was the decision to go public in 1983. At the time, the company was generating the second-most revenue of any newspaper chain in the United States at $1.4 billion, trailing only Gannett's $1.5 billion. The Tribune Company was still profitable, having earned $98.3 million in profits in 1982, though this was down from $139.8 million in 1979. The Canadian newsprint operations remained a valuable part of the company's overall holdings and, in fact, were at the time worth more than the company's broadcast properties. In 1982, the newsprint manufacturing operations were valued at $488.1 million while the broadcasting and cable properties were worth $321.7 million. The newspapers themselves were in 1982 valued at some $756.7 million.[37]

This proved to be a pivotal moment for the company in many ways. In 1984, Robert McCormick Schmon, the son of Arthur Schmon, stepped down as chief executive officer of the Tribune's Canadian newsprint subsidiary, effectively ending a long line of family management of those operations. *Tribune* publisher Robert McCormick himself had no children, and the Schmon connection had been as close as his corporation came to having the kind of long-term family control that would mark the media properties owned by the Ochs/Sulzberger and Bancroft families. The loss of these personal connections to the McCormick era of the paper pointed toward what would be, before long, the end of the *Tribune's* corporate presence in Canada. The movement in this direction began in the 1980s, when the company found that it was using only 60 percent of the newsprint it produced in Thorold and Baie Comeau and selling the balance of its paper output to "outside customers." This created a change in the nature of the business. When the company used all the newsprint that it produced, it benefited from vertical integration in controlling its production costs. When it was not using all that production, it was forced to sell substantial qualities on the open market, where prices could fluctuate. As the company began to consider going public, it was faced with a looming question about how to handle operations that were losing money or at least not sufficiently profitable for Wall Street. In 1983, when asked whether the company was considering selling the newsprint manufacturing operations, Tribune Company vice president Joseph Hayes remarked cannily that executives were "always exploring the back end of the house to see what is the best mix."[38]

To read the Canadian newsprint subsidiary's internal reports in the 1980s and early 1990s is to read the chronicle of a firm struggling in a declining industry. The February 1984 *President's Report* noted that the "worst year in the Company's history is behind us." In 1989, the company noted that the "continuing downturn in market conditions this year means a bumpy ride is ahead for all Canadian newsprint manufacturers—and that includes our company." The poor economic climate of the George H. W. Bush administration continued to affect the company's operations. In November 1991, the company noted that the "continuing decline in the North American newsprint market" was "part of a dark economic picture for our industry that is growing worse as we head towards 1992." The Tribune Company's Canadian subsidiary projected a $25.5 million net loss and warned of even greater struggles moving forward. The main problem, the company noted, was "oversupply and low demand" in the newsprint market. This, the company claimed, had to do with the fact that newspapers at the time were getting smaller because of "declining advertising lineage during the recession," as well as the fact that circulation was at best "growing slowly in most areas." By February 1992, the company claimed that there was "no doubt the industry is facing its worst financial crisis since the Great Depression prior to the Second World War." Even with "continuing cost reductions" and increased operating efficiency, the company reported a net loss of $37 million in the first half of 1992 alone, part of a continued downward trend across the industry. Vertical integration, once a path to industrial self-sufficiency, had become a drag on overall corporate performance.[39]

Hoping to make a profitable exit from what it saw as a declining sector, the Tribune Company in 1993 took its Canadian newsprint operations (by then the seventh largest in North America) public and then began selling off its stake in the firm, a process that concluded in late 1995 when the paper sold the last of its shares for roughly $330 million, with Donohue Inc. acquiring the firm. The Tribune Company, which had built the industrial operations from scratch more than eighty years earlier, had decided, as one company official put it in 1994, to "take a new direction—to become an information and entertainment company." For the Tribune Company, a corporation that had revolutionized the business of the metropolitan newspaper in the twentieth century by vertically integrating its publishing operations, the move marked a decisive transition away from being the purveyor of a daily mass-produced industrial commodity and toward a more nebulously defined "information and entertainment company" circulating content in multiple media. A Wall Street analyst hailed the decision, remarking that it was a "positive step. . . . I see no real point in tying up assets in a forest products company."[40]

In a move that was emblematic of the shift from being an industrial newspaper toward being a publicly traded "information and entertainment company," John Madigan took over as the company's CEO in mid-1995. Madigan had been an investment banker before becoming a Tribune executive in 1975, and he worked his way up the ranks on the business side of the operations prior to becoming publisher of the *Chicago Tribune* in 1993. When Madigan took over, as Ontario Paper Company chairman John Houghton told me, he came with the idea that "here you have an industry in Canada that they really didn't need. Originally, the Colonel, he needed it. . . . John knew that we're a capital intensive business and we make a commodity product and we can't control the pricing. . . . John figured that was a business that he didn't need, and he was right." Though he was dismayed to see the Canadian subsidiary sold, Houghton understood the decision. "I didn't disagree" with Madigan's course of action, John Houghton remarked, "but I didn't like the outlook." Another executive, Adrian Barnet, had a similar view about Madigan selling the Canadian subsidiary. While it was "actually a brilliant strategy" from the perspective of the company's overall trajectory, Barnet told me in 2015, "it wasn't the strategy of the old Chicago Tribune."[41]

Under Madigan's leadership, the Tribune Company continued its evolution away from both the company's older industrial strategy and its primary focus on printed newspapers. Not long after taking over the company, Madigan engineered what became the "biggest deal" in the company's history, and one that "did not include a single newspaper," in the $1.13 billion purchase of Renaissance Communications. The acquisition of Renaissance's six television stations meant that, for the first time in the Tribune Company's history, newspapers would no longer constitute the largest revenue-generating share of the business. The deal, the trade press noted, "serves notice that Tribune intends to be in the top tier of broadcasters," and the company's sixteen total television stations meant that it reached an estimated 33.4 percent of American homes. The Tribune Company's original flagship television station, WGN, had become a staple offering of basic cable systems in the United States in the late 1970s, and it remained central to the Tribune's new broadcast-heavy focus. The company's aggressively acquisitive and expansionist streak continued in the following years, and it continued to add television stations and newspapers in major deals. In 2000, the Tribune Company paid $8 billion for the Times-Mirror Corporation, a transaction that was the "largest newspaper merger ever" and one that gave the company thirteen daily newspapers and twenty-two television stations.[42]

As the Tribune Company continued expanding its television and newspaper holdings, the company also began moving into the emerging digital sector. The Tribune Company was an early investor in both the grocery delivery company Peapod and the internet firm America Online (AOL). The latter investment proved particularly lucrative for the company, as it purchased a $6 million stake in 1992 that it sold for some $260 million in a series of transactions between 1995 and 1999. In 1999, one Tribune executive estimated that the total value of the Tribune Company's "Internet stock holdings represents about 5 percent of the company's $10 billion total market capitalization."[43]

In addition to investing in other emerging digital firms, the Tribune Company also began trying to adapt its own content to the internet. In 1997, the company began, as one report noted, "recruiting an army of reporters, editors and photographers this summer to reinforce the paper's position as the dominant force in local news, both online and in print." The company planned to hire what it forecast would be the "largest staff of any online newspaper." No other newspaper, the report noted, had "announced expansion plans as ambitious as the *Tribune*'s," and it claimed that the company was the first "to begin developing a corps of reporters with a primary allegiance to the Web site rather than the ink-on-paper product."[44]

Though active in developing web operations and seeking to adapt its print-based model to the computer screen, the Tribune experienced some difficulties in developing a staff and infrastructure for the web. In 1999, the company created what it called Tribune Interactive as an "Internet division completely separate from the print operations," and within months more than 15 percent of the employees had left the company. Some left simply because there were better salaries available at internet firms than there were at newspapers trying to adapt to the internet. As one account noted, the "brain drain at one of the nation's top media companies vividly illustrates the difficulties faced by traditional media in this new-media era." A number of Tribune employees quit in order to take "prime Internet jobs at the likes of Yahoo! and Broadcast.com, top companies that can lure away even happy employees." Others left out of dissatisfaction with the way the Tribune conceptualized and managed the new operations. Some blamed the company for "de-emphasizing writing and journalism" and for creating a poor workplace environment. One employee reportedly blasted the company in a resignation memo by inviting it to "have fun slipping into the cold sea on the *Titanic* because you don't know jack shit about the Internet."[45]

Overall, the Tribune Company's performance record in the twenty-first century has been checkered, at best. Some reports noted an increasingly strained and dysfunctional corporate culture even prior to real estate mogul Sam Zell's ultimately disastrous acquisition of the company in 2007. Zell's managerial style outraged many employees, and the deal's complicated financial structure became increasingly precarious as the broader economy staggered, and the company was eventually forced to declare bankruptcy. After a "tumultuous four-year bankruptcy process" ended in early 2013, the Tribune Company struggled to find its way to profitability. As it did so, it employed a strategy that had become a "popular model for big media companies" in spinning off its publishing assets into a publicly traded company separate from its other nonprint businesses. Rupert Murdoch had done this in creating News Corp. as a newspaper firm completely separated from 21st Century Fox, the corporation holding his other media properties. The creation of the new Tribune Publishing was soon followed by the Gannett Company doing the same thing. As one analyst noted, these firms were "separating the secularly declining newspapers from TV stations and digital" as the result of "huge pressure from investors over the years."[46]

In September 2015, the *New York Times* noted of the Tribune's newspaper operations that "even by the low standards of a troubled industry, Tribune Publishing has struggled." Subsequent efforts did little to assuage investor or public confidence. In June 2016, the company announced that it would change the name of Tribune Publishing to the purposely lowercase "tronc," a move that puzzled many. As one marketing professional noted, "I think it's sort of truncated in a way, pun intended. It feels like a longer word that got cut off." In a press release, the company claimed that its new mission would be to become a "content curation and monetization company focused on creating and distributing premium, verified content across all channels." Few were clear on what this meant in practice, and the shift in the company's purpose and name was widely mocked and derided. One newspaper columnist remarked, "There is a chance that the executive who approved that description knows exactly what it was supposed to mean. There is no chance anyone else does." As of this writing, tronc's future remains unclear.[47]

During the twentieth century, the Tribune Company did much to expand the boundaries of the newspaper business. This was the case in the ways that it crossed an international border to vertically integrate its industrial manufacturing, and it was the case in the ways that it expanded its "content" business into other media beyond print. Sometimes these efforts in the overall media business were hugely successful, as was the case with WGN on radio and television, and sometimes they were not, as was the case with facsimile broadcasting. But throughout the twentieth century, the Tribune Company did much to expand, evolve, and push the boundaries of what it meant to be in the newspaper business. In the twenty-first century, those boundaries have contracted significantly. The Tribune Company no longer owns or operates any newsprint mills in Canada, and the print business that had generated the revenues and content on which the company's multimedia expansion was built were spun off into a separate corporation essentially designed to quarantine it from the relatively healthier broadcast properties. To those in the newspaper business, it remains unclear what firms like the Tribune Company, which had long been organized around selling information and advertisements printed on paper, are supposed to do or be in an era in which declining numbers of people have been wanting to exchange money for newsprint.

Conclusion

In April 1980, several weeks before the Cable News Network (CNN) began broadcasting, its founder Ted Turner remarked that he believed that CNN would "compete directly with newspapers" in providing Americans with daily news and information. "The basic problem with newspapers," Turner claimed, was that "they use trees—lots of them. And it takes lots of energy to keep the trees moving to the finished product and the product to the customer." Not all media executives were worried about this potential competition from cable news. Joe Dealey, president of the *Dallas Morning News*, noted rather presciently in 1980 that for the "indeterminate future I believe newspaper readers will want to hold hard copy in their hands. The time is a long way off before they will rely on the tube for all the news. Now, if somewhere down the road, say 20 or 25 years from now, I'm proved wrong, newspapers which are in the business of collecting and disseminating news will still continue to collect it although they might distribute it in a different way." The newspaper trade press agreed with Dealey and piled criticism on Ted Turner, with *Editor & Publisher* calling him a "sports impresario" from the worlds of "yachting and baseball" and chiding him for ignoring newspapers' long record of successful adaptation to new media. "We in the newspaper business have been hearing this gloomy dirge sung periodically for the last 50 years," the magazine claimed in 1980. "First it was radio, then it was television, that was going to kill off newspapers." And now, the magazine stated, "Turner and others like him" thought cable might finally "do the job," as they believed that the "high cost and the problem of supply of newsprint will pay a major role" in initiating the newspaper's final farewell. This would not be the case, *Editor & Publisher* noted. "Broadcasting by whatever method," the magazine noted, "will always suffer from the lack of a hard copy to read at leisure." It was not television that posed the threat of substituting another mediated experience for newspaper reading, but rather home printing. "If an efficient method can be found to produce a hard copy in the home to equal the volume of advertising and news now published in the daily newspaper," the magazine claimed, that would present meaningful competition. As of 1980, there was yet "no way" that "any broadcast device" could "deliver the modern newspaper at the same price" when printed in the home instead of being purchased on the street.[48]

Roughly twenty years later, or about the time that Joe Dealey suggested, newspaper publishers did in fact find an "efficient method" of getting their products to readers through digital distribution. Through the internet, publishers had solved the problem of paper in a way that facsimile in the 1930s and 1940s and screen-based efforts like Viewtron had not in the 1970s and 1980s. The unintended consequence, however, was that their twenty-first-century solution disrupted the idea of what a "modern newspaper" was, especially in terms of the way that it delivered advertising, and this has created tremendous commercial challenges that have reshaped—some might say killed—the business of newspaper publishing.

What *Editor & Publisher* meant in 1980 by "modern newspaper" was the industrial model that this book has traced. This was a factory-produced good relying on massive quantities of newsprint manufactured from forest products, and it is what many have come to call "dead tree media." Many of these printed newspapers still circulate, though in declining quantities. In 2016, for example, daily newspaper circulation was still 34.6 million, but this was down significantly from 1995, when national circulation was 58.2 million. At the same time, newspaper advertising has plummeted from $36.3 billion in 1995 to $18.3 billion in 2016. The share of revenues from digital advertising has increased from 17 percent in 2011 to 29 percent in 2016, but this rise has done little to stem the overall loss in revenue. Simply put, newspapers increasingly have been confronted with a huge and unfilled gap in revenue, something that many observers identify as the industry's primary challenge. As John Houghton, the president of the Tribune Company's Canadian newsprint operations, told me in 2015, "The thing to me that's killing the newspapers is . . . advertising."[49]

What this reminds us is that, as we talk about the decline or "death" of the newspaper, we are talking about two related but distinct trends: one having to do with the public appetite for news, and the other having to do with spending on newspapers, both by the public and, perhaps more importantly, by advertisers. In many ways, there are two intertwined narratives that make up the history of the newspaper business, one tracing a history of journalism and the other a history of advertising. One must be careful not to conflate the declining financial health of firms selling printed newspapers with a declining demand for news. In many ways, the key challenge facing newspapers in the twenty-first century is figuring out how to reimagine a commercial market for news without a physical paper object as the primary commodity of production and exchange.

In September 2016, the Newspaper Association of America, a group founded in 1887, dropped the word "Newspaper" as it chose its new name, "News Media Alliance." *New York Times* columnist Jim Rutenberg reported that this was because of the fact that "the word 'newspaper' has become meaningless in reference to many of the group's members." As News Media Alliance head David Chavern claimed, "'Newspaper' is not a big enough word to describe the industry anymore. . . . The future of this industry is much broader." Michael Klingensmith, the *Minneapolis Star-Tribune* publisher and News Media Alliance vice chairman, added that the "name change to me isn't about not being paper anymore—it's really just about expanding opportunities." Whether there is realistic cause for this optimism remains to be seen.[50]

Until very recently, the newspaper business was organized around paper, both in terms of selling individual paper copies to readers and in terms of selling space on that paper to advertisers. In the twentieth century, publishers used paper within industrial publishing enterprises, and the daily business of managing newspapers involved attention to the costs and quantities of rolls of newsprint on one end and the distribution of physical copies on the other end. These were problems of paper supply and distribution. The problem of paper in twenty-first century newspaper publishing has shifted to one of demand, which has declined precipitously among both readers and advertisers. For newspapers in the twenty-first century, the problem of paper is that not enough people want to buy it and not enough advertisers want to buy space on it.

Media Infrastructures, Old and New

In the twentieth century, the financial success of the US newspaper had direct effects on communities across Canada where pulp and paper mills were built to supply the export market. Motivated by policies put into place in the early 1910s, newsprint manufacturers in Canada built an industry that provided US newspapers with much of the paper that they relied on to print news and advertising. High and steady US demand for paper contributed to the creation and expansion of cities like Baie Comeau and Thorold. In the twentieth century, government policies in the United States and Canada created the incentives and infrastructures to serve a continental market for newsprint, and the US newspaper was a direct beneficiary of policies relating to tariffs, forestry, hydroelectric power generation, and canal building. These various policies, though often hidden from the view of US readers, ultimately served to subsidize both the US newspaper business and regional and industrial development in Canada. In the twenty-first century, the interdependence of US newspaper publishing and Canadian newsprint manufacturing has created significant challenges in these Canadian communities. The disappearance of 23.6 million in annual newspaper circulation and the interrelated $18 billion decline in advertising revenue since 1995 are indications of a troubled industry, and the US newspaper's declining fortunes have had negative consequences not just in American newsrooms but also all the way through the industrial supply chains to Canadian paper mills.

Across Canada, newsprint mills have curtailed production and shut down over the past two decades, including in Thorold, where the Tribune Company had operated since 1912. In 1985, mayor Don McMillan remarked, "For 73 years, Ontario Paper has been Thorold's number one industry, under the early guidance of Colonel Robert McCormick, then that of Arthur Schmon." The company had always been "outstanding and generous" in its "municipal involvement," McMillan claimed, and from the top down. "In the past," he noted in 1985, "there have been so many instances when Ontario Paper's carpenters, plumbers, electricians and maintenance people have given a helping hand to our churches, fire halls, library, arenas and service clubs." As the Tribune began scaling back manufacturing operations there in subsequent years, and as newsprint production declined more

generally, the Thorold area became part of the cross-border Rust Belt. As McMillan told me in 2015, in the 1990s newsprint production "started to fade and now
there's nothing, really." At the Tribune Company's mill, formerly a key to the city's
economy, "they only run one or two, one shift part time or something. It's sad . . .
the whole area is tough. There used to be, you know, working, several families
would go through the Ontario Paper . . . the whole area was really pulp and
paper. . . . Now there's not much."[1]

While Thorold and nearby St. Catharines show some signs of economic diversification and revitalization in other sectors, perhaps most notably as a result of
Brock University, the consequences of declining paper production have been devastating in other cities dominated by paper mills. In Dalhousie, New Brunswick,
for example, AbitibiBowater's closure of its mill in January 2008 decimated the
town. As one twenty-nine-year veteran of the mill stated, "Walking feels as if
we've got no legs. They took the legs from underneath us. We counted so much
on the mill for our lives, now it's turmoil. . . . It's the last nail in the casket for a
lot of us people, a big big downfall. I wish to God something comes here for our
people." Tony Tremblay, a professor at St. Thomas University in New Brunswick
and a native of Dalhousie, added that the mill was "really everything in Dalhousie, to the extent that people who didn't work in the mill, in a way worked in the
mill." For example, he noted, since the workers and their families spent so much
of their wages in Dalhousie, "there were indirect and direct influences everyone
who lived in the town felt." In many ways, as the industrial newspaper declined,
it has contributed to deindustrialization experienced across the continent.[2]

But if the printed newspaper's decline is related to patterns of North American deindustrialization, it is important to remember that the newspaper remains
very much a product of industrial supply chains, though in different ways and in
new geographic contexts. We are living through a historical moment in media
history defined not by de-objectification or de-materialization, but rather by re-
objectification and re-materialization. Newspaper reading in the United States
retains a connection to international supply chains producing material objects,
but today the most important sites of industrial production are no longer North
American newsprint mills and newspaper printing plants, but rather Asian factories manufacturing computers and portable electronic devices. There is a common thread across this longer media history, and that is that the material goods
providing daily news and information have been and remain parts of industrial
supply chains crossing international borders.

Moreover, it is not just the electronic devices that are part of a continuous industrial media economy tracing back to newsprint production and the realm of
trees and rivers; there is also the fact that much of the power used to run the servers

that undergird the internet is drawn from nature in the form of hydroelectric dams. The so-called "cloud" is in reality a distinctly terrestrial phenomenon, as the *Times* (London) reported in 2012. "The giants of the web are spreading the idea that everything that can be digitised can be stored, often for free, in an electronic 'cloud.' And yet the cloud is not a cloud. It is an extraordinary proliferation of huge, heavy machinery fed night and day with electricity and watched over by anxious technicians on whose expertise the new global economy depends." In 2012, a *New York Times* report claimed that the massive collections of servers around the world created a situation in which the "foundation of the information industry is sharply at odds with its image of sleek efficiency and environmental friendliness. Most data centers, by design, consume vast amounts of energy in an incongruously wasteful manner." In one example of these longer-term connections in the media business, Microsoft built a data center on 75 acres in Quincy, a town of seven thousand people in central Washington, with hydroelectric power coming from the Columbia River. Though relatively less remote from US media consumers than was Baie Comeau, the town was still distant from the "software meccas of Northern California or Seattle" to which it was connected. As paper companies had been attracted to Canada not only by timber but also by cheap power, twenty-first-century digital firms are finding a similar attraction in readily available hydroelectricity. The Columbia River dams, whose cheap power made the region a "magnet for large agricultural operations and heavy industries like aluminum, steel, paper and chemical plants," provided the same inducement for computer companies. And like the newsprint manufacturers a hundred years before, which were drawn to Canada through tariff policy, corporations like Microsoft benefited from generous government subsidies, often in the form of tax breaks, to come to these remote areas and build data centers.[3]

Though many would like to draw a clear line between the era of "dead tree media" and the digital age, in many areas the construction of data centers has directly followed the geography of the pulp and paper industry, as cheap power remains a crucial factor for the industrial plants undergirding the digital media business. As one 2011 report noted, the internet "now consumes two to three per cent of the world's electricity. If the Internet was a country, it would be the planet's fifth-biggest consumer of power, ahead of India and Germany." Around the world, digital media firms have responded to this need for power by employing some of the same infrastructures that pulp and paper companies had. In the past several years, for example, Facebook built a $750 million data center in Luleå, Sweden, where the nearby Lule River generates double the power of the Hoover Dam. In Hamina, Finland, Google spent $265 million converting a paper mill into a massive server facility. In North America, the availability of cheap hydroelectric power attracted data center construction along the Columbia River in both the United States and Canada.[4]

In addition to using massive amounts of electricity, data centers also generate a tremendous amount of heat and need to be cooled, so northern climates provide an additional incentive to build there because of naturally cooler temperatures. As had been the case in the international market for newsprint, Canada found itself competing with Scandinavian countries in the digital age to attract investment from leading digital firms. As a correspondent noted in the *Globe and Mail* in 2012, the "combination of cheap power and cold weather puts Canada in a similar league with Sweden and Finland, which have recently become the hosts of huge data centres built by Facebook and Google, respectively."[5]

Across Canada, provinces are competing to attract data centers as they used to try to appeal to newsprint manufacturers. As the *Telegraph-Journal* in New Brunswick noted, "Data centres are the new-economy equivalent of pulp mills," and it encouraged the provincial government to "accept that access to affordable power is the starting point for industrial growth." As was the case with the newsprint business of the twentieth century, the growth of the internet in the twenty-first century in many places is being driven by power concessions from the state. When we consider electronics manufacturing, and when we consider the size and scope of international data centers, it is clear that today's digital media remain, as was the case throughout the twentieth century, firmly embedded in global networks of both industrial production and natural resources, and in many instances the infrastructures are the same.[6]

Ultimately, the question of what this all means for the printed newspaper in the twenty-first century remains open. Considered over the longer term, one of the most striking things about the printed newspaper's history is how resilient the medium has been. Futurists have been predicting the death of the newspaper for nearly a century, and until very recently their proclamations were consistently overeager and incorrect. In the twentieth century, radio, facsimile, and television were each, in turn, hailed as the new media innovation that would render the printed newspaper obsolete. None of them succeeded. It remains to be seen whether newspaper firms can find ways of developing new business models to take advantage of the fact that increasing numbers of readers are accessing their content digitally rather than on newsprint. It is wise to remain open to the possibility that they can.

Archival Abbreviations

BANQ: Bibliothèque et Archives nationales du Québec, Centre d'archives de Québec, Quebec City, QC
  MDF: Maurice Duplessis Fonds, ZC44

LAC: Library and Archives Canada, Ottawa, ON
  Borden Fonds: Sir Robert Borden Fonds, R6113-0-X-E
  Grey Fonds: Albert Henry George Grey, 4th Earl Grey Fonds, R4689-0-6-E
  King Fonds: William Lyon Mackenzie King Fonds, R10383-0-6-E
  Laurier Fonds: Sir Wilfrid Laurier Fonds, R10811-0-X-E
  Meighen Fonds: Arthur Meighen Fonds, R14423-0-6-E
  Pringle: Commission to Inquire into and Report upon the Manufacture, Sale, Price and Supply of News Print Paper within Canada Fonds, R1133-0-6-E
  QOPC: Quebec and Ontario Paper Company Fonds, R6120, acquisition no. 2006-00392-5

LFCASC: Lake Forest College Archives and Special Collections, Lake Forest College, Lake Forest, IL
  JMP: Joseph Medill Patterson Papers

LOC: Library of Congress, Washington, DC
  Pinchot Papers: Gifford Pinchot Papers
  Pulitzer Papers: Joseph Pulitzer Papers
  Taft Papers: William Howard Taft Papers

NACP: National Archives and Records Administration, College Park, MD
  FS: Records of the Forest Service, Record Group 95
  OIAA: Records of the Office of Inter-American Affairs, 1918–51, Record Group 229
  USFAA: Records of US Foreign Assistance Agencies, Record Group 469

NSA: Nova Scotia Archives, Halifax, NS
  WSF: W. S. Fielding Fonds, MG2

NYPL: Manuscripts and Archives Division, The New York Public Library, Astor, Lenox and Tilden Foundations, New York, NY
  NYT: New York Times Company Records, General Files 1836–2000, MSS 17802
  Sulzberger Papers: New York Times Company Records, Arthur Hays Sulzberger Papers, MSS 17782

SHCN: Société Historique de la Côte-Nord, Baie Comeau, QC
  Dupuy: Fonds René Dupuy, Collection P88
  PTBC: Fonds Petit Théâtre de Baie-Comeau, Collection P85
  QNS: Fonds Quebec North Shore, Collection P26

SHSW: State Historical Society of Wisconsin, Madison, WI
  Hedges Papers: William Saxby Hedges Papers

Tribune Archives: Colonel Robert R. McCormick Research Center, First Division Museum at Cantigny, Wheaton, IL
  RRMB: Series I-60, Robert R. McCormick Business Correspondence, 1927–55

Introduction · What Was a Newspaper?

1. John Perry Barlow, "A Declaration of the Independence of Cyberspace," Feb. 8, 1996, www.eff.org/cyberspace-independence.

2. I have adapted the term "industrial newspaper" from Scott Casper, Jeffrey Groves, Stephen Nissenbaum, and Michael Winship, eds., *A History of the Book in America*, vol. 3, *The Industrial Book, 1840–1880* (Chapel Hill: University of North Carolina Press, 2007); and John Nerone, *The Media and Public Life: A History* (New York: Polity, 2015).

3. Royal Kellogg, *Newsprint Paper in North America* (New York: Newsprint Service Bureau, 1948), 78.

4. "Trees to Tribunes," *Chicago Tribune*, Feb. 3, 1929, D6; "Trees to Tribunes," *Chicago Tribune*, Feb. 10, 1929, E8.

5. "Trees to Tribunes," *Chicago Tribune*, Mar. 3, 1929, E11; "Trees to Tribunes," *Chicago Tribune*, Mar. 10, 1929, E10; "Trees to Tribunes," *Chicago Tribune*, Apr. 13, 1930, C4; "Trees to Tribunes," *Chicago Tribune*, Apr. 27, 1930, C4; "Trees to Tribunes," *Chicago Tribune*, May 11, 1930, C4.

6. Arthur Evans, "Thorold Docks with Big Cargo of Newsprint," *Chicago Tribune*, June 25, 1930, 5; James O'Donnell Bennett, "Tribune Plant Lures Throngs of Sightseers," *Chicago Tribune*, Aug. 30, 1931, 6.

7. In addition to these articles in the newspaper, the Tribune Company routinely used the title "Trees to Tribunes" in other venues. A short film with the title became a part of the visitors' tour of the Tribune Tower in the 1930s. See Bennett, "Tribune Plant Lures Throngs of Sightseers." A version of the film is available at www.archive.org/details/TreestoT1937. A modified but more prominent use of the phrase came in Carl Wiegman, *Trees to News: A Chronicle of the Ontario Paper Company's Origin and Development* (Toronto: McClelland & Stewart, 1953). This book is the only significant monograph on the Tribune's Canadian papermaking activities, and it is a fascinating portrait of the industrial production of the printed newspaper. Many scholars analyzing the history of newsprint production cite it as authoritative. Wiegman's knowledge of the Tribune Company is sound, and, indeed, I rely significantly on his research and chronology in my own work. But Wiegman's book does require a brief comment, as it needs to be treated with some care as a historical source. As a work of nonfiction, it is a portrait of the Tribune Company which is in some respects as selective and evasive as the "Trees to Tribunes" articles. Wiegman was a longtime *Chicago Tribune* writer and editor (he was on the staff from 1933 to 1971), and he wrote *Trees to News* with the strong guidance of company leaders. As executive Arthur Schmon wrote to Tribune Company president Robert McCormick in 1952, Wiegman's book had "value not only in giving a history and a background to the employees of our own companies but also in public relations. It will be very valuable in Canadian public relations in informing important people of Canada about you and the Company. . . . There is one thing which all sections of the story do have in common, and which is equally applicable to and understandable by any type of reader, and that is a record of pioneering and, regardless of what section of what chapter any one may happen to glance at, it is, as a public relations man emphasized to me, a pioneering story." Arthur Schmon to Robert McCormick, Dec. 9, 1952, 1–2 (emphasis in original), vol. 25353 (acquisition no. 2006-00392-5, box 64), folder Trees to News, pt. 2, QOPC, LAC. Tribune Company executives dictated not only the focus of Wiegman's book but also its title. Arthur Schmon suggested, in addition to the ultimately used *Trees to News*, a series of others. Some fell rather flat, for example, *The Forest and the*

Press, Wood in News, and *Wood Pulp and White Water.* Others were more evocative, for example, *Newsprint Trail, Wilderness Empire,* and *In Cartier's Footsteps.* Schmon did suggest one title, *Behind Metropolitan Newspapers,* that was ultimately discarded but in many respects captures the aim of my own book. Schmon and Wiegman were interested in going "behind" the production of the *Tribune* as long as the story cast the Tribune Company in a positive light, and *Trees to News* reflects this desire to present history in the form of public relations, at the expense of considering some of the broader and important political contexts of that history, as I will do here. See Arthur Schmon, "Proposed List of Titles for Carl Wiegman's History of the Ontario Paper Company," n.d., 1–2, vol. 25353 (acquisition no. 2006-00392-5, box 64), folder Trees to News, pt. 2, ibid.

8. In that sense, this book follows the work of Pietra Rivoli in seeking to explain commodity production in the global economy from natural setting to finished product. See Pietra Rivoli, *The Travels of a T-Shirt in the Global Economy: An Economist Examines the Markets, Power, and Politics of World Trade,* 2nd ed. (Hoboken: John Wiley & Sons, 2009).

9. Dwayne Winseck, "Reconstructing the Political Economy of Communication for the Digital Media Age," *Political Economy of Communication* 4, no. 2 (2016): 81–82.

10. Andrew Pettegree, *The Invention of News: How the World Came to Know about Itself* (New Haven, CT: Yale University Press, 2014), 5–11, 75, 182–83; Andrew Pettegree, *The Book in the Renaissance* (New Haven, CT: Yale University Press, 2010), 130–36; Charles E. Clark, *The Public Prints: The Newspaper in Anglo-American Culture, 1665–1740* (New York: Oxford University Press, 1994), 3, 15–25; Kevin Barnhurst and John Nerone, *The Form of News: A History* (New York: Guilford, 2001).

11. William Scott, *Scientific Circulation Management for Newspapers* (New York: Ronald, 1915), 64; "Seattle Times Jumps Its Prices," *Editor & Publisher* 52, no. 16 (Sept. 18, 1919): 7.

12. For excellent analyses of the ways that early twentieth-century newspapers provided readers with guides for urban life through content and advertising, see Richard Abel, *Menus for Movieland: Newspapers and the Emergence of American Film Culture, 1913–1916* (Oakland: University of California Press, 2015); and Julia Guarneri, *Newsprint Metropolis: City Papers and the Making of Modern Americans* (Chicago: University of Chicago Press, 2017).

13. "New York Newspaper Circulations, Total Pages, Total Advertising, Dry Goods and Foreign Advertising, and Circulation Rates, Jan. 1–Aug. 31, 1922 and 1921," *Editor & Publisher* 55, no. 18 (Sep. 30, 1922): 12; Susan B. Carter, Scott Sigmund Gartner, Michael R. Haines, Alan L. Olmstead, Richard Sutch, and Gavin Wright, eds., *Historical Statistics of the United States, Earliest Times to the Present, Millennial Edition* (New York: Cambridge University Press, 2006), tables De482–515 and Dg287–92.

14. John Pomfret, "The Best of Times, the Worst of Times," *Washington Post,* Dec. 1, 2006, C1; Joel Achenbach, "I Really Need You to Read This Article, Okay?," *Washington Post,* Aug. 19, 2007, B3.

15. Advertisement, Chicago Tribune Company, *Editor & Publisher* 53, no. 38 (Feb. 19, 1921): cover; advertisement, Great Northern Paper Company, *Editor & Publisher* 95, no. 9 (Mar. 3, 1962): 25.

16. "A Solution of Paper Problem," editorial, *Editor & Publisher* 7, no. 23 (Nov. 30, 1907): 4; Jason Rogers, "Factory Idea Best for Newspapers," *Editor & Publisher* 51, no. 39 (Mar. 8, 1919): 14.

17. Robert Park, "The Natural History of the Newspaper," *American Journal of Sociology* 29, no. 3 (Nov. 1923): 273–74; "Trees and Newspapers," editorial, *Editor & Publisher* 56, no. 22 (Oct. 27, 1923): 40.

18. Kellogg, *Newsprint Paper in North America*, 29, 73; Gifford Pinchot, foreword to *Forest Resources of the World*, by Raphael Zon and William Sparhawk, 2 vols. (New York: McGraw-Hill, 1923), 1:vii.

19. US Senate, 66th Cong., 2nd Sess., *Newsprint Paper Industry: Hearing before a Subcommittee of the Committee on Manufactures, Pursuant to S. Res. 164* (Washington, DC: Government Printing Office, 1920), 20, 213; Chicago Tribune Company, *From Trees to Tribunes* (Chicago: Chicago Tribune Company, 1922), 52; "Page and Paper Sizes and Other Mechanical Data of 1,935 Dailies," *Editor & Publisher* 61, no. 23 (Oct. 27, 1928): 54, 62, 64, 74.

20. Don Seitz, *Training for the Newspaper Trade* (Philadelphia: J. B. Lippincott, 1916), 40.

21. William Cronon, *Nature's Metropolis: Chicago and the Great West* (New York: Norton, 1991), 155–59; Mark Kuhlberg, *In the Power of the Government: The Rise and Fall of Newsprint in Ontario, 1894–1932* (Toronto: University of Toronto Press, 2015), 27–30.

22. Advertisement, Hearst Newspapers, *Editor & Publisher* 68, no. 13 (Aug. 10, 1935): 20–21.

23. John Guthrie, *The Newsprint Paper Industry: An Economic Analysis* (Cambridge, MA: Harvard University Press, 1941), 18–23; Jack P. Oden, "Charles Holmes Herty and the Birth of the Southern Newsprint Paper Industry, 1927–1940," *Journal of Forest History* 21, no. 2 (Apr. 1977): 76–89; Canadian Pulp and Paper Association, *A Handbook of the Canadian Pulp and Paper Industry* (Montreal: Canadian Pulp and Paper Association, 1920), 4.

24. Michael Hart, *A Trading Nation: Canadian Trade Policy from Colonialism to Globalization* (Vancouver: UBC Press, 2002), 37, 57, 65–68, 96; John Bartlet Brebner, *North Atlantic Triangle: The Interplay of Canada, the United States, and Great Britain* (1945; repr., New York: Columbia University Press, 1958).

25. Canadian Pulp and Paper Association, *From Watershed to Watermark: The Pulp and Paper Industry of Canada* (Montreal: Canadian Pulp and Paper Association, 1950), 10–11, 14, 53.

26. Ibid., 7; Hugh G. J. Aitken, "The Changing Structure of the Canadian Economy, with Particular Reference to the Influence of the United States," in *The American Economic Impact on Canada*, ed. Hugh G. J. Aitken et al. (Durham, NC: Duke University Press, 1959), 13, 15.

27. Earle Clapp and Charles Boyce, *How the United States Can Meet Its Present and Future Pulpwood Requirements*, US Department of Agriculture, bulletin no. 1241 (Washington, DC: US Department of Agriculture, 1924), 82; Newsprint Association of Canada, *Newsprint Data, 1951* (Montreal: Newsprint Association of Canada, 1951), 6, 11, 21.

28. Royal Kellogg, "The International Movement of Pulpwood," *Pulp & Paper Magazine of Canada* 40 (Oct. 1939): 648.

29. Richard R. John, *Spreading the News: The American Postal System from Franklin to Morse* (Cambridge, MA: Harvard University Press, 1995).

30. "Prof. H. A. Innis Noted as Historian, Author," *Globe and Mail*, Nov. 10, 1952, 10; Harold Innis, "The Decline in the Efficiency of Instruments Essential in Equilibrium," *American Economic Review* 43, no. 1 (Mar. 1953): 21.

31. Innis, "Decline in the Efficiency of Instruments," 21–22.

32. Charles Acland and William Buxton, "Introduction: Harold Innis: A Genealogy of Contesting Portraits," in *Harold Innis in the New Century: Reflections and Refractions*, ed. Charles Acland and William Buxton (Montreal and Kingston: McGill-Queen's University Press, 1999), 6–10.

33. Marshall McLuhan, "The Later Innis," *Queen's Quarterly* 60 (Jan. 1953): 385, 390. Donald Creighton's 1957 biography similarly promoted the interpretation of Innis's career as being divided into separate and largely discontinuous phases, though Creighton grouped these phases into intervals defined by the decades of the 1920s, 1930s, and 1940s. Creighton also focused primarily on Innis's work on staples and devoted very little attention to Innis's consideration of communications. See Donald Creighton, *Harold Adams Innis: Portrait of a Scholar* (Toronto: University of Toronto Press, 1957), chap. 3.

34. For other works emphasizing the continuities between Innis's early and later works, see Daniel Czitrom, *Media and the American Mind: From Morse to McLuhan* (Chapel Hill: University of North Carolina Press, 1982), chap. 6; W. T. Easterbrook, "Harold Adams Innis," *American Economic Review* 43, no. 1 (Mar. 1953): 8–12; Graeme Patterson, *History and Communications: Harold Innis, Marshall McLuhan, the Interpretation of History* (Toronto: University of Toronto Press, 1990); Paul Heyer, *Harold Innis* (Lanham: Rowman & Littlefield, 2003); and Menahem Blondheim and Rita Watson, "Introduction: Innis, McLuhan and the Toronto School," in *The Toronto School of Communication Theory: Interpretations, Extensions, Applications*, ed. Rita Watson and Menahem Blondheim (Toronto and Jerusalem: University of Toronto Press and Hebrew University Magnes Press, 2007), 7–26.

35. Alexander John Watson, *Marginal Man: The Dark Vision of Harold Innis* (Toronto: University of Toronto Press, 2006); W. T. Easterbrook and Hugh G. J. Aitken, *Canadian Economic History* (1956; repr., Toronto: University of Toronto Press, 2008), 538, 540, 546; Harold Innis, *Empire and Communications* (1950; repr., Lanham: Rowman & Littlefield, 2007).

36. Harold Innis, *The Fur Trade in Canada* (1930; repr., Toronto: University of Toronto Press, 2001), 384–85.

37. Harold Innis, "Pulp-and-Paper Industry," in *The Encyclopedia of Canada*, ed. W. Stewart Wallace (Toronto: University Associates of Canada Limited, 1937), 5:179, 184.

38. For reviews, see Harold Innis, review of *The Newsprint Paper Industry: An Economic Analysis*, by John A. Guthrie, *Journal of Political Economy* 50, no. 4 (Aug. 1942): 624–25; Harold Innis, review of *The Newsprint Paper Industry: An Economic Analysis*, by John A. Guthrie; *The Background and Economics of American Papermaking*, by Louis Tillotson Stevenson; *Canada Gets the News*, by Carlton McNaught; *AP: The Story of News*, by Oliver Gramling; and *The Human Interest Story*, by Helen MacGill Hughes, *Canadian Journal of Economics and Political Science* 7, no. 4 (Nov. 1941): 578–83; Harold Innis, review of *The Chicago Tribune: Its First Hundred Years*, vol. 1, *1847–1865*, by Philip Kinsley, *Journal of Economic History* 4, no. 1 (May 1944): 100; and Harold Innis, review of *The Chicago Tribune: Its First Hundred Years*, vol. 2, *1865–1880*, by Philip Kinsley, *Journal of Economic History* 7, no. 1 (May 1947): 121–22. For scholarly articles, see, e.g., Harold Innis, introduction to *The Canadian Economy and Its Problems*, ed. Harold Innis and A. F. W. Plumptre (Toronto: Canadian Institute of International Affairs, 1934), 19–21; Harold Innis, "The Newspaper in Economic Development," *Journal of Economic History* 2, no. S1 (Dec. 1942): S15–16; Harold Innis, "On the Economic Significance of Culture," *Journal of Economic History* 4, no. S1 (Dec. 1944): S90; Harold Innis, "The English Publishing Trade in the Eighteenth Century," *Manitoba Arts Review* 4, no. 4 (Winter 1945): 14; and Harold Innis, "The English Press in the Nineteenth Century: An Economic Approach," *University of Toronto Quarterly* 15, no. 1 (Oct. 1945): 37–39.

39. Harold Innis, "The Penetrative Powers of the Price System," *Canadian Journal of Economics and Political Science* 4, no. 3 (Aug. 1938): 313.

40. Harold Innis, *Political Economy in the Modern State* (Toronto: Ryerson, 1946), ix–x; Harold Innis, *The Press: A Neglected Factor in the Economic History of the Twentieth Century* (London: Oxford University Press, 1949), 36–37.

41. Harold Innis, *The Bias of Communication* (1951; repr., Toronto: University of Toronto Press, 2006), 77–78, 187–90.

42. Donald Innis, comment on Harold Innis, "The Decline in the Efficiency of Instruments Essential in Equilibrium," *American Economic Review* 43, no. 1 (Mar. 1953): 23–25.

43. For other explorations of this connection, see James Carey, "Space, Time, and Communications: A Tribute to Harold Innis," in *Communication as Culture: Essays on Media and Society* (1989; repr., New York: Routledge, 1992), 142–72; and Jody Berland, *North of Empire: Essays on the Cultural Technologies of Space* (Durham, NC: Duke University Press, 2009), chap. 2.

44. Harold Innis, "Great Britain, the United States and Canada," in *Changing Concepts of Time* (1954; repr., Lanham: Rowman & Littlefield, 2004), 106.

45. For examples of exceptional works that have considered aspects of the relationship between the United States and Canada, see Seymour Martin Lipset, *Continental Divide: The Values and Institutions of the United States and Canada* (New York: Routledge, 1991); Caren Irr, *The Suburb of Dissent: Cultural Politics in the United States and Canada during the 1930s* (Durham, NC: Duke University Press, 1998); Rachel Adams, *Continental Divides: Remapping the Cultures of North America* (Chicago: University of Chicago Press, 2009); and Benjamin H. Johnson and Andrew R. Graybill, eds., *Bridging National Borders in North America: Transnational and Comparative Histories* (Durham, NC: Duke University Press, 2010).

46. Kelly Cryderman, "Oil Prices Surge on Plan to Tackle Global Glut," *Globe and Mail*, Apr. 13, 2016, B1; US Energy Information Administration, "U.S. Total Crude Oil and Products Imports," www.eia.gov/dnav/pet/pet_move_impcus_a2_nus_ep00_im0_mbbl_m.htm.

47. George A. Gonzalez, *American Empire and the Canadian Oil Sands* (New York: Palgrave MacMillan, 2016).

48. Melville Watkins, "A Staple Theory of Economic Growth," *Canadian Journal of Economics and Political Science* 29, no. 2 (May 1963): 141.

49. Immanuel Wallerstein, *The Modern World-System I: Capitalist Agriculture and the Origins of the European World-Economy in the Sixteenth Century* (1974; repr., Berkeley: University of California Press, 2011), 42.

50. For rubber, see Greg Grandin, *Fordlandia: The Rise and Fall of Henry Ford's Forgotten Jungle City* (New York: Metropolitan Books, 2009). For cotton, see Sven Beckert, *Empire of Cotton: A Global History* (New York: Alfred A. Knopf, 2014). For aluminum, see Mimi Sheller, *Aluminum Dreams: The Making of Light Modernity* (Cambridge, MA: MIT Press, 2014). For guano, see Gregory Cushman, *Guano and the Opening of the Pacific World: A Global Ecological History* (New York: Cambridge University Press, 2013).

51. Robert U. Brown, "Average Daily's Expenses Increased 28% in 1947," *Editor & Publisher* 81, no. 17 (Apr. 17, 1948): 18; Robert U. Brown, "Costs Outgain Revenues for Third Year in Row," *Editor & Publisher* 83, no. 16 (Apr. 15, 1950): 5; Robert U. Brown, "Revenue Outpaces Expense," *Editor & Publisher* 93, no. 16 (Apr. 16, 1960): 11.

52. "Business Is Only an Incidental of Journalism, Says McCormick," *Editor & Publisher* 63, no. 27 (Nov. 22, 1930): 11.

53. Arthur Schmon, statement to the Select Committee of the Legislature of the Province of Ontario, May 7, 1940, printed in "The Status of the Ontario Paper Company, Limited, Including Its Wholly Owned Subsidiary Quebec North Shore Paper Company, as a Non-

commercial Newsprint Manufacturer," June 5, 1940, 15–16, vol. 25808 (acquisition no. 2006-00392-5, box 519), folder Status of the Ontario Paper Company Limited 1940, QOPC, LAC; Arthur Schmon, "Our Paper Mills," address at the Chicago Tribune Circulation Department Centennial Prevue, Chicago, IL, Oct. 9, 1946, 10, vol. 25799 (acquisition no. 2006-00392-5, box 510), folder Addresses by Arthur A. Schmon, ibid.

54. The "American Paper for Americans" slogan appeared regularly on the *Tribune*'s masthead from the mid-1940s to the early 1970s. The "World's Greatest Newspaper" slogan has been in regular usage in the paper since 1906.

55. The three best works on the subject are Richard Kaplan, *Politics and the American Press: The Rise of Objectivity, 1865–1920* (New York: Cambridge University Press, 2002); David Mindich, *Just the Facts: How "Objectivity" Came to Define American Journalism* (New York: New York University Press, 1998); and Michael Schudson, *Discovering the News: A Social History of American Newspapers* (New York: Basic Books, 1978).

56. In their comparative analysis of national media systems, Daniel Hallin and Paolo Mancini cite Robert McCormick and Time-Life owner Henry Luce as the two important exceptions to these trends in twentieth-century American journalism history. Daniel Hallin and Paolo Mancini, *Comparing Media Systems: Three Models of Media and Politics* (New York: Cambridge University Press, 2004), 219.

57. On Robert McCormick's life and politics, see Frank C. Waldorp, *McCormick of Chicago: An Unconventional Portrait of a Controversial Figure* (Englewood Cliffs, NJ: Prentice-Hall, 1966); Jerome Edwards, *The Foreign Policy of Col. McCormick's Tribune, 1929–1941* (Reno: University of Nevada Press, 1971); and Richard Norton Smith, *The Colonel: The Life and Legend of Robert R. McCormick* (New York: Houghton Mifflin, 1997).

58. Jack Alexander, "The Duke of Chicago," *Saturday Evening Post* 214, no. 3 (July 19, 1941): 70, 74–75.

59. A. J. Liebling, "The Wayward Press," *New Yorker* 25 (Jan. 7, 1950): 52, 56.

Chapter 1 · The Making of Industrial Print Culture

1. Alexis de Tocqueville, *Democracy in America*, trans. and ed. Harvey Mansfield and Delba Winthrop (Chicago: University of Chicago Press, 2000), 493.

2. "A. S. Ochs Tells How He Made the *New York Times*," *Editor & Publisher* 49, no. 2 (June 24, 1916): 3, 24–25; advertisement, *Editor & Publisher* 58, no. 31 (Dec. 26, 1925): 21.

3. William Scott, *Scientific Circulation Management for Newspapers* (New York: Ronald, 1915), 172.

4. Gunther Barth, *City People: The Rise of Modern City Culture in Nineteenth-Century America* (New York: Oxford University Press, 1980), chap. 3; John Nerone, *The Media and Public Life: A History* (New York: Polity, 2015), chap. 4; David Paul Nord, "The Victorian City and the Urban Newspaper," in *Making News: The Political Economy of Journalism in Britain and America from the Glorious Revolution to the Internet*, ed. Richard R. John and Jonathan Silberstein-Loeb (New York: Oxford University Press, 2015), 73–106, chap. 4; Michael Schudson, *Discovering the News: A Social History of American Newspapers* (New York: Basic Books, 1978), chap. 1.

5. Elizabeth Eisenstein, *The Printing Press as an Agent of Change: Communications and Cultural Transformations in Early-Modern Europe* (New York: Cambridge University Press, 1979).

6. Lucien Febvre and Henri-Jean Martin, *The Coming of the Book: The Impact of Printing 1450–1800*, trans. David Gerard (1958; repr., New York: Verso, 1997), 30; Juraj Kittler,

"From Rags to Riches: The Limits of Early Paper Manufacturing and Their Impact on Book Print in Renaissance Venice," *Media History* 21 no. 1 (Feb. 2015): 8–22; Anna Melograni, "Manuscript Materials: Cost and the Market for Parchment in Renaissance Italy," in *Trade in Artists' Materials: Markets and Commerce in Europe to 1700*, ed. Jo Kirby, Susie Nash, and Joanna Cannon (London: Archetype, 2010), 199–219.

7. Dard Hunter, *Papermaking: The History and Technique of an Ancient Craft* (1947; repr., New York: Dover, 1978), chap. 1. For other accounts of the long-term history of paper, see Nicholas Basbanes, *On Paper: The Everything of Its Two-Thousand-Year History* (New York: Alfred A. Knopf, 2013); Lothar Müller, *White Magic: The Age of Paper*, trans. Jessica Spengler (New York: Polity, 2014); and Ian Sansom, *Paper: An Elegy* (New York: William Morrow, 2012).

8. Hunter, *Papermaking*, chap. 2.

9. Febvre and Martin, *Coming of the Book*, 30–33; Hunter, *Papermaking*, chap. 14; Judith McGaw, *Most Wonderful Machine: Mechanization and Social Change in Berkshire Paper Making, 1801–1885* (Princeton, NJ: Princeton University Press, 1987), 39–49.

10. Susan Strasser, *Waste and Want: A Social History of Trash* (New York: Henry Holt, 1999), 80–90.

11. McGaw, *Most Wonderful Machine*, 96–103; Hunter, *Papermaking*, chap. 12; Alfred McClung Lee, *The Daily Newspaper in America: The Evolution of a Social Instrument* (New York: Macmillan, 1947), 100.

12. David C. Smith, *History of Papermaking in the United States (1691–1969)* (New York: Lockwood, 1970), 121–26; Febvre and Martin, *Coming of the Book*, 36.

13. L. Ethan Ellis, *Print Paper Pendulum: Group Pressures and the Price of Newsprint* (New Brunswick, NJ: Rutgers University Press, 1948), 3–5; Canadian Pulp and Paper Association, *From Watershed to Watermark: The Pulp and Paper Industry of Canada* (Montreal: Canadian Pulp and Paper Association, 1950), 11, 14–16.

14. McGaw, *Most Wonderful Machine*, 201; "When Did Newspapers Begin to Use Wood Pulp?," *Bulletin of the New York Public Library* 33, no. 10 (Oct. 1929): 743–45. This research was done at the New York Public Library, where archivists examined their holdings of physical newspapers to ascertain the materials that were used in printing.

15. In 1879, US paper manufacturers had used 22,570 tons of wood pulp and 200,005 tons of rags as raw materials. These figures increased respectively to 2,018,764 and 294,552 tons in 1904. By 1919, paper manufacturers used 4,019,696 tons of wood pulp and only 277,849 tons of rags as their raw materials. Earle Clapp and Charles Boyce, *How the United States Can Meet Its Present and Future Pulpwood Requirements*, US Department of Agriculture, bulletin no. 1241 (Washington, DC: US Department of Agriculture, 1924), 71.

16. Royal Kellogg, *Newsprint Paper in North America* (New York: Newsprint Service Bureau, 1948), 29.

17. John Guthrie, *The Newsprint Paper Industry: An Economic Analysis* (Cambridge, MA: Harvard University Press, 1941), 18–23, 48–51.

18. Advertisement, *Editor & Publisher* 48, no. 27 (Dec. 11, 1915): cover; Royal Kellogg quoted in "Eleven Billion Newspapers," *Pulp & Paper Magazine of Canada* 40 (Nov. 1939): 711.

19. James Moran, *Printing Presses: History and Development from the Fifteenth Century to Modern Times* (Berkeley: University of California Press, 1973); Adrian Johns, *The Nature of the Book: Print and Knowledge in the Making* (Chicago: University of Chicago Press, 1998); Jeffrey Pasley, *The Tyranny of Printers: Newspaper Politics in the Early Republic* (Charlottesville: University of Virginia Press, 2001), 24–28.

20. Lee, *Daily Newspaper in America*, 113–19.

21. James Carey, "Technology and Ideology: The Case of the Telegraph," in *Communication as Culture: Essays on Media and Society* (1989; repr., New York: Routledge, 1992): 201–30; "The Mechanical Side," editorial, *Editor & Publisher* 63, no. 23 (Oct. 25, 1930): 36.

22. "Fine New Newspaper Plants Inadequate under Stress of Rapid Delivery," *Editor & Publisher* 58, no. 44 (Mar. 27, 1926): 3; "Detroit News Published from New Home; $2,000,000 Plant Finest in United States," *Editor & Publisher* 50, no. 19 (Oct. 20, 1917): pt. 2, 1; "Giant Visible Presses with 430,000 16-Page Capacity," *Editor & Publisher* 50, no. 19 (Oct. 20, 1917): pt. 2, 6; "Free Press Occupies New $4,000,000 Plant," *Editor & Publisher* 58, no. 27 (Nov. 28, 1925): 6; "Construction Work Started on New $2,500,000 Detroit Times Plant," *Editor & Publisher* 61, no. 44 (Mar. 23, 1929): 26. On newspaper buildings generally, see Aurora Wallace, *Media Capital: Architecture and Communications in New York City* (Urbana: University of Illinois Press, 2012).

23. "Tribune Deserts Park Row for New Modern Plant Uptown," *Editor & Publisher* 55, no. 47 (Apr. 21, 1923): 14; "New Easton Express Home Embodies Latest in Small Plant Design," *Editor & Publisher* 56, no. 22 (Oct. 27, 1923): 10–11; "Matching Strides with Long Beach, Cal.," *Editor & Publisher* 57, no. 42 (Mar. 14, 1925): 6, 32; "Milwaukee Journal Plant Cost $2,000,000," *Editor & Publisher* 57, no. 26 (Nov. 22, 1924): 6–7; "A New Idea for Metropolitan Newspapers," *Editor & Publisher* 57, no. 22 (Oct. 25, 1924): 5, 45.

24. "Fine New Newspaper Plants," 4; "Philadelphia Inquirer in Palatial Home," *Editor & Publisher* 58, no. 8 (July 18, 1925): 9.

25. "Towering New Homes for Many Papers Built at Cost of $100,000,000," *Editor & Publisher* 58, no. 18 (Sept. 26, 1925): 3; "Los Angeles Times New $4,000,000 Home Combines Beauty and Efficiency," *Editor & Publisher* 68, no. 13 (Aug. 10, 1935): sec. 2, 2, 6, 17.

26. William Allen White, "How Free Is the Press?," *Collier's* 103 (Apr. 8, 1939): 16, 88.

27. Silas Bent, *Ballyhoo: The Voice of the Press* (New York: Horace Liveright, 1927), 249; Susanne Freidberg, *Fresh: A Perishable History* (Cambridge, MA: Belknap Press of Harvard University Press, 2009), 198, 203, 224.

28. The exception to this in the daily newspaper market was postal delivery of newspapers, which could take several days to get the paper to the reader. For some, subscribing to a distant urban paper was a way to keep up on important national trends in culture, politics, and the economy, while others liked to read about ongoing events in hometowns in which they no longer resided. For most newspapers, mail circulation was a small part of the business. For the *New York Times* in 1915, for example, only 23,000 of the 300,000 daily papers sold were to mail subscribers. Scott, *Scientific Circulation Management*, 86, 225. Sunday newspapers occupied a somewhat different segment of the market, as they had many features that were designed to be more like magazines and kept around for perusal throughout the week.

29. Don Seitz, *Training for the Newspaper Trade* (Philadelphia: J. B. Lippincott, 1916), 137–39.

30. Scott, *Scientific Circulation Management*, 17, 108.

31. Alexander Fleisher, *The Newsboys of Milwaukee* (Milwaukee: Milwaukee Bureau of Economy and Efficiency, 1911), 67; Scott, *Scientific Circulation Management*, 115–16.

32. Scott, *Scientific Circulation Management*, 90, 94, 96; Donald Abramoske, "The *Chicago Daily News*: A Business History, 1875–1901" (PhD diss., University of Chicago, 1963), 24–25.

33. Fleisher, *Newsboys of Milwaukee*, 69; Federation of Chicago Settlements, *Newsboy Conditions in Chicago* (Chicago: Federation of Chicago Settlements, 1907), 6.

34. Federation of Chicago Settlements, *Newsboy Conditions in Chicago*, 5, 10 19, 21. On newsboy labor in Chicago more generally, see Jon Bekken, "Crumbs from the Publishers' Golden Tables: The Plight of the Chicago Newsboy," *Media History* 6, no. 1 (2000): 45–57.

35. Fleisher, *Newsboys of Milwaukee*, 61; Scott, *Scientific Circulation Management*, 111–12; US Department of Labor, Children's Bureau, *Children Engaged in Newspaper and Magazine Selling and Delivering* (Washington, DC: Government Printing Office, 1935), 20–22, 32; Anthony Smith, *Goodbye Gutenberg: The Newspaper Revolution of the 1980s* (New York: Oxford University Press, 1980), 73–74.

36. "Five Buffalo Dailies Double Selling Price," *Editor & Publisher* 49, no. 25 (Dec. 2, 1916): 10.

37. Stephen Lacy and Todd Simon, *The Economics and Regulation of United States Newspapers* (Norwood, NJ: Ablex, 1993), 17–19.

38. Richard Kaplan, "Press, Paper, and the Public Sphere: The Rise of the Cheap Mass Press in the USA, 1870–1910," *Media History* 21, no. 1 (2015): 42–54.

39. "Cost and Price," editorial, *Editor & Publisher* 5, no. 23 (Nov. 25, 1905): 4; "Some Hope for Free Paper," editorial, *Editor & Publisher* 7, no. 42 (Apr. 18, 1908): 4.

40. "The Paper Pulp Problem," editorial, *Brooklyn Daily Eagle*, Sept. 5, 1904, 4.

Chapter 2 · Forests, Trade, and Empire

1. Frederick Jackson Turner, "The Significance of the Frontier in American History," in *Frontier and Section: Selected Essays of Frederick Jackson Turner* (Englewood Cliffs, NJ: Prentice-Hall, 1961), 37, 62.

2. William Cronon, *Nature's Metropolis: Chicago and the Great West* (New York: Norton, 1991), 32.

3. An excellent recent attempt to do this is Kristin Hoganson, "Meat in the Middle: Converging Borderlands in the U.S. Midwest, 1865–1900," *Journal of American History* 98, no. 4 (Mar. 2012): 1025–51.

4. Cronon, *Nature's Metropolis*, 169, 175, 206.

5. L. Ethan Ellis, *Print Paper Pendulum: Group Pressures and the Price of Newsprint* (New Brunswick, NJ: Rutgers University Press, 1948), 7.

6. Michael Hart, *A Trading Nation: Canadian Trade Policy from Colonialism to Globalization* (Vancouver: UBC Press, 2002), 179–80; Dani Rodrik, *The Globalization Paradox: Democracy and the Future of the World Economy* (New York: Norton, 2011), chaps. 2–3.

7. César Ayala, *American Sugar Kingdom: The Plantation Economy of the Spanish Caribbean, 1898–1934* (Chapel Hill: University of North Carolina Press, 1999), chap. 4; Stephen Scheinberg, "Invitation to Empire: Tariffs and American Economic Expansion in Canada," *Business History Review* 47, no. 2 (Summer 1973): 218–38.

8. US Tariff Commission, *Reciprocity with Canada: A Study of the Arrangement of 1911* (Washington, DC: Government Printing Office, 1920), 22, 24–26; Hart, *Trading Nation*, 62–68.

9. Douglas A. Irwin, *Clashing over Commerce: A History of US Trade Policy* (Chicago: University of Chicago Press, 2017), chaps. 5–6; Richard Bensel, *The Political Economy of American Industrialization, 1877–1900* (New York: Cambridge University Press, 2000), 107, 183; F. W. Taussig, *The Tariff History of the United States*, 8th ed. (1931; repr., New York: Augustus M. Kelley, 1967), 172–75; L. Ethan Ellis, *Reciprocity 1911: A Study in Canadian-American Relations* (New Haven, CT: Yale University Press, 1939), 1, 8.

10. Ellis, *Print Paper Pendulum*, 20–24.

11. Gifford Pinchot testimony, May 19, 1907, US Congress, House of Representatives, 60th Cong., 2nd Sess., *Pulp and Paper Investigation Hearings*, document no. 1502 (Washington, DC: Government Printing Office, 1909), 2:1357.

12. E. H. Bronson to Wilfrid Laurier, Nov. 1, 1898, 2, reel C760, vol. 90, Laurier Fonds, LAC.

13. *Report of the Royal Commission on Pulpwood*, 14–15 George V, Sessional Paper no. 310 (Ottawa: Printer to the King's Most Excellent Majesty, 1924), 92–93; H. V. Nelles, *The Politics of Development: Forests, Mines, and Hydroelectric Power in Ontario, 1849–1941*, 2nd ed. (1974; repr., Montreal and Kingston: McGill-Queen's University Press, 2005), 13–14, 63, 73–74, 87; Constant Southworth, "The American-Canadian Newsprint Paper Industry and the Tariff," *Journal of Political Economy* 30, no. 5 (Oct. 1922): 691–92; Lomer Gouin, Feb. 16, 1911, Quebec, Legislative Assembly, *Débats de l'Assemblée législative*, 12th Legislature, 3rd Sess., 1:347.

14. Mark Kuhlberg, *In the Power of the Government: The Rise and Fall of Newsprint in Ontario, 1894–1932* (Toronto: University of Toronto Press, 2015), 34–35, 75–81; Mark Kuhlberg, "'Pulpwood Is the Only Thing We Do Export': The Myth of Provincial Protectionism in Ontario's Forest Industry, 1890–1930," in *Smart Globalization: The Canadian Business and Economic History Experience*, ed. Andrew Smith and Dimitry Anastakis (Toronto: University of Toronto Press, 2014), 59–91; Earle Clapp and Charles Boyce, *How the United States Can Meet Its Present and Future Pulpwood Requirements*, US Department of Agriculture, bulletin no. 1241 (Washington, DC: US Department of Agriculture, 1924), 84.

15. Quoted in "The Great Pulp War," *Editor & Publisher* 2, no. 11 (Sept. 6, 1902): 1–2.

16. Emerson Bristol Biggar, *Canada's Approaching Peril* (Toronto: Biggar-Wilson, 1908), 5, 8–9, 12–13.

17. Richard R. John, "Proprietary Interest: Merchants, Journalists, and Antimonopoly in the 1880s," in *Media Nation: The Political History of News in Modern America*, ed. Bruce Schulman and Julian Zelizer (Philadelphia: University of Pennsylvania Press, 2017), 10–35.

18. Thomas Heinrich, "Product Diversification in the U.S. Pulp and Paper Industry: The Case of International Paper, 1898–1941," *Business History Review* 75, no. 3 (Autumn 2001): 469–71; "Makes Its Own Paper," *Editor & Publisher* 5, no. 23 (Nov. 25, 1905): 1; "A Tax upon Intelligence," editorial, *New York World*, Dec. 29, 1904, 6.

19. American Newspaper Publishers Association, brief in favor of free paper and free pulp, Dec. 1898, 1, 4, reel C761, vol. 94, Laurier Fonds, LAC; John Norris to Wilfrid Laurier, Jan. 24, 1899, reel C762, vol. 98, ibid.; John Norris to Wilfrid Laurier, Nov. 13, 1901, reel C789, vol. 212, ibid.; Wilfrid Laurier to John Norris, Nov. 20, 1901, reel C789, vol. 212, ibid.

20. "What Derb Says," *Paper Mill and Wood Pulp News* 27, no. 15 (Apr. 9, 1904): 1; Chester Lyman to John Adair, Feb. 14, 1908, reprinted in US Congress, *Pulp and Paper Investigation Hearings*, 2:1210; "Champ Clark on Pulp Duty," *Editor & Publisher* 3, no. 41 (Apr. 2, 1904): 3.

21. US Congress, *Pulp and Paper Investigation Hearings*, 1:67–76. These are representative examples taken from a list of over two hundred newspapers reporting price increases during this period.

22. "War on Paper Trust," *Editor & Publisher* 7, no. 13 (Sept. 21, 1907): 1–2; "Publishers Are United," editorial, *Editor & Publisher* 7, no. 43 (Apr. 25, 1908): 6.

23. H.R. 344, 60th Cong., 1st Sess., reprinted in US Congress, *Pulp and Paper Investigation Hearings*, 3:1619–20; Frank MacLennan testimony, May 6, 1908, ibid., 1:504–14; John Norris testimony, Apr. 25, 1908, ibid., 1:5.

24. "Report of Select Committee on Pulp and Paper Investigation," printed in US Congress, *Pulp and Paper Investigation Hearings*, 5:3314; Ellis, *Print Paper Pendulum*, 55–56.

25. A collection of remarks from newspapers around the country is found in "The Tariff," *Editor & Publisher* 8, no. 39 (Mar. 27, 1909): 13.

26. "Pungent Protest," *Editor & Publisher* 8, no. 37 (Mar. 13, 1909): 1; "Norris Attacks," *Paper Mill and Wood Pulp News* 32, no. 17 (Apr. 24, 1909): 1.

27. Taussig, *Tariff History of the United States*, 362–63, 375; "The Paper Trust," editorial, *New Yorker Staats-Zeitung*, Mar. 16, 1908, reprinted in US Congress, *Pulp and Paper Investigation Hearings*, 2:1206.

28. Taussig, *Tariff History of the United States*, 362–63, 409; Ellis, *Reciprocity 1911*, 29, 31.

29. Taussig, *Tariff History of the United States*, 410–15.

30. US Tariff Commission, *Reciprocity with Canada*, 33–38; William Howard Taft, "Reciprocity with Canada," in Western Economic Society, *Reciprocity with Canada* (Chicago: University of Chicago Press, 1911), 125; Patrice Dutil and David MacKenzie, *Canada 1911: The Decisive Election That Shaped the Country* (Toronto: Dundurn, 2011), 84.

31. US Tariff Commission, *Reciprocity with Canada*, 39; US Senate, 61st Cong., 3rd Sess., document no. 787, *Canadian Reciprocity* (Washington, DC: Government Printing Office, 1911), 2.

32. US Senate, document no. 787, *Canadian Reciprocity*, iv–vii.

33. Albert Beveridge, "The March of the Flag," 7–9, 16, pamphlet reprint, https://archive.org/details/marchofflagbegin00beve.

34. 46 Cong. Rec. S2181 (daily ed. Feb. 9, 1911).

35. 46 Cong. Rec. H125 (appendix Feb. 14, 1911).

36. 46 Cong. Rec. H2902 (daily ed. Feb. 19, 1911); 47 Cong. Rec. H404 (daily ed. Apr. 19, 1911).

37. Herman Ridder to ANPA members, Jan. 27, 1911, reprinted in US Congress, Senate, 62nd Cong., 1st Sess., *Reciprocity with Canada: Hearings before the Committee on Finance of the United States Senate on H.R. 4412*, document no. 56 (Washington, DC: Government Printing Office, 1911), 2:1224; 47 Cong. Rec. S3139 (daily ed. July 21, 1911).

38. Hugh Graham to W. S. Fielding, Jan. 5, 1911, vol. 521, folder 72, WSF, NSA. I thank Elsbeth Heaman for bringing this document to my attention.

39. The *Chicago Tribune* polled some 10,000 newspaper representatives in the West and Midwest, and of the roughly 43% who responded, 3,313 favored reciprocity and 1,127 did not. "West Favors Reciprocity 2 ½ to 1," *Chicago Tribune*, June 3, 1911, 1.

40. "The Real Issue of Reciprocity," editorial, *Chicago Tribune*, Feb. 10, 1911, 10.

41. US Tariff Commission, *Reciprocity with Canada*, 77–78; Taussig, *Tariff History of the United States*, 414; William Howard Taft to Charles Taft, July 23, 1911, 1, reel 508, ser. 8, vol. 28, Taft Papers, LOC.

42. "Reciprocity at Last," editorial, *Chicago Tribune*, July 22, 1911, 10.

43. Albert Grey, memorandum of meeting with Wilfrid Laurier, Jan. 9, 1905, 3, reel 2, vol. 1, file 4, Grey Fonds, LAC.

44. Dutil and MacKenzie, *Canada 1911*, 9.

45. Dominion of Canada, House of Commons, Official Report of Debates, 11th Parl., 3rd Sess., Jan. 26, 1911, 99:2501.

46. Dominion of Canada, House of Commons, Official Report of Debates, 11th Parl., 3rd Sess., Feb. 16, 1911, 99:3744–45.

47. Dominion of Canada, House of Commons, Official Report of Debates, 11th Parl., 3rd Sess., July 28, 1911, 102:10535; Gordon Stewart, *The American Response to Canada since 1776* (East Lansing: Michigan State University Press, 1992), 113–16; Canadian National League, *Reciprocity with the United States* (Toronto: Canadian National League, 1911), 16, 2, reel C4197, vol. 2, Borden Fonds, LAC; 46 Cong. Rec. H2457 (daily ed. Feb. 13, 1911); 46 Cong. Rec. H2520 (daily ed. Feb. 14, 1911).

48. "Canada's Verdict," editorial, *Chicago Tribune*, Sept. 23, 1911, 8; "The 'Annexation' Bogey in Ottawa and London," editorial, *New York Herald*, Feb. 23, 1911, 8; "Annexation the End; Reciprocity Is Merely the Means," editorial, *Montreal Star*, Sept. 16, 1911, 30; James Bryce to Lord Grey, Aug. 27, 1911, reel 2, vol. 11, file 470, Grey Fonds, LAC.

49. Dominion of Canada, House of Commons, Official Report of Debates, 11th Parl., 3rd Sess., Feb. 21, 1911, 100:3993.

50. W. S. Fielding, to the electors of the counties of Shelburne and Queen's, Aug. 17, 1911, 4–5, vol. 518, folder 62, WSF, NSA; Dutil and MacKenzie, *Canada 1911*, 126, 161–63; "Reciprocity Up to People," *Chicago Tribune*, July 30, 1911, A1.

51. Ellis, *Reciprocity, 1911*, 182; Rudyard Kipling, "The White Man's Burden," in *Rudyard Kipling's Verse, Inclusive Edition* (Garden City: Doubleday, Page, 1922), 371; "It Is Her Own Soul That Canada Risks To-Day," *Montreal Star* (Sept. 7, 1911), 1.

52. Dutil and MacKenzie, *Canada 1911*, 251–53; "Reciprocity Is Repudiated," *Montreal Star*, Sept. 22, 1911, 1.

53. Michael Hart, "The Road to Free Trade," in *Free Trade: Risks and Rewards*, ed. L. Ian MacDonald (Montreal and Kingston: McGill-Queen's University Press, 2000), 11.

54. Ellis, *Reciprocity 1911*, ix; Lewis Gould, *Four Hats in the Ring: The 1912 Election and the Birth of Modern American Politics* (Lawrence: University Press of Kansas, 2008), 3, 8–12, 80–81, 177–78.

55. US Tariff Commission, *Reciprocity with Canada*, 39–40; Ellis, *Reciprocity, 1911*, 191–92; Nathan Reich, "National Problems of Canada: The Pulp and Paper Industry in Canada," McGill University Economic Studies no. 7 (Toronto: Macmillan Company of Canada, for the Department of Economics and Political Science, McGill University, Montreal, 1926), 64–65.

56. A. Scott Berg, *Wilson* (New York: Putnam, 2013), 292–99; John Milton Cooper, *The Warrior and the Priest: Woodrow Wilson and Theodore Roosevelt* (Cambridge, MA: Belknap Press of Harvard University Press, 1983), 232–33; Ajay Mehrotra, *Making the Modern American Fiscal State: Law, Politics, and the Rise of Progressive Taxation, 1877–1929* (New York: Cambridge University Press, 2013), chap. 5; "What Free Paper Means," editorial, *Editor & Publisher* 12, no. 43 (Apr. 12, 1913): 10; "Norris' Fight Ended," *Editor & Publisher* 13, no. 17 (Oct. 11, 1913): 319.

57. E. B. Biggar, "More Paper Plants," *Toronto Globe*, Jan. 2, 1913, 13.

58. Samuel P. Hays, *Conservation and the Gospel of Efficiency: The Progressive Conservation Movement, 1890–1920* (Cambridge, MA: Harvard University Press, 1959); Mark Kuhlberg, *One Hundred Rings and Counting: Forestry Education and Forestry in Toronto and Canada, 1907–2007* (Toronto: University of Toronto Press, 2009).

59. Gifford Pinchot, *A Primer of Forestry, Part II—Practical Forestry* (Washington, DC: Government Printing Office, 1905), 38; Gifford Pinchot, *The Fight for Conservation* (1910; repr., Seattle: University of Washington Press, 1967), 110; Gifford Pinchot, address to the School of Pedagogy, St. Louis, MO, Mar. 1908, 3, 7, box 771, folder Forestry, Pinchot Papers,

LOC; Char Miller, *Gifford Pinchot and the Making of Modern Environmentalism* (Washington, DC: Island, 2001), 4–5.

60. Nelles, *Politics of Development*, 39, 200, 182; "Pulpwood Supply," *Editor & Publisher* 5, no. 32 (Jan. 27, 1906): 9.

61. Herman Ridder, letter to ANPA members, Jan. 27, 1911, reprinted in US Congress, *Reciprocity with Canada*, 2:1224; Gifford Pinchot, "Out of Sight, Out of Mind," speech given in Altoona, PA, Sept. 24, 1921, 2, box 775, folder Forestry Talk—Out of Sight, Out of Mind, Pinchot Papers, LOC; Gifford Pinchot, quoted in Ben Mellon, "Wasted Forests and Newsprint Shortage," *Editor & Publisher* 53, no. 18 (Oct. 2, 1920): 5.

62. Ian Tyrrell, *Crisis of the Wasteful Nation: Empire and Conservation in Theodore Roosevelt's America* (Chicago: University of Chicago Press, 2015), 31–35; "Save the Tree," editorial, *Editor & Publisher* 57, no. 47 (Apr. 18, 1925): 46.

63. Quoted in "Views of Publishers," *Editor & Publisher* 7, no. 14 (Sept. 28, 1907): 1–2; "Mr. Roosevelt and Pulp," editorial, *New York Sun*, Mar. 26, 1908, 6.

Chapter 3 · The Continental Newsprint Market and the Perils of Dependency

1. Earle Clapp and Charles Boyce, *How the United States Can Meet Its Present and Future Pulpwood Requirements*, US Department of Agriculture, bulletin no. 1241 (Washington, DC: US Department of Agriculture, 1924), 82.

2. William Pape, quoted in "Congress Unlikely to Put New Duty on Paper Imports from Europe," *Editor & Publisher* 53, no. 37 (Feb. 12, 1921): 7.

3. "Reforestation," editorial, *Editor & Publisher* 53, no. 9 (July 31, 1920): 70.

4. "Newfoundland Pulp," *Editor & Publisher* 7, no. 30 (Jan. 18, 1908): 1–2; W. J. Reader, *Bowater: A History* (Cambridge: Cambridge University Press, 1981), 27–28.

5. Raphael Zon, *The Forest Resources of the World* (Washington, DC: Government Printing Office, 1910), 32–33; James Hiller, "The Origins of the Pulp and Paper Industry in Newfoundland," *Acadiensis: Journal of the History of the Atlantic Region* 11, no. 2 (Spring 1982): 42–68.

6. "Plan Activities in Canada," *Editor & Publisher* 11, no. 7 (Aug. 12, 1911): 1; "Another Invasion of Canada," *Editor & Publisher* 12, no. 13 (Sept. 14, 1912): 1; Thomas Heinrich, "Product Diversification in the U.S. Pulp and Paper Industry: The Case of International Paper, 1898–1941," *Business History Review* 75, no. 3 (Autumn 2001): 475–76; "Less News Print Paper," *Editor & Publisher* 13, no. 29 (Jan. 3, 1914): 553.

7. "No U.S. Paper Famine," *Editor & Publisher* 14, no. 8 (Aug. 8, 1914): 155.

8. "Paper Reserve Ample," *Editor & Publisher* 14, no. 9 (Aug. 15, 1914): 171; "Paper Conservation," *Editor & Publisher* 14, no. 11 (Aug. 29, 1914): 201–2.

9. Royal Kellogg, *Newsprint Paper in North America* (New York: Newsprint Service Bureau, 1948), 50; "Government Planning to Help Paper Trades," *Editor & Publisher* 48, no. 39 (Mar. 4, 1916): 1174; G. W. Harris, "Paper Will Not Be Cheaper, Says P. T. Dodge," *Editor & Publisher* 48, no. 52 (June 3, 1916): 1643.

10. Peter Fearon, *War, Prosperity and Depression: The U.S. Economy 1917–1945* (Lawrence: University Press of Kansas, 1987), 9–11; US Federal Trade Commission, *Report of the Federal Trade Commission on the News-Print Paper Industry* (Washington, DC: Government Printing Office, 1917), 122.

11. Robert Barrett, "How to Offset the High Cost of Paper," *Editor & Publisher* 49, no. 7 (July 29, 1916): 6.

12. "Washington Papers Cutting Down," *Editor & Publisher* 49, no. 9 (Aug. 12, 1916): 7; "New York City Papers Decrease Size to Help Conserve News Print Supply," *Editor & Publisher* 49, no. 8 (Aug. 5, 1916): 3, 22; "Co-operation of Publishers for the Common Good," editorial, *Editor & Publisher* 49, no. 9 (Aug. 12, 1916): 20; Jason Rogers, "Newspaper Making," *Editor & Publisher* 49, no. 28 (Dec. 23, 1916): 20; Michael Stamm, "The Space for News: Ether and Paper in the Business of Media," *Media History* 21, no. 1 (2015): 55–73.

13. "Department of Justice Not to Be Hampered," *Editor & Publisher* 49, no. 38 (Mar. 3, 1917): Special Newsprint Supplement, 1–3; "Philip T. Dodge Explains the Attitude of the Manufacturers," *Editor & Publisher* 49, no. 8 (Aug. 5, 1916): News Print Conference Supplement, 4–5.

14. E. O. Merchant, "The Government and the News-Print Manufacturers," *Quarterly Journal of Economics* 32, no. 2 (Feb. 1918): 238–56; "News Print Price Set at $2.50 at the Mill," *Editor & Publisher* 49, no. 39 (Mar. 10, 1917): 9, 10, 28; "A Victory for the Square Deal," editorial, *Editor & Publisher* 49, no. 39 (Mar. 10, 1917): 20.

15. "Seven News Print Manufacturers Indicted," *Editor & Publisher* 49, no. 44 (Apr. 14, 1917): 7, 44; untitled editorial, *Editor & Publisher* 49, no. 46 (Apr. 28, 1917): pt. 1, 28; "Strengthening the Trade Commission," editorial, *Editor & Publisher* 49, no. 46 (Apr. 28, 1917): pt. 1, 28; Jeff Nichols, "Propaganda, Chicago Newspapers, and the Political Economy of Newsprint during the First World War," *Journalism History* 43, no. 1 (Spring 2017): 21–31.

16. US Federal Trade Commission, *Report of the Federal Trade Commission*, 11–13, 81, 127–28; "News Print Manufacturers Pay Fines and Agree to Federal Control of Prices," *Editor & Publisher* 50, no. 25 (Dec. 1, 1917): 5–9; Merchant, "Government and the News-Print Manufacturers," 249–50; "Commission Names $3.10 per Hundred Price for News Print in Car Lots during War," *Editor & Publisher* 51, no. 2 (June 22, 1918): 5–6; "Publishers Ask New Hearing on Cost of News Print," *Editor & Publisher* 51, no. 27 (Dec. 14, 1918): 11.

17. Robert D. Cuff, *The War Industries Board: Business-Government Relations during World War I* (Baltimore: Johns Hopkins University Press, 1973); "Newspapers of Principal American Cities Are Meeting Requirements of War Board," *Editor & Publisher* 51, no. 12 (Aug. 31, 1918): 5; "U.S. Takes Control of News Print," *Editor & Publisher* 51, no. 21 (Nov. 2, 1918): 8; "Government Pushes for Full Control of News Print," *Editor & Publisher* 51, no. 22 (Nov. 9, 1918): 8.

18. "Paper Manufacturers Refuse to Sign Agreement," *Editor & Publisher* 51, no. 23 (Nov. 16, 1918): 15; "Newspapers to Remain on 'War Basis' until February 1," *Editor & Publisher* 51, no. 24 (Nov. 23, 1918): 30.

19. "Advocating Economic Madness," editorial, *Editor & Publisher* 49, no. 22 (Nov. 11, 1916): 18; "News Print Men Ready to Abide by Price Set by Federal Commission," *Editor & Publisher* 49, no. 48 (May 12, 1917): 21; "Canadian Mills Must Sell Paper for $2.50," *Editor & Publisher* 50, no. 1 (June 16, 1917): 10; "Canada Holds $2.85 News Print Price," *Editor & Publisher* 51, no. 4 (July 6, 1918): 36; "$69 per Ton the New Canadian Price," *Editor & Publisher* 51, no. 16 (Sept. 28, 1918): 42.

20. E. H. Macklin to R. A. Pringle, Nov. 4, 1918, 1–2, box 4, folder Newsprint Commission—Newsprint Inquiry Correspondence—F2, Pringle, LAC.

21. R. A. Pringle to Henry Drayton, Dec. 12, 1919, 1, box 4, folder Newsprint Commission—Newsprint Inquiry Correspondence—F3, Pringle, LAC; "U.S. Dollars Hurt Canada Newspapers," *Detroit Free Press*, Jan. 10, 1920, 2; "Western Newspapers May Have to Suspend for Lack of Newsprint; Company's Stock May Be Seized," *Ottawa Journal*, Jan. 12,

1920, 12; "Newsprint Cut Off by Canada," *Detroit Free Press*, Jan. 21, 1920, 7; "News Famine in Winnipeg," *Editor & Publisher* 52, no. 34 (Jan. 22, 1920): 9.

22. "Canadian Officials Split over Embargo—Legal Action Threatened," *Editor & Publisher* 52, no. 34 (Jan. 22, 1920): 9; Tom Traves, *The State and Enterprise: Canadian Manufacturers and the Federal Government, 1917–1931* (Toronto: University of Toronto Press, 1979), 40–45.

23. "To the Honourable, the Members of the Newsprint Committee of the Government of Canada," undated, 2–3, attachment to Vernon Knowles to Arthur Meighen, Mar. 22, 1920, reel C3217, vol. 10, Meighen Fonds, LAC.

24. Gaston Maillet to Robert Borden, June 12, 1920, 1–2, reel C3217, vol. 10, Meighen Fonds, LAC.

25. Herbert Ridout, "Lord Riddell Sees No Relief Near from World Paper Dearth," *Editor & Publisher* 53, no. 3 (June 19, 1920): 9; "The Menace to Newsprint," editorial, *Editor & Publisher* 53, no. 11 (Aug. 14, 1920): 26; "British Designs on Canadian Newsprint," editorial, *Editor & Publisher* 53, no. 12 (Aug. 21, 1920): 26; "Small Papers Already Feel Pinch of Tight Newsprint Market," *Editor & Publisher* 52, no. 12 (Aug. 21, 1919): 18, 23; J. S. Lewis, "Southwest Papers Are Threatened with Extinction," *Editor & Publisher* 52, no. 25 (Nov. 20, 1919): 5–6.

26. Frank Munsey, May 1, 1920, US Senate, 66th Cong., 2nd Sess., *Newsprint Paper Industry: Hearing before a Subcommittee of the Committee on Manufactures, Pursuant to S. Res. 164* (Washington, DC: Government Printing Office, 1920), 1:134, 130; US Senate, 66th Cong., 2nd Sess., *Newsprint Paper Industry Investigation*, report no. 662, June 2, 1920, printed in ibid., 2:4, 7.

27. Thomas Heinrich and Bob Batchelor, *Kotex, Kleenex, Huggies: Kimberly-Clark and the Consumer Revolution in American Business* (Columbus: Ohio State University Press, 2004), chap. 2.

28. D. W. Stevick, "Places Bulk of Blame on Jobber for High Newsprint Price," *Editor & Publisher* 52, no. 41 (Mar. 13, 1920): 10; "Form Canadian I.P. Co.," *Editor & Publisher* 53, no. 31 (Jan. 1, 1921): 32; Mark Kuhlberg, *In the Power of the Government: The Rise and Fall of Newsprint in Ontario, 1894–1932* (Toronto: University of Toronto Press, 2015), 196–99, 276–77.

29. Owen McGillicuddy, "The Paper and Pulpwood Situation," *North American Review* 219 (May 1924): 617.

30. "Newsprint—a Warning!," editorial, *Editor & Publisher* 61, no. 3 (June 9, 1928): 32; Barry E. C. Boothman, "High Finance / Low Strategy: Corporate Collapse in the Canadian Pulp and Paper Industry, 1919–1932," *Business History Review* 74, no. 4 (Winter 2000): 621.

31. Kuhlberg, *In the Power of the Government*, chap. 11; "Montreal Newsprint Meeting Ends without Promised Price Statement," *Editor & Publisher* 61, no. 34 (Jan. 9, 1929): 18; Charles Vining, "Newsprint Saved by Sensible Selling," *Pulp & Paper Magazine of Canada* 40 (Jan. 1939): 8–9.

32. American Newspaper Publishers Association, *Newsprint Now and in the Next Decade* (New York: American Newspaper Publishers Association, 1951), 7; M. O. Neilson, "A Design for Newsprint," *Pulp & Paper Magazine of Canada* 40 (Jan. 1939): 14; E. A. Forsey, "The Pulp and Paper Industry," *Canadian Journal of Economics and Political Science* 1, no. 3 (Aug. 1935): 501.

33. "Newsprint Price Fight Continues," *Editor & Publisher* 67, no. 24 (Oct. 27, 1934): 7; Canadian Pulp and Paper Association, *From Watershed to Watermark: The Pulp and Paper Industry of Canada* (Montreal: Canadian Pulp and Paper Association, 1950), 51; "Newsprint after 1938," editorial, *Editor & Publisher* 70, no. 46 (Nov. 13, 1937): 24.

34. John Loomis, "Who Owns the Daily Press?," *Nation* 128 (Apr. 17, 1929): 446.

35. "News Print at $35 a Ton from U.S. Forests," *Editor & Publisher* 49, no. 41 (Mar. 24, 1917): 24; "New Forest for Paper Shortage in 20 Years," *Editor & Publisher* 50, no. 9 (Aug. 11, 1917): 26; US Department of Agriculture, Office of Information, memorandum, "Newsprint Situation Summarized," July 30, 1919, 2, box 113, folder Newsprint Situation Summarized, Records of the Division of Information and Education, Series General Correspondence, 1905–41, Entry Pl-18-49, FS, NACP.

36. George H. Manning, "Alaska Pulp Resources Shown by U.S.," *Editor & Publisher* 63, no. 11 (Aug. 2, 1930): 18; "Protests Alaskan Pulp Development," *Editor & Publisher* 64, no. 12 (Aug. 8, 1931): 7, 16; George H. Manning, "Alaskan Newsprint Project Held Up," *Editor & Publisher* 65, no. 6 (June 25, 1932): 45; Stuart Chambers to Joseph Pulitzer, July 25, 1947, 1, reel 82, container 101, folder Saugerties Mill and Sheffield Mill 1946–48, Pulitzer Papers, LOC.

37. "South Is Considering Building Paper Mills," *Editor & Publisher* 49, no. 16 (Sept. 30, 1916): 12; Jack P. Oden, "Charles Holmes Herty and the Birth of the Southern Newsprint Industry, 1927–1940," *Journal of Forest History* 21, no. 2 (Apr. 1977): 76–89; William Boyd, *The Slain Wood: Papermaking and Its Environmental Consequences in the American South* (Baltimore: Johns Hopkins University Press, 2015).

38. "Foreign Newsprint Control Menaces U.S. Press Freedom Says Francis P. Garvan," *Editor & Publisher* 67, no. 5 (June 16, 1934): 12; "S.N.P.A. Finds Pine Paper Practical; Will Underwrite Cost Survey," *Editor & Publisher* 67, no. 7 (June 30, 1934): 8; "Will Start at Once on Southern Mill," *Editor & Publisher* 67, no. 12 (Aug. 4, 1934): 8; "Showdown on Newsprint Expected as Companies Hold Up Statements," *Editor & Publisher* 67, no. 25 (Nov. 3, 1934): 6; "Impasse Confronts Southern Newsprint," *Editor & Publisher* 69, no. 47 (Nov. 21, 1936): 10.

39. "Southern Newsprint Group Makes Plans for $5,000,000 Texas Plant," *Editor & Publisher* 70, no. 6 (Feb. 6, 1937): 8; "RFC Loans $3,425,000 to Southern Newsprint Mill," *Editor & Publisher* 71, no. 49 (Dec. 3, 1938): 9; "First Southern Pine Newsprint Made at Lufkin," *Editor & Publisher* 73, no. 3 (Jan. 20, 1940): 5.

40. "Capehart Offers Bill to Legalize Paper Pooling," *Editor & Publisher* 80, no. 16 (Apr. 12, 1947): 8; Robert L. Smith, June 28, 1950, US House of Representatives, 81st Cong., 2nd Sess., *Study of Monopoly Power of the Committee on the Judiciary: Hearing before the Subcommittee on Study of Monopoly Power*, serial no. 14, pt. 6A (Washington, DC: Government Printing Office, 1950), 580.

41. R. M. Fowler, "Pulp and Paper," address given at Town Hall, New York, Mar. 24, 1953, 3, box 5, folder Country File, Canada—General, Department of State, International Cooperation Administration, Office of the Deputy Director for Technical Services, Office of Industrial Resources, Industrial Specialist Division, Minerals and Processing Industries Branch, Series Country Files Related to Paper Products, 1948–56, Entry P-179D, US-FAA, NACP.

Chapter 4 · *The Local Newspaper as International Corporation*

1. Lloyd Wendt, *Chicago Tribune: The Rise of a Great American Newspaper* (Chicago: Rand McNally, 1979), 41–42, 46–50, 63–65, 222–23, 243–47, 265–67.

2. The quote appears in Joseph A. Reaves, "Press Runs Ended at Tribune Tower," *Chicago Tribune*, Sept. 19, 1982, 3; and Wendt, *Chicago Tribune*, 488.

3. Megan McKinney, *The Magnificent Medills: America's Royal Family of Journalism during a Century of Turbulent Splendor* (New York: HarperCollins, 2011); Amanda Smith,

Newspaper Titan: The Infamous Life and Monumental Times of Cissy Patterson (New York: Alfred A. Knopf, 2011).

4. Jack Alexander, "The Duke of Chicago," *Saturday Evening Post* 214, no. 3 (July 19, 1941): 10, 71.

5. "Tribune's 100th," editorial, *Editor & Publisher* 80, no. 25 (June 14, 1947): 42.

6. On the connection between newspapers and urban development in Chicago, see David Paul Nord, "The Public Community: The Urbanization of Journalism in Chicago," in *Communities of Journalism: A History of American Newspapers and the Readers* (Urbana: University of Illinois Press, 2001), 108–32.

7. Donald Miller, *City of the Century: The Epic of Chicago and the Making of America* (New York: Simon & Schuster, 1996).

8. Chicago Tribune Company, *Trees to Tribunes* (Chicago: Chicago Tribune Company, 1949), 5; Charles Leavelle, "Paper Thunders Underground to Tribune Presses," *Chicago Tribune*, Oct. 8, 1942, 3; "Chicago's Tribune," *Editor & Publisher* 1, no. 35 (Feb. 22, 1902): 14; "Chicago Tribune's New Home," *Editor & Publisher* 1, no. 45 (May 3, 1902): 3.

9. "Publishers' Census," *Editor & Publisher* 2, no. 3 (July 12, 1902): 7.

10. Walter Strong, address to the 22nd Annual Convention of the National Association of Building Owners and Managers, Montreal, Quebec, June 10, 1929, 3, 16–17, box 2, folder 3—Speeches, 1929–61, Hedges Papers, SHSW.

11. "Electrical Control Dominates Plant of Chicago Tribune," *Editor & Publisher* 54, no. 24 (Nov. 12, 1921): 13; Walter H. Wood, "The Chicago Tribune's New Home Latest in Plant Construction," *Editor & Publisher* 53, no. 1 (June 5, 1920): 36, 38.

12. The other of the three highest-circulating papers, William Randolph Hearst's *Chicago American*, sold 82% of its papers in the city and suburbs. "'ABC' Reports Summarized for the Space Buyer," *Editor & Publisher* 53, no. 7 (July 17, 1920): sec. 2, 7.

13. Advertisement, *Editor & Publisher* 53, no. 8 (July 24, 1920): cover; James O'Donnell Bennett, "Chicagoland's Shrines: A Tour of Discoveries," *Chicago Tribune*, July 27, 1926, 1; Julia Guarneri, *Newsprint Metropolis: City Papers and the Making of Modern Americans* (Chicago: University of Chicago Press, 2017), 166–67, 179–88.

14. Advertisement, *Editor & Publisher* 11, no. 28 (Jan. 6, 1912): 5.

15. "Excise Tax on Paper Recommended by Attorney General's Aide," *Editor & Publisher* 52, no. 50 (May 15, 1920): 6, 26; Chicago Tribune Company, *From Trees to Tribunes* (Chicago: Chicago Tribune Company, 1922), 52; Leavelle, "Paper Thunders Underground," 3.

16. Richard Norton Smith, *The Colonel: The Life and Legend of Robert McCormick, 1880–1955* (New York: Houghton Mifflin, 1997), 12–13, 77–82, 113–14; Jerome Edwards, *The Foreign Policy of Col. McCormick's Tribune, 1929–1941* (Reno: University of Nevada Press, 1971), 5–10.

17. Wendt, *Chicago Tribune*, 366–68, 380–81; Edwards, *Foreign Policy of Col. McCormick's Tribune*, 10–11; McKinney, *Magnificent Medills*, 97–98.

18. "The Last Straw," editorial, *Chicago Tribune*, Aug. 22, 1945, 12.

19. "Hearst's Big Paper Deal," *Editor & Publisher* 3, no. 42 (Apr. 9, 1904): 2; John Norris testimony, May 24, 1911, US Congress, Senate, 62nd Cong., 1st Sess., *Reciprocity with Canada: Hearings before the Committee on Finance of the United States Senate on H.R. 4412*, document no. 56 (Washington, DC: Government Printing Office, 1911), 2:1206.

20. Medill McCormick testimony, May 1, 1908, US Congress, House of Representatives, 60th Cong., 2nd Sess., *Pulp and Paper Investigation Hearings*, document no. 1502 (Washington, DC: Government Printing Office, 1909), 1:290–92; "Newspaper Hearing," *Editor &*

Publisher 7, no. 45 (May 9, 1908): 3; Robert McCormick, "Memoirs: XI," *Chicago Tribune*, Nov. 16, 1952, 24.

21. John Norris testimony, Apr. 25, 1909, US Congress, *Pulp and Paper Investigation Hearings*, 1:38; Royal Kellogg, *Newsprint Paper in North America* (New York: Newsprint Service Bureau, 1948), 76–77.

22. Arthur Sulzberger to Henry Luce, Aug. 5, 1960, 1, box 219, folder 14—Newsprint, 1960–68, Sulzberger Papers, NYPL; "Newsprint Strategy," attachment to D. K. Fletcher and H. J. Wilson to J. C. Goodale and W. E. Mattson, Nov. 14, 1977, 1, box 178, folder 5—Newsprint, 1966–96, NYT, NYPL; "N.Y. Times Group Visits Canadian Mill," *Editor & Publisher* 61, no. 18 (Sept. 22, 1928): 22; Kellogg, *Newsprint Paper in North America*, 76–77; Mark Kuhlberg, *In the Power of the Government: The Rise and Fall of Newsprint in Ontario, 1894–1932* (Toronto: University of Toronto Press, 2015), 238–47; Thomas Heinrich and Bob Batchelor, *Kotex, Kleenex, Huggies: Kimberly-Clark and the Consumer Revolution in American Business* (Columbus: Ohio State University Press, 2004), 71–73.

23. Wendt, *Chicago Tribune*, 387.

24. Carl Wiegman, *Trees to News: A Chronicle of the Ontario Paper Company's Origin and Development* (Toronto: McClelland & Stewart, 1953), 15; Robert McCormick, speech at Ontario Paper Company Twenty-Fifth Anniversary Dinner, Sept. 11, 1938, 9, box 64, folder OPC—Anniversaries 1938–53, RRMB, Tribune Archives; Eugene Griffin, "Tribune Mill Marks 50 Years of Paper Making," *Chicago Tribune*, July 28, 1963, 3; Wendt, *Chicago Tribune*, 387–88; memorandum with respect to tax matters, Nov. 16, 1932, 1, vol. 25354 (acquisition no. 2006-00392-5, box 65), folder Amalgamation—Annexation Town of Thorold with Thorold Township, pt. 3, QOPC, LAC.

25. Quebec and Ontario Paper Company, *Q&O: Our Story* (St. Catharines: Quebec and Ontario Paper Company, 1988), 9.

26. Wiegman, *Trees to News*, 352; memorandum of agreement between the Ontario Paper Company Ltd. and the Tribune Company, Dec. 4, 1915, vol. 25793 (acquisition no. 2006-00392-5, box 504), folder Historical Documents: Organization and Construction, The Ontario Paper Company, pt. 1, QOPC, LAC; Arthur Schmon, quoted in transcript of proceedings, folder Ontario Paper Company Twenty-Fifth Anniversary Dinner, Sept. 11, 1938, 6, vol. 25798 (acquisition no. 2006-00392-5, box 509), folder Ontario Paper Company Twenty-Fifth Anniversary Dinner 1937, ibid.

27. Greg Grandin, *Fordlandia: The Rise and Fall of Henry Ford's Forgotten Jungle City* (New York: Metropolitan Books, 2009), 106.

28. Program, "Chicago Tribune 29th Annual Spring Luncheon," Apr. 24, 1951, 2, vol. 25180, QOPC, LAC; Robert McCormick, "Memoirs: XVII," *Chicago Tribune*, Feb. 15, 1953, 20; Arthur Schmon to Robert McCormick, Sept. 13, 1935, 1, vol. 25292 (acquisition no. 2006-00392-5, box 3), folder Land Purchase Negotiations Manicouagan / Baie Comeau, pt. 16, QOPC, LAC; I. D. Bird, "Woodlands," May 27, 1974, 1, attachment to notes for Mr. Schmon for Investment Bankers Meeting, June 4, 1974, vol. 25737 (acquisition no. 2006-00392-5, box 448), folder Investment Bankers (A-O General), ibid.

29. Questionnaire, "Heron Bay," Apr. 16, 1962, vol. 25793 (acquisition no. 2006-00392-5, box 504), folder Ontario Paper Company Historical and Background Material, pt. 1, QOPC, LAC; "The Heron Bay Barking Plant," *Pulp & Paper Magazine of Canada* 40 (Aug. 1939): 537–39.

30. Questionnaire, "Franquelin," Apr. 16, 1962, 1, vol. 25793 (acquisition no. 2006-00392-5, box 504), folder Ontario Paper Company Historical and Background Material, pt.

1, QOPC, LAC; questionnaire, "Shelter Bay," 1, Apr. 16, 1962, vol. 25793 (acquisition no. 2006-00392-5, box 504), folder Ontario Paper Company Historical and Background Material, pt. 1, ibid.; M. C. Martin to S. E. Thomason, Feb. 13, 1923, 2, vol. 25291 (acquisition no. 2006-00392-5, box 2), folder Land Purchase Negotiations Manicouagan/Baie Comeau, pt. 1, ibid.; "Memorandum Relating to Ontario Paper Company Limited, for Use in Connection with Problems Presented with Respect to the Manicouagan Development," May 4, 1936, 4, vol. 25307 (acquisition no. 2006-00392-5, box 18), folder Government, pt. 1, ibid.; Glyn Osler to Arthur Schmon, Dec. 27, 1937, 4, vol. 25308 (acquisition no. 2006-00392-5, box 19), folder Government, pt. 3, ibid.

31. T. E. Siegerman to M. C. Martin, 1, Feb. 25, 1935, vol. 25307 (acquisition no. 2006-00392-5, box 18), folder Government, pt. 1, QOPC, LAC; Arthur Schmon to Robert McCormick, Sept. 13, 1935, 1–2, vol. 25292 (acquisition no. 2006-00392-5, box 3), folder Land Purchase Negotiations Manicouagan/Baie Comeau, pt. 16, ibid.; Wiegman, *Trees to News*, 115.

32. Honoré Mercier, *Les forêts et les forces hydrauliques de la province de Québec* (Québec: s.n., 1923), 33, 43, 53.

33. Bernard Vigod, *Quebec before Duplessis: The Political Career of Louis-Alexandre Taschereau* (Montreal and Kingston: McGill-Queen's University Press, 1986), 11–16, 119–20; Herbert Quinn, *The Union Nationale: Quebec Nationalism from Duplessis to Lévesque*, 2nd ed. (Toronto: University of Toronto Press, 1979), 31–32.

34. "A. A. Schmon of Tribune Company Dies," *Chicago Tribune*, Mar. 19, 1964, 3; Eleanor Page, "Mrs. Chesser Campbell Will Be Wed to Arthur A. Schmon," *Chicago Tribune*, Nov. 26, 1963, B1; "Tribune Names New Chief of Ontario Paper," *Chicago Tribune*, June 28, 1984, sec. 2, 5.

35. Robert McCormick to Warren Curtis, Feb. 6, 1924, 1, vol. 25300 (acquisition no. 2006-00392-5, box 11), folder Manicouagan and Outardes Bible—Baie Comeau Development, pt. 1, QOPC, LAC; H. G. Acres, "The Outardes Falls Power Development," undated (ca. 1937), 2, vol. 25788 (acquisition no. 2006-00392-5, box 499), folder Publicity Manicouagan, pt. 4, ibid.; "Memorandum Relating to Ontario Paper Company Limited," 3, 7–8; "The Tribune's New Power and Paper Mill Development," *Chicago Tribune*, Aug. 2, 1931, F6.

36. Louis Garnier, *Dog Sled to Airplane: A History of the St. Lawrence North Shore*, trans. Hélène Nantais and Robert Nantais (Quebec: s.n., 1949), 147–48; Pierre Frenette, *Histoire de la Côte-Nord* (Quebec City: Laval University Press, 1996), 361–65; W. J. Reader, *Bowater: A History* (Cambridge: Cambridge University Press, 1981), 27–28; "Quebec Paper Mills for Harmsworths," *Editor & Publisher* 53, no. 11 (Aug. 14, 1920): 32.

37. "Government Is Attacked for Timber Deals," *Montreal Gazette*, Mar. 15, 1924, 1, 17; Alfred Duranleau, Jan. 21, 1925, Quebec, Legislative Assembly, *Débats de l'Assemblée législative*, 16th Legislature, 2nd Sess., 1:131.

38. Arthur Schmon to Robert McCormick, Sept. 13, 1935, 2; Arthur Schmon to J. D. Gilmour, July 13, 1929, 1, vol. 25300 (acquisition no. 2006-00392-5, box 11), folder Agreement between Ontario Paper Co. and Anglo-Canadian Co. re financing, QOPC, LAC; Timber Land Holdings in Quebec of Quebec North Shore Paper Co., undated (ca. 1943), 21, vol. 25293 (acquisition no. 2006-00392-5, box 4), folder Land Purchase Negotiations Manicouagan/Baie Comeau, pt. 24, ibid.

39. Robert Parisé, *Géants de la Côte-Nord* (Quebec: Éditions Garneau, 1974), 77; Wiegman, *Trees to News*, 158–59; Arthur Schmon to E. M. Antrim, Apr. 21, 1928, 1, vol. 25291 (acquisition no. 2006-00392-5, box 2), folder Land Purchase Negotiations Manicouagan/Baie Comeau, pt. 1, QOPC, LAC.

40. E. M. Antrim to Arthur Schmon, July 9, 1929, vol. 25290 (acquisition no. 2006-00392-5, box 1), folder Wharf Contracts and Negotiations Manicouagan, pt. 5, QOPC, LAC; Arthur Schmon to E. M. Antrim, Apr. 21, 1928, 3.

41. Memorandum, "Pulpwood Movement," undated, vol. 25347 (acquisition no. 2006-00392-5, box 58), folder House Organ, pt. 8, QOPC, LAC.

42. W. G. Cates, "History of the Welland Canal Recounts Period of Industrial Development of the District," *Financial Post*, Nov. 14, 1924, 29, 31; "Gateway Towns Are Prosperous in Industries," *Financial Post*, Nov. 14, 1924, 29.

43. Al Sykes and Skip Gillham, *Pulp and Paper Fleet: A History of the Quebec and Ontario Transportation Company* (St. Catharines: Stonehouse, 1988), 1; M. Stephen Salmon, "'This Remarkable Growth': Investment in Canadian Great Lakes Shipping, 1900–1959," *Northern Mariner* 15, no. 3 (July 2005): 29–35; Harry Laird, "Chicago Tribune Again Pioneers in Ship Construction," *Chicago Tribune*, June 23, 1935, D6; "Tribune Builds Boats to Carry Paper from Mills," *Chicago Tribune*, May 20, 1928, 18; "Tribune Victor in Lake Race; N.Y. News Is 2D," *Chicago Tribune*, Apr. 16, 1932, 15.

44. See, e.g., "Steamer Brings 6,791 Tons of Tribune Paper," *Chicago Tribune*, Apr. 20, 1933, 5; "Big Newsprint Shipment on Way to Tribune," *Chicago Tribune*, Apr. 14, 1938, 29; "Boats Speed Paper to Tribune," *Chicago Tribune*, Apr. 27, 1940, 2; "First Tribune Ship of Year Arrives in City," *Chicago Tribune*, Apr. 22, 1947, 11; "First Tribune Ship of Season Bringing Paper," *Chicago Tribune*, Apr. 3, 1951, 14; and "Tribune Waits 1st Newsprint of '59 Season," *Chicago Tribune*, Apr. 8, 1959, A2.

45. F. J. Byington, "Shipping Operations: Tribune Company and Subsidiaries," June 28, 1950, 1–3, 6–7, 10, 14–18, vol. 25796 (acquisition no. 2006-00392-5, box 507), folder Quebec & Ontario Transportation Co. 1985–87, QOPC, LAC; Sykes and Gillham, *Pulp and Paper Fleet*, 25.

46. "Heron Bay Barking Plant," 540; Byington, "Shipping Operations," 2.

47. "Heron Bay Barking Plant," 540, 543, 548; Carl Wiegman, "A Day in the Life of a Lumberjack," *Chicago Tribune*, Feb. 15, 1953, F16.

48. Wiegman, *Trees to News*, 92–93; "Trees to Tribunes," *Chicago Tribune*, Feb. 3, 1929, D6; Chicago Tribune Company, *Trees to Tribunes* (Chicago: Chicago Tribune Company, 1930), 13–14; "Trees to Tribunes," *Chicago Tribune*, Apr. 13, 1930, C4.

49. Jack Alexander, "The World's Greatest Newspaper," *Saturday Evening Post* 214, no. 4 (July 26, 1941): 82, 84; Jack Alexander, "Vox Populi-II," *New Yorker* 14 (Aug. 13, 1938): 21.

50. Alexander, "World's Greatest Newspaper," 86; John Chapman, *Tell It to Sweeney: The Informal History of the New York Daily News* (Garden City, NJ: Doubleday, 1961), 16–18.

51. Alfred McClung Lee, *The Daily Newspaper in America: The Evolution of a Social Instrument* (New York: Macmillan, 1947), 274; "Announce New York Illustrated News," *Editor & Publisher* 52, no. 3 (June 19, 1919): 39.

52. Chapman, *Tell It to Sweeney*, 59, 61; Alexander, "Vox Populi-II," 19; "What Is the Lure of the Tabloid Press?," *Editor & Publisher* 57, no. 9 (July 26, 1924): 7; John W. Perry, "N.Y. News, Now 15, Holds Grip on Masses," *Editor & Publisher* 67, no. 7 (June 30, 1934): 5; advertisement, *Editor & Publisher* 59, no. 44 (Mar. 26, 1927): cover; Silas Bent, *Ballyhoo: The Voice of the Press* (New York: Horace Liveright, 1927), 198.

53. Advertisement, *Editor & Publisher* 54, no. 20 (Oct. 15, 1921): cover; Walter E. Schneider, "Fabulous Rise of N.Y. Daily News due to Capt. Patterson's Genius," *Editor & Publisher* 72, no. 25 (June 24, 1939): 5–7; Perry, "N.Y. News," 5.

54. Alexander, "Vox Populi-II," 23; "Modern Newspaper Plant Is Ready for New York Daily News," *Editor & Publisher* 53, no. 44 (Apr. 2, 1921): 11; Philip Schuyler, "N.Y. News Breaks Record at New Plant," *Editor & Publisher* 59, no. 41 (Mar. 5, 1927): 7; Chapman, *Tell It to Sweeney*, 267–68; "N.Y. Daily News Plans Brooklyn Plant," *Editor & Publisher* 58, no. 44 (Mar. 27, 1926): 4.

55. Aurora Wallace, *Media Capital: Architecture and Communications in New York City* (Urbana: University of Illinois Press, 2012), 92, 96–106; "N.Y. Daily News Plant to Cost $10,000,000," *Editor & Publisher* 61, no. 39 (Feb. 16, 1929): 9; John F. Roche and Allan Delafons, "Branch Plants Are Solving Publication Problems of Metropolitan Dailies," *Editor & Publisher* 61, no. 46 (Apr. 6, 1929): 6.

56. Wallace, *Media Capital*, 98.

57. "Chicago Newspapers Increase Price to Two Cents to Meet Economic Conditions," *Editor & Publisher* 49, no. 48 (May 12, 1917): 6; "An Impossible Business Situation," editorial, *Editor & Publisher* 49, no. 49 (May 19, 1917): 18; "Plan Conservation 'That All May Live' at Special A.N.P.A. Convention," *Editor & Publisher* 52, no. 24 (Nov. 13, 1919): pt. 2, 2; advertisement, *Editor & Publisher* 53, no. 38 (Feb. 19, 1921): cover.

58. Robert McCormick to R. A. Pringle, May 19, 1917, 1, box 13, folder Statements Filed by Companies, Pringle, LAC; Warren Curtis to R. A. Pringle, Dec. 4, 1917, box 6, folder Newsprint Commission—Newsprint Inquiry Correspondence—O, ibid.

59. Ontario Paper Company to R. A. Pringle, Nov. 2, 1918, box 6, folder Newsprint Commission—Newsprint Inquiry Correspondence—O, Pringle, LAC. The Canadian government ignored these protests and continued to solicit the Tribune Company's help in providing newsprint to other Canadian newspapers over the next year. The company complied, though in late 1919 one official reiterated that "we desire to do all in our power to aid you in every way possible, but nevertheless are able to furnish paper to the Canadian trade only at a tremendous sacrifice." M. C. Martin to R. A. Pringle, Nov. 15, 1919, box 6, folder Newsprint Commission—Newsprint Inquiry Correspondence—O, ibid.

60. Quoted in "Dodge Decries High Newsprint Prices," *Editor & Publisher* 52, no. 51 (May 22, 1920): 6.

61. "Corner Stone of Tribune's N.Y. Paper Mill Laid," *Chicago Tribune*, Nov. 3, 1923, 8; Wiegman, *Trees to News*, 103–10; J. Herbert Hodgins, "We Are Depleting Our Population and Forests Because Pulpwood Embargo Is Delayed," *MacLean's Magazine* 38, no. 4 (Feb. 15, 1925): 11; "International to Buy Tribune Paper Mill," *Editor & Publisher* 63, no. 51 (May 9, 1931): 15; "International Paper Co. Buys Tonawanda Mill," *Chicago Tribune*, June 22, 1931, 25.

62. Philip Schuyler, "Newsprint Men Financing Newspaper Deals," *Editor & Publisher* 61, no. 31 (Dec. 22, 1928): 5; "Drops Suit against International," *Editor & Publisher* 62, no. 25 (Nov. 9, 1929): 10; Warren Curtis to Robert McCormick, May 14, 1924, vol. 25300 (acquisition no. 2006-00392-5, box 11), folder Manicouagan and Outardes Bible—Baie Comeau Development, pt. 1, QOPC, LAC; John Stadler to Arthur Schmon, Jan. 27, 1931, vol. 25300 (acquisition no. 2006-00392-5, box 11), folder Manicouagan and Outardes Bible—Baie Comeau Development, pt. 4, ibid.

63. Robert McCormick to Arthur Schmon, May 24, 1929, vol. 25300 (acquisition no. 2006-00392-5, box 11), folder Layouts, QOPC, LAC; Robert McCormick to Arthur Schmon, June 4, 1929, vol. 25300 (acquisition no. 2006-00392-5, box 11), folder Layouts, ibid.; Arthur Schmon to Elbert Antrim, May 9, 1929, 1, box 1, Localisation MN//1/1.24, QNS, SHCN;

Wiegman, *Trees to News*, 125–28; "Memorandum Relating to Ontario Paper Company Limited," 11.

Chapter 5 · Robert McCormick and the Politics of Planning

1. "Government by Inquisition and for the Commissars," editorial, *Chicago Tribune*, Apr. 1, 1936, 14.

2. Kim Phillips-Fein, *Invisible Hands: The Businessmen's Crusade against the New Deal* (New York: Norton, 2010), 3–6, 13–15, 19–22; Richard Norton Smith, *The Colonel: The Life and Legend of Robert R. McCormick, 1880–1955* (Boston: Houghton Mifflin, 1997), 345–48.

3. Jason Scott Smith, *Building New Deal Liberalism: The Political Economy of Public Works, 1933–1956* (Cambridge: Cambridge University Press, 2006), 3; "Constitutional Obstructions," editorial, *Chicago Tribune*, Sept. 11, 1936, 14. On New Deal planning and its critics, see Alan Brinkley, *The End of Reform: New Deal Liberalism in Recession and War* (New York: Vintage, 1995), 229–35.

4. "Socialism in the Tennessee Valley," editorial, *Chicago Tribune*, Aug. 17, 1939, 14; "Mr. Roosevelt's Popular Front," editorial, *Chicago Tribune*, Oct. 26, 1936, 12.

5. William Leuchtenburg, "Roosevelt, Norris, and the 'Seven Little TVA's,'" *Journal of Politics* 14, no. 3 (Aug. 1952): 418–41; Charles McCarthy, "TVA and the Tennessee Valley," *Town Planning Review* 21, no. 2 (July 1950): 116–30; Thomas McCraw, *TVA and the Power Fight, 1933–1939* (Philadelphia: J. B. Lippincott, 1971); Philip Selznick, *TVA and the Grass Roots: A Study in the Sociology of Formal Organization* (1949; repr., New York: Harper, 1966), 19.

6. "To Be Licked by Their Own Mistakes?," editorial, *Chicago Tribune*, May 3, 1933, 12. The paper felt so strongly about this particular editorial that it reprinted it verbatim three years later "by request," it claimed. "To Be Licked by Their Own Mistakes?," editorial, *Chicago Tribune*, Feb. 22, 1936, 12.

7. Norris was one of what New Dealers hoped would be a series of well-planned and federally constructed communities, others being Greenbelt, Maryland, and Greendale, Wisconsin. Mostly because of political opposition, the broader project never extended to the scope to which its proponents aspired. On the New Deal's planned towns and their critics, see Daniel Rodgers, *Atlantic Crossings: Social Politics in a Progressive Age* (Cambridge, MA: Belknap Press of Harvard University Press, 1998), 454–73.

8. Margaret Crawford, *Building the Workingman's Paradise: The Design of American Company Towns* (New York: Verso, 1995), 195–99.

9. "Pillars of the New Deal," editorial, *Chicago Tribune*, Oct. 10, 1938, 12.

10. "The Rubber Yardstick," editorial, *Chicago Tribune*, Feb. 6, 1936, 12. The spelling of the word "burocratic" was purposeful. McCormick's *Tribune* used a number of strategies to shorten words it found some particular fault with, for example, printing "thoroly" instead of "thoroughly." In early 1934, the paper explained this practice in a pair of front-page stories claiming that it was promoting what it called "sane trends toward simpler spelling of the English language" by making spelling more efficient through the elimination of "superfluous letters." In this linguistic modernization project, these letters were those that could be "dropped without affecting the pronunciation of the words, without blurring their derivation, and without giving them a grotesque appearance." See James O'Donnell Bennett, "Tribune Adopts Saner Spelling of Many Words," *Chicago Tribune*, Jan. 28, 1934, 1, 4; and James O'Donnell Bennett, "Tribune Adds 18 Words to Sane Spelling List," *Chicago Tribune*, Feb. 11, 1934, 1, 10.

11. "A Power Program Made in Russia," editorial, *Chicago Tribune*, Sept. 14, 1936, 10; "The Entering Wedge of the No Deal," editorial, *Chicago Tribune*, Nov. 4, 1934, 16 (emphasis in original).

12. Bob McStay, "Carving City from Wilds," *Toronto Star Weekly*, Oct. 10, 1936, sec. 3, 6.

13. Ontario Paper Company, "Properties, Allowable Annual Cuts, and Merchantable Volumes of Timber in Ontario and Quebec," Nov. 1975, 1, vol. 25397 (acquisition no. 2006-00392-5, box 108), folder Forest Properties in Ontario & Quebec—Allowable Annual Cuts, QOPC, LAC.

14. Smith, *Colonel*, 55–61.

15. Jerome Edwards, *The Foreign Policy of Col. McCormick's Tribune, 1929–1941* (Reno: University of Nevada Press, 1971), 8–9; Crawford, *Building the Workingman's Paradise*, 155.

16. Louis P. Cain, *Sanitation Strategy for a Lakefront Metropolis: The Case of Chicago* (DeKalb: Northern Illinois University Press, 1978); *The Sanitary District of Chicago: Proceedings of the Board of Trustees, 1910* (Chicago: Chicago Sanitary District, 1910), 78–80.

17. Leonard Schlemm, memorandum in connection with plans for a townsite at Comeau Bay, P.Q., June 10, 1931, vol. 25292 (acquisition no. 2006-00392-5, box 3), folder Land Purchase Negotiations Manicouagan / Baie Comeau, pt. 2, QOPC, LAC; Arthur Schmon to Hugh Stewart, Dec. 27, 1934, 1–2, vol. 25290 (acquisition no. 2006-00392-5, box 1), folder Wharf Contracts and Negotiations Manicouagan, pt. 13, ibid.

18. Arthur Robb, "Newspaper Groups Doubled in Decade," *Editor & Publisher* 66, no. 40 (Feb. 17, 1934): 11; "Memorandum Relating to Ontario Paper Company Limited, for Use in Connection with Problems Presented with Respect to the Manicouagan Development," May 4, 1936, 8, vol. 25307 (acquisition no. 2006-00392-5, box 18), folder Government, pt. 1, QOPC, LAC; "The Status of the Ontario Paper Company, Including Its Wholly Owned Subsidiary Quebec North Shore Paper Company, as a Non-commercial Newsprint Manufacturer," June 5, 1940, 57, vol. 25808 (acquisition no. 2006-00392-5, box 519), folder Status of the Ontario Paper Company Limited 1940, ibid.

19. Arthur Schmon to Robert McCormick, Sept. 13, 1935, box 82, folder OPC—Schmon, Arthur A., Southern Pine, 1934–37, RRMB, Tribune Archives; Robert Nelson to Robert McCormick, Jan. 7, 1936, box 82, folder OPC—Schmon, Arthur A., Southern Pine, 1934–37, ibid.; Germaine M. Reed, "Realization of a Dream: Charles H. Herty and the South's First Newsprint Mill," *Forest & Conservation History* 39, no. 1 (Jan. 1995): 7–8; Arthur Schmon to Robert McCormick, Feb. 7, 1936, 2, box 53, folder 5, JMP, LFCASC; Arthur Schmon to Robert McCormick, Mar. 20, 1935, box 53, folder 4, ibid.

20. Arthur Schmon to Robert McCormick, May 18, 1929, 2–3, vol. 25300 (acquisition no. 2006-00392-5, box 11), folder Manicouagan and Outardes Bible—Baie Comeau Development, pt. 2, QOPC, LAC.

21. Joseph Patterson to Robert McCormick, Jan. 24, 1936, box 53, folder 5, JMP, LFCASC; Joseph Patterson to Robert McCormick, Mar. 5, 1935, box 53, folder 4, ibid.; Robert McCormick to Joseph Patterson, Mar. 6, 1935, box 53, folder 4, ibid.; Robert McCormick to Arthur Schmon, Mar. 31, 1934, box 53, folder 3, ibid.

22. Interview with Patricia Whitelaw, Apr. 14, 2013, transcript in author's possession.

23. Arthur Schmon to Robert McCormick, June 24, 1931, vol. 25292 (acquisition no. 2006-00392-5, box 3), folder Land Purchase Negotiations Manicouagan / Baie Comeau, pt. 3, QOPC, LAC.

24. M. C. Martin to Robert McCormick, May 7, 1938, 1, vol. 25346 (acquisition no. 2006-00392-5, box 57), folder Subsidiary Companies, pt. 2, QOPC, LAC.

25. Ibid.

26. Bernard Vigod, *Quebec before Duplessis: The Political Career of Louis-Alexandre Taschereau* (Montreal and Kingston: McGill-Queen's University Press, 1986), 182–83, 240–41; Herbert Quinn, *The Union Nationale: Quebec Nationalism from Duplessis to Lévesque*, 2nd ed. (Toronto: University of Toronto Press, 1979), 63–66.

27. Noel Dorion to Maurice Duplessis, Feb. 14, 1936, 1, reel ZC44/1, Entry Correspondence 1936, MDF, BANQ; Arthur Schmon to Robert McCormick, May 8, 1936, box 1, Localisation MN//1/1.34, QNS, SHCN.

28. Louis-Philippe Cote, "Faisons la lumière," *La Terre de Chez Nous*, July 15, 1936, 12; E. C. LaRose to Arthur Schmon, May 2, 1938, 1, vol. 25346 (acquisition no. 2006-00392-5, box 57), folder Subsidiary Companies, pt. 2, QOPC, LAC.

29. M. C. Martin to Robert McCormick, May 7, 1938, 1.

30. Robert McCormick to Arthur Schmon, Aug. 21, 1936, box 53, folder 6, JMP, LFCASC; Arthur Schmon to Robert McCormick, Oct. 10, 1936, 1, vol. 25307 (acquisition no. 2006-00392-5, box 18), folder Government, pt. 2, QOPC, LAC; Jonathan Robinson to Arthur Schmon, Jan. 13, 1937, 2, box 1, Localisation MN//1/1.34, QNS, SHCN.

31. Arthur Schmon to Robert McCormick, Apr. 15, 1937, 3–4, box 1, Localisation MN//1/1.34, QNS, SHCN; Arthur Schmon to Robert McCormick, Jan. 28, 1937, 1, box 1, Localisation MN//1/1.34, ibid.

32. Quinn, *Union Nationale*, 76, 81; M. C. Martin to Robert McCormick, May 7, 1938, 2–3.

33. M. C. Martin to Robert McCormick, May 7, 1938, 3; Arthur Schmon to D. M. Deininger, May 26, 1938, vol. 25346 (acquisition no. 2006-00392-5, box 57), folder Subsidiary Companies, pt. 2, QOPC, LAC; Robert McCormick to Joseph Patterson, Aug. 6, 1938, box 53, folder 9, JMP, LFCASC.

34. Arthur Schmon to Robert McCormick, May 28, 1938, vol. 25346 (acquisition no. 2006-00392-5, box 57), folder Subsidiary Companies, pt. 2, QOPC, LAC.

35. John Dickinson and Brian Young, *A Short History of Quebec*, 3rd. ed. (Montreal and Kingston: McGill-Queen's University Press, 2003), 289–96; Conrad Black, *Duplessis* (Toronto: McClelland & Stewart, 1977), 583, 616–20, 623.

36. "Memorandum Relating to Ontario Paper Company Limited," 12–14.

37. James Scott, *Seeing Like a State: How Certain Schemes to Improve the Human Condition Have Failed* (New Haven, CT: Yale University Press, 1998), 4; James Scott, "High Modernist Social Engineering: The Case of the Tennessee Valley Authority," in *Experiencing the State*, ed. Lloyd Rudolph and John Kurt Jacobsen (New York: Oxford University Press, 2006), 21.

38. Greg Grandin, *Fordlandia: The Rise and Fall of Henry Ford's Forgotten Jungle City* (New York: Metropolitan Books, 2009), 22–28, 106, 156–57, 220–30, 273–74, 298–99, 353.

39. Lloyd Wendt, *Chicago Tribune: The Rise of a Great American Newspaper* (Chicago: Rand McNally, 1979), 421–22; Robert McCormick, "Memoirs: No. 27," *Chicago Tribune*, Sept. 13, 1953, 20; "Ford's Rubber Farm in Brazil Gets Clean Bill," *Chicago Tribune*, May 1, 1929, 26; L. Palacios Galvez, "Presto Chango! And Ford Grows Rubber in Jungle," *Chicago Tribune*, Dec. 15, 1929, 24; M. M. Corpening, "Clearing Carved in Jungle Begins Rubber Venture," *Chicago Tribune*, Apr. 10, 1939, 7; Anthony Patric, "Ford's Men Whip Amazon Jungle; Turn Out Rubber," *Chicago Tribune*, June 8, 1941, 19.

40. Lewis Mumford, *The Culture of Cities* (1938; repr., New York: Harcourt Brace Jovanovich, 1970), 11, 183, 272, 392. On company towns in a global perspective, see Oliver J.

Dinius and Angela Vergara, eds., *Company Towns in the Americas: Landscape, Power, and Working-Class Communities* (Athens: University of Georgia Press, 2011); and John S. Garner, ed., *The Company Town: Architecture and Society in the Early Industrial Age* (New York: Oxford University Press, 1992).

41. Stanley Buder, *Pullman: An Experiment in Industrial Order and Community Planning* (New York: Oxford University Press, 1967), 72, 92–96, 137, 168, 191–94, 228–29.

42. Hardy Green, *The Company Town: The Industrial Edens and Satanic Mills That Shaped the American Economy* (New York: Basic Books, 2010), 118–20, 128, 130–32.

43. Anne Mosher, *Capital's Utopia: Vandergrift, Pennsylvania, 1855–1916* (Baltimore: Johns Hopkins University Press, 2004); Margot Opdycke Lamme and Lisa Mullikin Parcell, "Promoting Hershey: The Chocolate Bar, the Chocolate Town, the Chocolate King," *Journalism History* 38, no. 4 (Winter 2013): 198–208; D. M. Deininger to Arthur Schmon, Feb. 19, 1936, box 1, Localisation MN//1/1.30, QNS, SHCN.

44. Arthur Sulzberger to W. L. G. Joerg, Feb. 15, 1929, box 191, folder Kapuskasing, Ontario, 1927–29, 1944–61 (191.16), Sulzberger Papers, NYPL; W. F. J. Sensenbrenner to Arthur Hays Sulzberger, Sept. 28, 1925, 1, box 248, folder Spruce Falls Power and Paper Company, 1922–25 (248.1), ibid.; George F. Hardy to W. F. J. Sensenbrenner, Mar. 24, 1927, box 247, folder Spruce Falls Power and Paper Company, 1927–29 (247.8), ibid.; Mark Kuhlberg, *In the Power of the Government: The Rise and Fall of Newsprint in Ontario, 1894–1932* (Toronto: University of Toronto Press, 2015), 240–47; Royal Kellogg, *Newsprint Paper in North America* (New York: Newsprint Service Bureau, 1948), 76–77.

45. Robert McCormick to Arthur Schmon, Nov. 3, 1946, vol. 25293 (acquisition no. 2006-00392-5, box 4), folder Land Purchase Negotiations Manicouagan / Baie Comeau, pt. 29, QOPC, LAC; Ontario Bureau of Municipal Affairs, *Report re Housing for 1921* (Toronto: Clarkson W. James, 1922), 7–8.

46. On the midcentury understandings of these kinds of company towns in Canada, see Ira Robinson, *New Industrial Towns on Canada's Resource Frontier*, Program of Education and Research in Planning Research Paper no. 4, Department of Geography Research Paper no. 73 (Chicago: University of Chicago, Department of Geography, 1962); and Rex A. Lucas, *Minetown, Milltown, Railtown: Life in Canadian Communities of Single Industry* (Toronto: University of Toronto Press, 1971).

47. Arthur Schmon to Robert McCormick, Jan. 10, 1936, 1–3, box 70, folder Ontario Paper Company, Schmon, Arthur A., 1936 Jan.–May, RRMB, Tribune Archives.

48. Robert McCormick to Arthur Schmon, Jan. 15, 1936, box 70, folder Ontario Paper Company, Schmon, Arthur A., 1936 Jan.–May, RRMB, Tribune Archives; Arthur Schmon to Robert McCormick, Jan. 20, 1936, 2, box 1, Localisation MN//1/1.30, QNS, SHCN.

49. "Incorporation of Baie Comeau Bill Strongly Opposed," *Quebec Chronicle-Telegraph,* Apr. 6, 1937, 1; Quebec, Legislative Assembly, May 7, 1937, *Débats de l'Assemblée législative,* 20th Legislature, 2nd Sess., 3:1050–52.

50. Quebec, Legislative Assembly, *An Act to Incorporate the Town of Baie Comeau,* May 20, 1937, 20th Legislature, 2nd Sess., chap. 120, 1 Geo. VI, 559–65.

51. C. A. Sankey, memorandum, "Total Labour Involved in Baie Comeau Project," Dec. 31, 1937, vol. 25789 (acquisition no. 2006-00392-5, box 500), folder Publicity Manicouagan, pt. 7, QOPC, LAC; promotional brochure, "Baie Comeau: A Modern Model City," undated, 1, 3, vol. 25302 (acquisition no. 2006-00392-5, box 13), folder Houses, pt. 4, ibid.

52. J. A. Marier, memorandum re housing requirements, Nov. 14, 1938, 2, vol. 25302 (acquisition no. 2006-00392-5, box 13), folder Houses, pt. 4, QOPC, LAC.

53. E. B. McGraw to Arthur Schmon, Mar. 30, 1944, vol. 25303 (acquisition no. 2006-00392-5, box 14), folder Houses, pt. 9, QOPC, LAC.

54. Arthur Schmon to Robert McCormick, Dec. 8, 1938, 2–3, vol. 25302 (acquisition no. 2006-00392-5, box 13), folder Houses, pt. 4, QOPC, LAC.

55. Arthur Schmon to Robert McCormick, Nov. 14, 1940, 1, vol. 25303 (acquisition no. 2006-00392-5, box 14), folder Houses, pt. 9, QOPC, LAC; R. E. Hayes, "Report on Baie Comeau Housing," undated, 2–3, vol. 25303 (acquisition no. 2006-00392-5, box 14), folder Houses, pt. 6, ibid.

56. H. A. Sewell to Maurice Duplessis, May 3, 1938, 1, 3, vol. 25308 (acquisition no. 2006-00392-5, box 19), folder Municipal Government, pt. 3, QOPC, LAC; H. A. Sewell to Arthur Schmon, May 5, 1938, 1, vol. 25308 (acquisition no. 2006-00392-5, box 19), folder Municipal Government, pt. 3, ibid.; Robert McCormick to Arthur Schmon, May 5, 1938, vol. 25308 (acquisition no. 2006-00392-5, box 19), folder Municipal Government, pt. 3, ibid.

57. H. A. Sewell to G. J. Lane, Jan. 14, 1939, vol. 25302 (acquisition no. 2006-00392-5, box 13), folder Camps and Company Buildings, pt. 3, QOPC, LAC; G. J. Lane to Arthur Schmon, Jan. 16, 1939, vol. 25302 (acquisition no. 2006-00392-5, box 13), folder Camps and Company Buildings, pt. 3, ibid.

58. "Paper Mill Opens June 11," *New York Times*, May 6, 1938, 41; "Premier Says Comeau Mill Will Be One of a Kind," *Montreal Gazette*, June 13, 1938, 1; "Baie Comeau Development," editorial, *Montreal Gazette*, June 13, 1938, 8.

59. L. Deschenes, "Baie Comeau—Baie des Miracles," *Le Progrès du Golfe*, June 24, 1938, 7; John R. Sturdy, "McCormick Fuels Hope," *Montreal Gazette*, June 13, 1938, 1.

60. J. B. Cote, "Reportage à la Baie Comeau," *L'Echo du Bas St-Laurent*, Aug. 13, 1937, 1; "A Lesson from Comeau Bay," editorial, *St. Maurice Valley Chronicle*, June 16, 1938, 3.

61. "U.S.-Owned Mill Argues against Being Prorated," *Globe and Mail*, May 8, 1940, 13; Lucile Keys to General J. L. Schley, Mar. 15, 1944, 1–2, box 649, folder Newsprint, Records of the Department of Transportation, Records Relating to Newsprint Shipments, 1941–45, Entry I-7-37, OIAA, NACP; Arthur Schmon to Robert McCormick, Sept. 26, 1947, 6, vol. 25310 (acquisition no. 2006-00392-5, box 21), folder Paper Handling and Storage, pt. 8, QOPC, LAC; "Canadian Trees Meet Adventure as NEWSprint," *New York Daily News*, Aug. 13, 1948, 7.

62. Timber Land Holdings in Quebec of Quebec North Shore Paper Co., undated (ca. 1943), 1–2, vol. 25293 (acquisition no. 2006-00392-5, box 4), folder Land Purchase Negotiations Manicouagan / Baie Comeau, pt. 24, QOPC, LAC.

63. M. C. Martin and Arthur Schmon, memorandum, Sept. 22, 1933, 1, vol. 25292 (acquisition no. 2006-00392-5, box 3), folder Land Purchase Negotiations Manicouagan / Baie Comeau, pt. 10, QOPC, LAC; Arthur Schmon to R. A. McInnis, Aug. 2, 1935, vol. 25290 (acquisition no. 2006-00392-5, box 1), folder Wharf Contracts and Negotiations Manicouagan, pt. 11, ibid.; Arthur Schmon to E. M. Little, Apr. 22, 1944, vol. 25294 (acquisition no. 2006-00392-5, box 5), folder Land Purchase Negotiations Manicouagan / Baie Comeau, pt. 32, ibid.; Arthur Schmon to Robert McCormick, Dec. 21, 1944, 1, vol. 25296 (acquisition no. 2006-00392-5, box 7), folder Land Purchase Negotiations Anglo Pulp and Paper Mills, pt. 6, ibid.; T. F. Flahiff to Onésime Gagnon, Oct. 23, 1951, vol. 25294 (acquisition no. 2006-00392-5, box 5), folder Wharf Contracts and Negotiations Manicouagan, pt. 43, ibid.; T. E. Siegerman to Arthur Schmon, Dec. 13, 1945, vol. 25296 (acquisition no. 2006-00392-5, box 7), folder Land Purchase Negotiations Anglo Pulp and Paper Mills, pt. 1, ibid.; Arthur Schmon to Robert McCormick, Sept. 27, 1945, vol. 25296 (acquisition no. 2006-00392-5,

box 7), folder Land Purchase Negotiations Anglo Pulp and Paper Mills, pt. 1, ibid.; "The Ontario Paper Company, Limited and Quebec North Shore Paper Company: The Company's Properties and Production Activities in Relation to the Papers' Newsprint Requirements," Aug. 13–17, 1948, 6, vol. 25316 (acquisition no 2006-00392-5, box 27), folder Baie Comeau 10th Anniversary, pt. 9, ibid.; memorandum, Woodlands, June 7, 1950, vol. 25294 (acquisition no. 2006-00392-5, box 5), folder Land Purchase Negotiations Manicouagan / Baie Comeau, pt. 35, ibid.

64. Eugene Griffin, "Col. McCormick Is Cited for His Help to Canada," *Chicago Tribune*, Sept. 25, 1950, 2.

65. Robert McCormick to Arthur Schmon, May 9, 1938, vol. 25317 (acquisition no. 2006-00392-5, box 28), folder Fences, Plaques, and Opening Ceremonies, pt. 3, QOPC, LAC; Robert McCormick to Arthur Schmon, May 12, 1938, vol. 25317 (acquisition no. 2006-00392-5, box 28), folder Fences, Plaques, and Opening Ceremonies, pt. 3, ibid.; Robert McCormick to Arthur Schmon, June 3, 1938, vol. 25317 (acquisition no. 2006-00392-5, box 28), folder Fences, Plaques, and Opening Ceremonies, pt. 8, ibid.

66. Arthur Schmon to Robert McCormick, Mar. 23, 1940, vol. 25308 (acquisition no. 2006-00392-5, box 19), folder Municipal Government, pt. 2, QOPC, LAC.

67. Arthur Schmon to J. E. Vallillee, Dec. 20, 1948, 1–2, vol. 25308 (acquisition no. 2006-00392-5, box 19), folder Municipal Government, pt. 8, QOPC, LAC; T. B. Fraser to Arthur Schmon, Nov. 22, 1954, 2, vol. 25302 (acquisition no. 2006-00392-5, box 13), folder Townsite, pt. 4, ibid.

68. Email correspondence with Doreen Scott, Oct. 2, 2013, in author's possession.

69. Corolyn Cox, "Memo—Baie Comeau—Col. Robert McCormick and Canada," 2–4, box 64, folder OPC—Baie Comeau—Cox Article, 1944–45, RRMB, Tribune Archives.

70. Ibid., 4–6, 8.

71. Ibid., 19–22.

72. Corolyn Cox to Robert R. McCormick, Feb. 1, 1945, box 64, folder OPC—Baie Comeau—Cox Article, 1944–45, RRMB, Tribune Archives; Robert McCormick to Corolyn Cox, Feb. 5, 1945, box 64, folder OPC—Baie Comeau—Cox Article, 1944–45, ibid.

Chapter 6 · *Work and Culture along the Newsprint Supply Chain*

1. "New Canadian Town Hails Its First Wedding," *Chicago Tribune*, Aug. 6, 1937, 4.

2. David Massell, *Quebec Hydropolitics: The Peribonka Concessions of the Second World War* (Montreal and Kingston: McGill-Queen's University Press, 2011), 6, 20, 24–25; Louis Garnier, *Dog Sled to Airplane: A History of the St. Lawrence North Shore*, trans. Hélène Nantais and Robert Nantais (Quebec: s.n., 1949), 228.

3. Arthur Schmon to James Hanrahan, May 3, 1929, vol. 25300 (acquisition no. 2006-00392-5, box 11), folder Colonization and Settlers Manicouagan / Baie Comeau, pt. 1, QOPC, LAC; Arthur Schmon to George Boisvert, Nov. 29, 1929, vol. 25300 (acquisition no. 2006-00392-5, box 11), folder Colonization and Settlers Manicouagan / Baie Comeau, pt. 1, ibid.; Arthur Schmon to James Hanrahan, June 20, 1929, 1, vol. 25300 (acquisition no. 2006-00392-5, box 11), folder Colonization and Settlers Manicouagan / Baie Comeau, pt. 1, ibid.; Arthur Schmon to George Boisvert, Feb. 27, 1930, 1, vol. 25300 (acquisition no. 2006-00392-5, box 11), folder Colonization and Settlers Manicouagan / Baie Comeau, pt. 1, ibid.

4. Chicago Tribune Company, *Trees to Tribunes* (Chicago: Chicago Tribune Company, 1930), 21–22; Arthur Schmon, "Our Paper Mills," speech at the Continental Hotel, Chicago, IL, Oct. 9, 1946, 10, vol. 25799 (acquisition no. 2006-00392-5, box 510), folder Ad-

dresses by Arthur A. Schmon, QOPC, LAC; Arthur Schmon to Harvey Smith, Dec. 16, 1958, 2, vol. 25353 (acquisition no. 2006-00392-5, box 64), folder Harvey Smith's Book, pt. 7, ibid.

5. John Menaugh, "How to Live in the Woods," *Chicago Tribune*, Dec. 9, 1951, C5, C10; Robert McCormick, foreword to *I Live in the Woods*, by Paul Provencher (Fredericton, NB: Brunswick Press, 1953).

6. James Scott, *Seeing Like a State: How Certain Schemes to Improve the Human Condition Have Failed* (New Haven, CT: Yale University Press, 1998), 311, 319.

7. J. H. Cunningham, "Return of Paper Making to U.S. Urged," *Editor & Publisher* 67, no. 30 (Dec. 8, 1934): 9.

8. Canadian Pulp and Paper Association, *The Pulpwood Harvest* (Montreal: Canadian Pulp and Paper Association, 1952), 14; Quebec North Shore Paper Company, report, "25th Anniversary, 1937–1962," 2, vol. 25317 (acquisition no. 2006-00392-5, box 28), folder Baie Comeau 25th Anniversary, pt. 4, QOPC, LAC; "50,900,000 Daily," editorial, *Editor & Publisher* 80, no. 9 (Feb. 22, 1947): 42.

9. Harvey Smith to "Mother and Father," Apr. 13, 1924, vol. 25353 (acquisition no. 2006-00392-5, box 64), folder Harvey Smith's Book, pt. 1, QOPC, LAC; Chicago Tribune Company, *Trees to Tribunes*, 25.

10. John Menaugh, "Winter Woodcraft in Canadian Wilds," *Chicago Tribune*, Dec. 30, 1951, G6–7.

11. Ian Radforth, *Bushworkers and Bosses: Logging in Northern Ontario, 1900–1980* (Toronto: University of Toronto Press, 1987), 3; Carl Wiegman, "A Day in the Life of a Lumberjack," *Chicago Tribune*, Feb. 15, 1953, F6, F16; James P. Hull, "The Second Industrial Revolution and the Staples Frontier in Canada: Rethinking Knowledge and History," *Scientia Canadensis: Canadian Journal of the History of Science, Technology and Medicine* 18, no. 1 (1994): 32–33.

12. Arthur Schmon to Harvey Smith, Dec. 16, 1958, 2.

13. Napoléon-Alexandre Labrie, "The Forest and the Social Problem in the Saguenay County," *Institut Social Populaire*, no. 430 (Jan. 1950): 15, 20.

14. Robert McCormick, "Memoirs: XVII," *Chicago Tribune*, Feb. 15, 1953, 20; Adelaide Leitch, "Along the St. Lawrence's North Shore," *New York Times*, June 10, 1956, 355.

15. John Menaugh, "Battling the Perils of the Canadian Wilderness," *Chicago Tribune*, Mar. 16, 1952, C14–15.

16. F. N. Greenleaf to Bess Vydra, Nov. 2, 1953, vol. 25345 (acquisition no. 2006-00392-5, box 56), folder Special Correspondence Colonel McCormick, pt. 21, QOPC, LAC; Gordon Godwin to F. N. Greenleaf, Nov. 10, 1953, vol. 25345 (acquisition no. 2006-00392-5, box 56), folder Special Correspondence Colonel McCormick, pt. 21, ibid.; F. N. Greenleaf to Robert McCormick, Nov. 12, 1953, vol. 25345 (acquisition no. 2006-00392-5, box 56), folder Special Correspondence Colonel McCormick, pt. 21, ibid.

17. F. J. Byington, "Shipping Operations: Tribune Company and Subsidiaries," June 28, 1950, 11, vol. 25796 (acquisition no. 2006-00392-5, box 507), folder Quebec & Ontario Transportation Co. 1985–87, QOPC, LAC; F. N. Greenleaf to Bess Vydra, Nov. 2, 1953; F. N. Greenleaf to Robert McCormick, Nov. 12, 1953.

18. "Diver Imprisoned beneath River in Entangled Lines," *Montreal Star*, Jan. 28, 1930, 1.

19. "Peter Trans Found Dead When Rescuers Reach Underwater Prison," *Montreal Star*, Jan. 30, 1930, 1–2.

20. "Safety and Welfare," *Pulp & Paper Magazine of Canada* 40 (Feb. 1939): 211; "Woods Accidents in Quebec," *Pulp & Paper Magazine of Canada* 40 (Mar. 1939): 260.

21. Frederic Babcock, "Liner Will Take Week-End Cruise to Baie Comeau," *Chicago Tribune*, Aug. 2, 1939, 18; Frederic Babcock, "Tribune's Paper Town in Canada on Cruise Tours," *Chicago Tribune*, May 15, 1940, 14.

22. Arthur Schmon to Leonard Schlemm, Oct. 10, 1941, 1, box 1, Localisation MN//1/1.43, QNS, SHCN; "Library Circulates 20,000 Books a Year," *Observation Post* 2, no. 1 (Jan. 1944): 2, vol. 25800 (acquisition no. 2006-00392-5, box 511), folder The Observation Post 1944, QOPC, LAC; "Comeau Children Have Big City Library Service," *Observation Post* 9, no. 3 (Mar. 1951): 11, 14, vol. 25801 (acquisition no. 2006-00392-5, box 512), folder The Observation Post 1951, ibid.; "Comeau Gets Films Soon after Big Cities," *Observation Post* 4, no. 12 (Dec. 1946): 5, vol. 25800 (acquisition no. 2006-00392-5, box 511), folder The Observation Post 1946, ibid.

23. W. S. Coolin to Arthur Schmon, July 21, 1955, vol. 25308 (acquisition no. 2006-00392-5, box 19), folder Municipal Government, pt. 6, QOPC, LAC; Town of Baie Comeau, Census, Jan. 1, 1953, vol. 25302 (acquisition no. 2006-00392-5, box 13), folder Townsite, pt. 3, ibid.; memorandum, "Town of Baie Comeau and District," undated, 1–5, vol. 25291 (acquisition no. 2006-00392-5, box 2), folder Wharf Contracts and Negotiations Manicouagan, pt. 18, ibid.

24. Town of Baie Comeau, Census, Jan. 1, 1953; Garnier, *Dog Sled to Airplane*, 16–43; J. A. Whitaker to Arthur Schmon, Jan. 15, 1938, vol. 25307 (acquisition no. 2006-00392-5, box 18), folder Roman Catholic Church, pt. 1, QOPC, LAC; Arthur Schmon to L. P. Gagne, Jan. 15, 1937, 1, vol. 25307 (acquisition no. 2006-00392-5, box 18), folder Roman Catholic Church, pt. 1, ibid.

25. Robert McCormick to Arthur Schmon, Sept. 1, 1937, vol. 25307 (acquisition no. 2006-00392-5, box 18), folder Roman Catholic Church, pt. 1, QOPC, LAC; Arthur Schmon to Robert McCormick, Aug. 21, 1939, vol. 25343 (acquisition no. 2006-00392-5, box 54), folder Special Correspondence Colonel McCormick, pt. 6, ibid.; L. P. Gagne to Arthur Schmon, Oct. 3, 1939, vol. 25307 (acquisition no. 2006-00392-5, box 18), folder Roman Catholic Church, pt. 2, ibid.; interview with Ian Sewell, Nov. 1, 2013, transcript in author's possession.

26. J. E. Vallillee to Arthur Schmon, Mar. 24, 1944, 1, vol. 25304 (acquisition no. 2006-00392-5, box 15), folder Schools, pt. 2, QOPC, LAC; J. E. Vallillee to Arthur Schmon, Dec. 28, 1945, 2, vol. 25307 (acquisition no. 2006-00392-5, box 18), folder Bishop Labrie, pt. 1, ibid.; Arthur Schmon, memorandum of conference with Bishop Labrie and Father Gagne, Oct. 15, 1945, 1, vol. 25307 (acquisition no. 2006-00392-5, box 18), folder Bishop Labrie, pt. 1, ibid.; Arthur Schmon to Robert McCormick, June 22, 1944, 1, vol. 25307 (acquisition no. 2006-00392-5, box 18), folder Roman Catholic Church, pt. 4, ibid.

27. Arthur Schmon to Robert McCormick, Jan. 4, 1946, vol. 25307 (acquisition no. 2006-00392-5, box 18), Bishop Labrie, pt. 1, QOPC, LAC; Robert McCormick to Arthur Schmon, Jan. 7, 1946, vol. 25307 (acquisition no. 2006-00392-5, box 18), Bishop Labrie, pt. 1, ibid.; N. A. LaBrie, speech at Baie Comeau, QC, Aug. 11, 1946, 1–2, vol. 25307 (acquisition no. 2006-00392-5, box 18), folder Bishop Labrie, pt. 2, ibid.; Garnier, *Dog Sled to Airplane*, 283.

28. Arthur Schmon, memorandum re school commission, Nov. 29, 1956, 2, vol. 25305 (acquisition no. 2006-00392-5, box 16), folder Schools, pt. 7, QOPC, LAC; memorandum, "School Commission," Nov. 21, 1956, vol. 25305 (acquisition no. 2006-00392-5, box 16),

folder Schools, pt. 7, ibid.; "Protestant Union Success, Baie Comeau Church Grows," *Montreal Gazette*, July 20, 1953, 3.

29. Program for a Baie Comeau Community Players version of "Petticoat Fever," by Mark Reed, Localisation P088/002/002, Dupuy, SHCN; J. A. Marier to Robert McCormick, Feb. 15, 1939, PTBC, SHCN; "First Eggs from Co-Op," *Observation Post* 1, no. 5 (Oct. 1943): 15, vol. 25800 (acquisition no. 2006-00392-5, box 511), folder The Observation Post 1943, QOPC, LAC; Larry Rue, "Weather Holds Col. McCormick at Baie Comeau," *Chicago Tribune*, Mar. 21, 1952, 8, 9; "Hails Faith Shown in Founding Quebec Paper Making Town," *Chicago Tribune*, July 17, 1952, 7.

30. *Town Crier* 1, no. 3 (May 1, 1941): 11, vol. 25789 (acquisition no. 2006-00392-5, box 500), folder Baie Comeau 1938–58, pt. 5, QOPC, LAC; Gilles Trepanier, "La Quebec North Shore Paper Company: Elément de progrès," editorial, *L'Aquilon*, Feb. 28, 1953, 3; "Pioneer Town in Quebec," *Times of London*, July 10, 1958, 11.

31. J. E. Vallillee to Arthur Schmon, Feb. 27, 1947, vol. 25347 (acquisition no. 2006-00392-5, box 58), folder House Organ, pt. 7, QOPC, LAC.

32. Gerald Zahavi, *Workers, Managers, and Welfare Capitalism: The Shoeworkers and Tanners of Endicott Johnson, 1890–1950* (Urbana: University of Illinois Press, 1988), 37.

33. United Paperworkers International Union, *100 Years of Challenge and Progress* (Nashville: United Paperworkers International Union, 1984), 2–4, 7–8; Robert Zieger, *Rebuilding the Pulp and Paper Workers' Union, 1933–1941* (Knoxville: University of Tennessee Press, 1984), 45–47; John Crispo, *International Unionism: A Study in Canadian-American Relations* (Toronto: McGraw-Hill, 1967), 20–22.

34. Quebec and Ontario Paper Company Ltd., "Our 75th Anniversary Album," 15, vol. 25809 (acquisition no. 2006-00392-5, box 520), folder Our 75th Anniversary Album, QOPC, LAC; United Paperworkers International Union, *100 Years of Challenge and Progress*, 8; The Ontario Paper Company Limited, Thorold Division, History of Operations, Sept. 10, 1963, 1, vol. 25364 (acquisition no. 2006-00392-5, box 75), folder Ontario Paper Co. Local 101, pt. 7, QOPC, LAC.

35. Matthew Burns, speech at Ontario Paper Company Twenty-Fifth Anniversary Dinner, Sept. 11, 1938, 17, 20, vol. 25798 (acquisition no. 2006-00392-5, box 509), folder Ontario Paper Company Twenty-Fifth Anniversary Dinner 1937, QOPC, LAC; Arthur Schmon, speech at Ontario Paper Company Twenty-Fifth Anniversary Dinner, Sept. 11, 1938, 3, 5, box 64, folder OPC—Anniversaries 1938–53, RRMB, Tribune Archives.

36. "Outlines Plot to Destroy U.S. Press Freedom," *Chicago Tribune*, Mar. 5, 1947, 4; transcript of speeches at the twenty-fifth-anniversary dinner of the Quebec North Shore Paper Co., July 28, 1962, 4–5, vol. 25790 (acquisition no. 2006-00392-5, box 501), folder Baie Comeau 25th Anniversary 1962, QOPC, LAC.

37. "Au Colonel Robert R. McCormick," *L'Echo du Bas St-Laurent*, Oct. 22, 1942, 3; Henri Dufresne, "Le bon génie qui allume des foyers," *La Patrie*, Nov. 17, 1946, 88–89; Kathleen Beebe to Robert McCormick, Feb. 24, 1947, 1, box 65, folder OPC—Baie Comeau—General, 1936–54, RRMB, Tribune Archives.

38. Alphonse Anctil to Robert McCormick, Oct. 18, 1939, vol. 25318 (acquisition no. 2006-00392-5, box 29), folder General, pt. 5, QOPC, LAC; Gabriel Lepage to Robert McCormick, undated, vol. 25343 (acquisition no. 2006-00392-5, box 54), folder Special Correspondence Colonel McCormick, pt. 8, ibid.; Arthur Schmon to Robert McCormick, July 30, 1941, vol. 25343 (acquisition no. 2006-00392-5, box 54), folder Special Correspondence Colonel McCormick, pt. 8, ibid.

39. Robert McCormick to Arthur Schmon, June 3, 1939, vol. 25343 (acquisition no. 2006-00392-5, box 54), folder Special Correspondence Colonel McCormick, pt. 6, QOPC, LAC; Robert McCormick to Arthur Schmon, July 9, 1940, vol. 25343 (acquisition no. 2006-00392-5, box 54), folder Special Correspondence Colonel McCormick, pt. 4, ibid.; Robert McCormick to Arthur Schmon, Feb. 6, 1955, vol. 25324 (acquisition no. 2006-00392-5, box 35), folder Ragueneau, ibid.

40. Memorandum, "Mill Office—Baie Comeau, Proposed Paint Schedule," Sept. 14, 1936, vol. 25302 (acquisition no. 2006-00392-5, box 13), folder Camps and Company Buildings, pt. 2, QOPC, LAC; J. A. Whitaker, memorandum re house building plan, fall and winter 1937–38, Sept. 22, 1937, 7, vol. 25302 (acquisition no. 2006-00392-5, box 13), folder Houses, pt. 1, ibid.; memorandum, "Possible Advantages and Savings Resulting from Duplication of Houses," Dec. 8, 1936, 1–2, vol. 25302 (acquisition no. 2006-00392-5, box 13), folder Houses, pt. 3, ibid.; Robert McCormick to Arthur Schmon, July 13, 1937, vol. 25302 (acquisition no. 2006-00392-5, box 13), folder Houses, pt. 5, ibid.

41. Robert McCormick, "Memoirs: XXIX," *Chicago Tribune*, Nov. 8, 1953, 16; Napoléon-Alexandre Comeau, *Life and Sport on the North Shore* (Quebec: Daily Telegraph Printing House, 1909), 7; Quebec North Shore Paper Company, "Building a Community: The First Ten Years at Baie Comeau, 1938–1948," 7–8, vol. 25316 (acquisition no. 2006-00392-5, box 27), folder Baie Comeau 10th Anniversary, pt. 1, QOPC, LAC; Carl Wiegman, *Trees to News: A Chronicle of the Ontario Paper Company's Origin and Development* (Toronto: McClelland & Stewart, 1953), 184–85; T. F. Flahiff to P. T. Ensor, Feb. 5, 1957, 1, vol. 25337 (acquisition no. 2006-00392-5, box 48), folder Townsite, QOPC, LAC; promotional brochure, "Baie Comeau: A Modern Model City," undated, vol. 25302 (acquisition no. 2006-00392-5, box 13), folder Houses, pt. 4, ibid.

42. Benedict Anderson, *Imagined Communities: Reflections on the Origin and Spread of Nationalism* (1983; repr., London: Verso, 2006), 26, 33.

Chapter 7 · The Diversified Newspaper Corporation

1. Program, *Chicago Tribune* Twenty-Fourth Annual Spring Luncheon, Apr. 23, 1946, vol. 25179, Scrapbook 1940–50, QOPC, LAC.

2. Paper plate from *Chicago Tribune* Twenty-Fourth Annual Spring Luncheon, Apr. 23, 1946, vol. 25179, Scrapbook 1940–50, QOPC, LAC; program, *Chicago Tribune* Twenty-Fourth Annual Spring Luncheon.

3. Program, *Chicago Tribune* Twenty-Fourth Annual Spring Luncheon.

4. Jerome Edwards, *The Foreign Policy of Col. McCormick's Tribune, 1929–1941* (Reno: University of Nevada Press, 1971), 28–29, 45–46, 53–58, 130–32.

5. Jack Alexander, "The Duke of Chicago," *Saturday Evening Post* 214, no. 3 (July 19, 1941): 71; Edwards, *Foreign Policy of Col. McCormick's Tribune*, 196–97.

6. "New Deal Uses Any Weapon to Smite Its Critics," *Chicago Tribune*, June 21, 1942, 3; Carl Wiegman, *Trees to News: A Chronicle of the Ontario Paper Company's Origin and Development* (Toronto: McClelland & Stewart, 1953), 213. The commercial competition and McCormick's attempt to block Field from getting membership in the Associated Press would eventually lead to the 1945 Supreme Court case *Associated Press v. United States*. In that proceeding, McCormick lost his fight with the government. See Margaret Blanchard, "The Associated Press Antitrust Suit: A Philosophical Clash over Ownership of First Amendment Rights," *Business History Review* 61, no. 1 (Spring 1987): 43–85.

7. Arthur Schmon to Robert McCormick, Dec. 22, 1939, vol. 25343 (acquisition no. 2006-00392-5, box 54), folder Special Correspondence Colonel McCormick, pt. 5, QOPC, LAC.

8. "One U.S. Newspaper That Might Be Dealt With," editorial, *Ottawa Journal*, May 28, 1940, 6; "Still Anti-British," editorial, *Ottawa Journal*, Jan. 9, 1941, 8; "Two Sides to This Drum," editorial, *Peterborough Examiner*, June 1, 1940, 4; "Has Been Using Canada Long Time," editorial, *Lindsay Post*, May 31, 1940, 2; "Using Our Pulp to Print Its Lies," editorial, *Lindsay Post*, May 30, 1940, 2; "The Chicago Tribune," editorial, *Winnipeg Free Press*, Feb. 4, 1941, 11; "*L'illustration Nouvelle* a lancé la première le cri d'alarme et avec raison," editorial, *L'Illustration Nouvelle*, May 31, 1940, 4; Edwards, *Foreign Policy of Col. McCormick's Tribune*, 195–96.

9. Dominion of Canada, House of Commons, Official Report of Debates, 19th Parl., 1st Sess., June 12, 1940, 222:721; Dominion of Canada, House of Commons, Sessional Paper no. 167, July 3, 1940, 2.

10. Edwards, *Foreign Policy of Col. McCormick's Tribune*, 195–96; Dominion of Canada, Senate, Official Report of Debates, 19th Parl., 1st Sess., June 6, 1940, 118–19; Dominion of Canada, Senate, Official Report of Debates, 19th Parl., 2nd Sess., Dec. 6, 1940, 50–51; Arthur Meighen to Maurice Dupre, June 18, 1940, 1, reel C3567, vol. 175, Meighen Fonds, LAC.

11. Arthur Schmon to Robert McCormick, June 20, 1940, 1–3, box 54, folder 2, JMP, LFCASC.

12. Confidential memorandum re "Chicago Tribune," May 30, 1940, 1, reel C4567, vol. 285, King Fonds, LAC; F. Charpentier and W. Eggleston to Pierre Casgrain, Feb. 13, 1941, reel C4860, vol. 300, ibid.; F. Charpentier and W. Eggleston, memorandum re "Scribner's Commentator," "Saturday Evening Post," "The Chicago Tribune," Feb. 13, 1941, 3, reel C4860, vol. 300, ibid.

13. "Good Leadership," editorial, *Editor & Publisher* 74, no. 7 (Feb. 15, 1941): 24; "Canadians Won't Interfere with Newsprint Flow," *Editor & Publisher* 74, no. 8 (Feb. 22, 1941): 45.

14. Claude Jagger, "Canada to Set Up Unified Rule for Newsprint," *Chicago Tribune*, Sept. 3, 1942, 25, 27; "Claims Canada Seeks to Close Paper Mill in Guise of War Need," *Chicago Tribune*, Sept. 6, 1942, 1.

15. Claude Jagger, "Newsprint Men Rap Growth of 'Cartelization,'" *Chicago Tribune*, Sept. 6, 1942, A9; Claude Jagger, "Newsprint Chief Studying Output Allocation Plan," *Chicago Tribune*, Sept. 9, 1942, 30; "Power Restored to Canadian Paper Mills," *Editor & Publisher* 77, no. 40 (Sept. 30, 1944): 9; "Freed Power to Prime Flow of Newsprint," *Chicago Tribune*, Sept. 27, 1944, 7; "We Have Kept the Faith," editorial, *Chicago Tribune*, Oct. 12, 1944, 16; Matthew Evenden, *Allied Power: Mobilizing Hydro-electricity during Canada's Second World War* (Toronto: University of Toronto Press, 2015), 113–14, 135–36.

16. James P. Hull, "The Second Industrial Revolution and the Staples Frontier in Canada: Rethinking Knowledge and History," *Scientia Canadensis: Canadian Journal of the History of Science, Technology and Medicine* 18, no. 1 (1994): 31–32; interview with Tony Yau, Oct. 1, 2015, transcript in author's possession.

17. Canadian Pulp and Paper Association, *The Pulpwood Harvest* (Montreal: Canadian Pulp and Paper Association, 1952), 11.

18. Hal Foust, "Lumber—Wasted Source of Material," *Chicago Tribune*, Feb. 14, 1943, E6.

19. "Tribune Makes Own Paper to Benefit World," *Chicago Tribune*, June 10, 1947, A10; Foust, "Lumber—Wasted Source of Material," E6; "Pioneer in Research," *St. Catharines*

Standard, July 26, 1963, 52; Ontario Paper Company, "Thorold Operations," 25, vol. 25791 (acquisition no. 2006-00392-5, box 502), folder Thorold Directors Visit, pt. 5, QOPC, LAC.

20. Hal Foust, "Huge Expansion Seen in Use of Forest Wastes," *Chicago Tribune*, June 20, 1943, 8.

21. "Can Make Rubber from By-Product of Paper Mills," *Editor & Publisher* 75, no. 34 (Aug. 22, 1942): 6; Robert McCormick, "Paper and By-Products," *Chicago Tribune*, June 20, 1943, 16. For an excellent account of these public-private dynamics of wartime industrial production in the United States, see Mark R. Wilson, *Destructive Creation: American Business and the Winning of World War II* (Philadelphia: University of Pennsylvania Press, 2016).

22. "Trib-buna Tire Leaves Imprint in Film Colony," *Chicago Tribune*, Oct. 18, 1943, 2; "Chi. Trib. Auto Using Synthetic Rubber and Fuel," *Editor & Publisher* 76, no. 35 (Aug. 28, 1943): 6.

23. Arthur Schmon, "Statement of the Ontario Paper Company with Respect to the Proposed Order Discontinuing the Supply of Electric Power to Its Thorold Mill," Aug. 31, 1942, 3–4, vol. 25793 (acquisition no. 2006-00392-5, box 504), folder Ontario Paper Company Historical and Background Material, pt. 1, QOPC, LAC; Ernest Mahler to Arthur Sulzberger, Oct. 24, 1942, box 247, folder Spruce Falls Power and Paper Company (247.2), Sulzberger Papers, NYPL; "Newsprint Producers Out to Beat 1943," *Financial Post*, Sept. 23, 1944, 21; Ontario Paper Company, "Thorold Operations," 11.

24. Arthur Schmon to Robert McCormick, Dec. 7, 1944, 1, vol. 25343 (acquisition no. 2006-00392-5, box 54), folder Special Correspondence Colonel McCormick, pt. 7, QOPC, LAC; "Big Expansion for 2 Tribune Paper Plants," *Chicago Tribune*, Aug. 21, 1945, 19; Ontario Paper Company, "Thorold Operations," 11; Arthur Schmon to Robert McCormick, Dec. 12, 1945, 1, vol. 25344 (acquisition no. 2006-00392-5, box 55), folder Special Correspondence Colonel McCormick, pt. 11, QOPC, LAC; "The Ontario Paper Company, Limited and Quebec North Shore Paper Company: The Company's Properties and Production Activities in Relation to the Papers' Newsprint Requirements," 11–12, vol. 25316 (acquisition no. 2006-00392-5, box 27), folder Baie Comeau 10th Anniversary, pt. 9, ibid.; Arthur Schmon, "Our Paper Mills," address at the Chicago Tribune Circulation Department Centennial Prevue, Continental Hotel, Chicago, IL, Oct. 9, 1946, 1, vol. 25799 (acquisition no. 2006-00392-5, box 510), folder Addresses by Arthur A. Schmon, ibid.

25. Donald J. Forgie to Arthur Kaufman, Dec. 8, 1959, 1, vol. 25759 (acquisition no. 2006-00392-5, box 470), folder Vanillin Plant Publicity, pt. 2, QOPC, LAC; "Tribune Makes Own Paper to Benefit World," A10; Lorne Merritt, "The Toxicology of Toluene—a Practical Approach," 1980, 1, vol. 25759 (acquisition no. 2006-00392-5, box 470), folder Vanillin Plant General, pt. 1, QOPC, LAC.

26. Steve Ettlinger, *Twinkie, Deconstructed* (New York: Hudson Street, 2007), 203–4; Tim Ecott, *Vanilla: Travels in Search of the Ice Cream Orchid* (New York: Grove, 2004), 2; Emilio Kourí, *A Pueblo Divided: Business, Property, and Community in Papantla, Mexico* (Stanford, CA: Stanford University Press, 2004), chap. 1.

27. Ettlinger, *Twinkie, Deconstructed*, 205–6; Patricia Rain, *Vanilla: The Cultural History of the World's Most Popular Flavor and Fragrance* (New York: Penguin, 2004), 153; US Tariff Commission, *Vanillin*, TC Publication 65, Aug. 1962, 3, vol. 25355 (acquisition no. 2006-00392-5, box 66), folder United States Tariff Commission Vanillin, QOPC, LAC.

28. "Plastics from Lignin," *Pulp & Paper Magazine of Canada* 40 (Oct. 1939): 638; Martin B. Hocking, "Vanillin: Synthetic Flavoring from Spent Sulfite Liquor," *Journal of*

Chemical Education 74, no. 9 (Sept. 1997): 1055–59; Bruce Gralow to Arthur Schmon, Jan. 18, 1947, vol. 25761 (acquisition no. 2006-00392-5, box 472), folder Vanillin Plant Flowsheets & Estimates, pt. 3, QOPC, LAC; memoranda for discussion at research meeting, Oct. 15, 1947, 1, vol. 25761 (acquisition no. 2006-00392-5, box 472), folder Vanillin Plant Flowsheets & Estimates, pt. 2, ibid.; Arthur Schmon to James MacKinnon, Jan. 10, 1948, 1–3, vol. 25344 (acquisition no. 2006-00392-5, box 55), folder Special Correspondence Colonel McCormick, pt. 13, ibid.

29. US Tariff Commission, *Vanillin*, 3–4.

30. C. A. Sankey, "Vanillin Story for Colonel McCormick—Draft of Detail Story," attachment to C. A. Sankey to B. B. Gralow, Apr. 17, 1953, vol. 25759 (acquisition no. 2006-00392-5, box 470), folder Vanillin Plant Publicity, pt. 1, QOPC, LAC; W. Fell to J. B. Jones, Dec. 11, 1956, vol. 25759 (acquisition no. 2006-00392-5, box 470), folder Vanillin Plant General, pt. 1, ibid.

31. "New Vanillin Process Uses Paper Wastes," *Chicago Tribune*, Jan. 3, 1952, B5.

32. Donald J. Forgie to Arthur Kaufman, Dec. 8, 1959, 1–2, vol. 25759 (acquisition no. 2006-00392-5, box 470), folder Vanillin Plant Publicity, pt. 2, QOPC, LAC; US Tariff Commission, *Vanillin*, 10–13; Bruce Gralow to Arthur Schmon, Dec. 28, 1951, vol. 25759 (acquisition no. 2006-00392-5, box 470), folder Vanillin Plant Negotiations with Dow Chemical of Canada Ltd., pt. 3, QOPC, LAC; "Brief for the Ontario Paper Company Limited before the United States Tariff Commission," June 29, 1962, vol. 25355 (acquisition no. 2006-00392-5, box 66), folder United States Tariff Commission Vanillin, ibid.; Ontario Paper Company, "Technical Vanillin Sales—Statement of 'Lioxin' Sales by Countries and Customers, Year 1961," vol. 25766 (acquisition no. 2006-00392-5, box 477), folder Vanillin Plant Sterling Drug Protest (Tariff Commission), pt. 5, ibid.

33. "Annual Payroll Economy Boost to Niagara Area," *St. Catharines Standard*, July 26, 1963, 46; "Ontario Paper Hikes Capacity for Vanillin," *Chicago Tribune*, Feb. 25, 1964, sec. 3, 5; Eugene Griffin, "Thorold Wages Pollution Fight," *Chicago Tribune*, Nov. 12, 1970, sec. 3, 9; "Brief for the Ontario Paper Company Limited before the United States Tariff Commission, Investigation under Section 201(a) of the Antidumping Act of 1921, as Amended, Technical Vanillin from Canada," 7–9, vol. 25767 (acquisition no. 2006-00392-5, box 478), folder Vanillin Customs Trial, pt. 3, QOPC, LAC; "Notes for Mr. Schmon for Investment Bankers Meeting," June 4, 1974, Marketing Section, 6, vol. 25737 (acquisition no. 2006-00392-5, box 448), folder Investment Bankers, ibid.; M. R. Jones, memorandum, "Vanillin Plant Expansion," Nov. 5, 1973, vol. 25467 (acquisition no. 2006-00392-5, box 168), folder Vanillin Plant Phase III Expansion, pt. 1, ibid.

34. Interview with Adrian Barnet, Sept. 30, 2015, transcript in author's possession; interview with Tony Yau, Oct. 1, 2015, transcript in author's possession; "Increased Newsprint Major Problem Now," *Editor & Publisher* 78, no. 35 (Aug. 25, 1945): 8.

35. Memorandum, "Manicouagan Power Development," Oct. 8, 1945, vol. 25452 (acquisition no. 2006-00392-5, box 163), folder Power Development Manicouagan, pt. 1, QOPC, LAC.

36. Harald S. Patton, "Hydro-electric Power Policies in Ontario and Quebec," *Journal of Land & Public Utility Economics* 3, no. 2 (May 1927): 132–36; Christopher Armstrong and H. V. Nelles, *Monopoly's Moment: The Organization and Regulation of Canadian Utilities, 1830–1930* (Philadelphia: Temple University Press, 1986), 154–56, 190–92, 296–97, 312–13.

37. Harald S. Patton, "Hydro-electric Power Policies in Ontario and Quebec," *Journal of Land & Public Utility Economics* 3, no. 3 (Aug. 1927): 226–27; John Dales, *Hydroelectricity*

and Industrial Development: Quebec 1898–1940 (Cambridge, MA: Harvard University Press, 1957), 30–31; Caroline Desbiens, *Power from the North: Territory, Identity, and the Culture of Hydroelectricity in Quebec* (Vancouver: UBC Press, 2013), 31–32.

38. Arthur Schmon to Robert McCormick, Dec. 13, 1945, 2, vol. 25295 (acquisition no. 2006-00392-5, box 6), folder Land Purchase Negotiations Project X, pt. 1, QOPC, LAC; memorandum, Sept. 10, 1948, vol. 25297 (acquisition no. 2006-00392-5, box 8), folder First Falls Manicouagan, pt. 12, ibid.

39. Arthur Schmon to T. F. Flahiff, Oct. 6, 1958, 1, vol. 25789 (acquisition no. 2006-00392-5, box 500), folder Baie Comeau 1938–58, pt. 4, QOPC, LAC.

40. Arthur Schmon to Robert McCormick, Sept. 3, 1948, 1, vol. 25307 (acquisition no. 2006-00392-5, box 18), folder Bishopric Amedee Township, pt. 3, QOPC, LAC.

41. N. A. Labrie to Maurice Duplessis, Aug. 31, 1948, 3, vol. 25307 (acquisition no. 2006-00392-5, box 18), folder Bishopric Amedee Township, pt. 3, QOPC, LAC.

42. Arthur Schmon to Robert McCormick, Sept. 11, 1948, vol. 25294 (acquisition no. 2006-00392-5, box 5), folder Land Purchase Negotiations Manicouagan / Baie Comeau, pt. 40, QOPC, LAC; Arthur Schmon to H. A. Sewell, Dec. 17, 1948, vol. 25307 (acquisition no. 2006-00392-5, box 18), folder Bishop Labrie, pt. 3, ibid.

43. N. A. Labrie, "Le Nouveau Quebec," *Le Soleil*, Dec. 3, 1948, 13.

44. Arthur Schmon to Robert McCormick, Dec. 17, 1948, 1–2, vol. 25307 (acquisition no. 2006-00392-5, box 18), folder Bishop Labrie, pt. 3, QOPC, LAC.

45. Maurice Duplessis, Mar. 10, 1949, Quebec Legislative Assembly, *Débats de l'Assemblée législative*, 23rd Legislature, 1st Sess., 491–92; T. F. Flahiff to Arthur Schmon, Jan. 11, 1954, 1–2, vol. 25295 (acquisition no. 2006-00392-5, box 6), folder Land Purchase Negotiations Manicouagan / Baie Comeau, pt. 51, QOPC, LAC.

46. Eugene Griffin, "Power Plant Will Harness New Niagara," *Chicago Tribune*, Jan. 26, 1951, 1, 6; Eugene Griffin, "Engineers Face Tough Problems to Create Big McCormick Dam," *Chicago Tribune*, Nov. 11, 1951, H8; transcript, "Dominion Magazine," CBC Dominion Network, June 4, 1951, 1, vol. 25188, Scrapbook, QOCP, LAC; "Theater of the Air to Salute New McCormick Dam," *Chicago Tribune*, May 25, 1951, C8; Eugene Griffin, "Huge McCormick Dam to Speed Development of Canada's North Shore Region," *Chicago Tribune*, Aug. 26, 1951, G6.

47. Eugene Griffin, "McCormick Dam, Tribune's Unique Power Project in Canada, Completed in Record Time," *Chicago Tribune*, June 22, 1952, D8; "Dedicate New McCormick Dam at Baie Comeau," *Chicago Tribune*, July 19, 1953, 3; "Power Plant Starts in Quebec," *New York Times*, July 19, 1953, 2; "Let Us Work Together," editorial, *Montreal Star*, July 21, 1953, 10; Chris McFarland, "Quebec's Resources 'Want' U.S. Capital, Says Duplessis," *Montreal Gazette*, July 20, 1953, 17; "Monarch of the Forest," *Time*, July 27, 1953, 24.

48. "McCormick Dam Sends 1st Power to Tribune Mill," *Chicago Tribune*, Dec. 7, 1952, 5.

49. Joy Parr, *Sensing Changes: Technologies, Environments, and the Everyday, 1953–2003* (Vancouver: UBC Press, 2010); Jamie Linton, *What Is Water? The History of a Modern Abstraction* (Vancouver: UBC Press, 2010), chap. 3; Daniel Macfarlane, *Negotiating a River: Canada, the US, and the Creation of the St. Lawrence Seaway* (Vancouver: UBC Press, 2014).

50. T. F. Flahiff, memorandum, May 23, 1955, 3, 5, vol. 25315 (acquisition no. 2006-00392-5, box 26), folder Power Consumers, pt. 6, QOPC, LAC; Arthur Schmon to R. H. Davis, Feb. 15, 1952, vol. 25314 (acquisition no. 2006-00392-5, box 25), folder Power Consumers, pt. 2, ibid.; Arthur Schmon to B. B. Gralow, Feb. 14, 1952, vol. 25314 (acquisition no. 2006-00392-5, box 25), folder Power Consumers, pt. 2, ibid.; Arthur Schmon to Robert McCor-

mick, Oct. 28, 1949, vol. 25298 (acquisition no. 2006-00392-5, box 9), folder First Falls Manicouagan, pt. 24, ibid.

51. Mimi Sheller, *Aluminum Dreams: The Making of Light Modernity* (Cambridge, MA: MIT Press, 2014), 1, 16–18, 37–47, 85; British Aluminium Company, *The History of the British Aluminium Company* (London: Norfolk House, 1955), 15–16; "Aluminum: Metal on the March," *Newsweek*, Jan. 21, 1957, 77.

52. David Massell, *Amassing Power: J. B. Duke and the Saguenay River, 1897–1922* (Montreal and Kingston: McGill-Queen's University Press, 2000), 56–58; David Massell, *Quebec Hydropolitics: The Peribonka Concessions of the Second World War* (Montreal and Kingston: McGill-Queen's University Press, 2011), 52–53; George David Smith, *From Monopoly to Competition: The Transformations of Alcoa, 1888–1986* (Cambridge: Cambridge University Press, 1988), 142–45.

53. Matthew Evenden, "Mobilizing Rivers: Hydro-electricity, the State, and World War II in Canada," *Annals of the Association of American Geographers* 99, no. 5 (2009): 847; Matthew Evenden, "Aluminum, Commodity Chains, and the Environmental History of the Second World War," *Environmental History* 16, no. 1 (Jan. 2011): 72–76.

54. F. J. Byington, "Shipping Operations: Tribune Company and Subsidiaries," June 28, 1950, 24–26, vol. 25796 (acquisition no. 2006-00392-5, box 507), folder Quebec & Ontario Transportation Co. 1985–87, QOPC, LAC; "Tribune Ships Help Canada Produce Aluminum," *Chicago Tribune*, Nov. 1, 1942, D2, D8; Evenden, *Allied Power*, 192.

55. British Aluminium Company, *The History of the British Aluminium Company*, 62; F. S. Smithers and Co., pamphlet, "The British Aluminium Company, Limited," 1956, 8–9, vol. 25338 (acquisition no. 2006-00392-5, box 49), folder Publicity, pt. 6, QOPC, LAC.

56. James Moxon, *Volta: Man's Greatest Lake*, rev. ed. (1969; repr., London: Andre Deutsch, 1984), 50, 59–61, 83–86; Stephen Miescher, " 'Nkrumah's Baby': The Akosombo Dam and the Dream of Development in Ghana, 1952–1966," *Water History* 6, no. 4 (Dec. 2014): 341–66.

57. T. F. Flahiff, memorandum, May 23, 1955, 15; Arthur Schmon to Robert McCormick, Nov. 1, 1954, 2, vol. 25315 (acquisition no. 2006-00392-5, box 26), folder Power Consumers, pt. 6, QOPC, LAC; Donald B. Macurda, "Progress Report on the British Aluminium Company, Limited," June 19, 1958, 1, vol. 25334 (acquisition no. 2006-00392-5, box 45), folder Financing, pt. 13, ibid.; Arthur Schmon to Maurice Duplessis, May 25, 1955, vol. 25332 (acquisition no. 2006-00392-5, box 43), folder Quebec Government, pt. 1, ibid.

58. "Position of Manicouagan Power Company during Period the Fifth Falls Reservoir Is Being Filled," Sept. 13, 1960, vol. 25325 (acquisition no. 2006-00392-5, box 36), folder First Falls Expansion, pt. 6, QOPC, LAC; "Canada to Expand Aluminum Output," *New York Times*, Oct. 17, 1955, 44; C. C. Milne, "New Tide of Industry Rolls on St. Lawrence," *Financial Post*, Sept. 1, 1956, 17; Eugene Griffin, "Baie Comeau's Aluminum Ingot Smelter Takes Shape," *Chicago Tribune*, Feb. 24, 1957, A9–10; C. Miller and W. G. Street, "Aluminium Reduction Plant at Baie Comeau, Que.," *Engineering Journal* 41, no. 7 (July 1958): 41–49; Milford Smith, "International Outpost at Baie Comeau Adds to Canada's Economic Strength," *Hamilton Spectator*, June 14, 1958, 23; "La Côte-Nord, richesse économique mais aussi richesse humaine," editorial, *L'Aquilon*, June 11, 1958, 4.

59. "A Master Builder's Monuments," *Chicago Tribune*, July 30, 1958, 16; Denis Masse, "Baie-Comeau: ville de l'aluminium," *Le Soleil*, June 16, 1958, 3; "Le premier ministre de la province inaugure à Baie Comeau sur la Côte Nord une aluminerie de $100,000,000," *L'Événement Journal*, June 16, 1958, 1, 8.

60. "At 25, Tribune Quebec Community Is Bustling," *Chicago Tribune*, July 29, 1962, 8; Eugene Griffin, "Baie Comeau Power Plans Lead Nation," *Chicago Tribune*, Jan. 14, 1963, C6.

61. Sean Mills, *The Empire Within: Postcolonial Thought and Political Activism in Sixties Montreal* (Montreal and Kingston: McGill-Queen's University Press, 2010), 26–27; Karl Froschauer, *White Gold: Hydroelectric Power in Canada* (Vancouver: UBC Press, 1999), 79–82.

62. T. F. Flahiff to Arthur Schmon, Apr. 5, 1960, vol. 25325 (acquisition no. 2006-00392-5, box 36), folder First Falls Expansion, pt. 3, QOPC, LAC; B. B. Gralow to Arthur Schmon, July 28, 1960, vol. 25325 (acquisition no. 2006-00392-5, box 36), folder First Falls Expansion, pt. 2, ibid.

63. B. B. Gralow to Arthur Schmon, June 10, 1960, 1, vol. 25325 (acquisition no. 2006-00392-5, box 36), folder First Falls Expansion, pt. 2, QOPC, LAC; "Big Power Plans," *Time*, Sept. 5, 1960, 10; Charles Miller, "Manicouagan Power Company Rights at First Falls and Water Regulation during Fill-Up Period of Duplessis Reservoir," Apr. 29, 1960, vol. 25325 (acquisition no. 2006-00392-5, box 36), folder First Falls Expansion, pt. 1, QOPC, LAC; "Position of Manicouagan Power Company during Period the Fifth Falls Reservoir Is Being Filled," 1–6; memorandum, "Manicouagan Power Company's Position during the Period When Hydro-Quebec Is Developing the Power Potential of the Manicouagan River," Aug. 20, 1962, vol. 25327 (acquisition no. 2006-00392-5, box 38), folder First Falls Expansion, pt. 24, ibid.

64. "Memorandum Presented by the Quebec North Shore Paper Company Concerning Problems to Immediate and Future Operations Created by the Flooding of Timber Limits and Crown Land along the Manicouagan River," May 4, 1960, 1–3, 9, vol. 25325 (acquisition no. 2006-00392-5, box 36), folder First Falls Expansion, pt. 1, QOPC, LAC.

65. Laurent Lauzier, "Problème épineux pour l'aménagement de la Manicouagan: trois millions de cordes de bois en quête d'acheteurs!," *La Presse*, Aug. 24, 1960, 1; Roger Champoux, "La forêt à saveur des eaux!," editorial, *La Presse*, Aug. 25, 1960, 4; "A Brief concerning Negotiations between the Quebec Government and Quebec North Shore Paper Company to Salvage and Use Pulpwood from the Manicouagan Watershed," Oct. 6, 1961, 3, vol. 25326 (acquisition no. 2006-00392-5, box 37), folder First Falls Expansion, pt. 15, QOPC, LAC.

66. Froschauer, *White Gold*, 81; Desbiens, *Power from the North*, 31–32.

67. T. F. Flahiff to Arthur Schmon, Apr. 3, 1962, 1–2, vol. 25326 (acquisition no. 2006-00392-5, box 37), folder First Falls Expansion, pt. 20, QOPC, LAC.

68. Arthur Schmon to Sir William Strath, Dec. 14, 1962, 1, vol. 25332 (acquisition no. 2006-00392-5, box 43), folder Quebec Government, pt. 2, QOPC, LAC.

69. H. T. Fisher, Diversification Report, Dec. 31, 1963, introduction, 1, vol. 25355 (acquisition no. 2006-00392-5, box 66), folder Diversification Report, QOPC, LAC.

70. H. T. Fisher, memorandum, Jan. 13, 1964, 1–2, vol. 25355 (acquisition no. 2006-00392-5, box 66), folder Diversification Report, QOPC, LAC.

Chapter 8 · The Industrial Newspaper and Its Legacies

1. Arthur Schmon to G. J. Lane, July 17, 1954, vol. 25345 (acquisition no. 2006-00392-5, box 56), folder Special Correspondence Colonel McCormick, pt. 23, QOPC, LAC; Eugene Griffin, "Col. McCormick Paid Tribute on 74th Birthday," *Chicago Tribune*, July 31, 1954, 3.

2. "Col. McCormick Rites Monday; Eisenhower Leads U.S. Tributes," *New York Daily News*, Apr. 2, 1955, 2.

3. "The Great Editors," *Time*, Apr. 11, 1955, 64; "The 'Tribune's' Colonel Dies," *Life*, Apr. 11, 1955, 47; "The Passing of the Giants," *Newsweek*, Apr. 11, 1955, 84.

4. "Aquilonnades," *L'Aquilon*, Apr. 15, 1955, 1; "Col. R. R. McCormick," editorial, *St. Catharines Standard*, Apr. 1, 1955, 4.

5. "Robert Rutherford McCormick," editorial, *Globe and Mail*, Apr. 2, 1955, 6; "McCormick of the Tribune," editorial, *Ottawa Journal*, Apr. 4, 1955, 6.

6. "Robert R. McCormick," editorial, *Editor & Publisher* 88, no. 16 (Apr. 9, 1955): 42.

7. "Major Papers Were Built Up in McCormick's Press Empire," *New York Times*, Apr. 1, 1955, 17.

8. John Nerone, *The Media and Public Life: A History* (New York: Polity, 2015), 184, 191; Daniel Hallin, "The Passing of the 'High Modernism' of American Journalism," *Journal of Communication* 42, no. 3 (Summer 1992): 14–25.

9. A. Kent MacDougall, "Trials at the Trib," *Wall Street Journal*, Feb. 25, 1971, 1.

10. "Col. McCormick Buys Washington Times-Herald," *Editor & Publisher* 82, no. 31 (July 23, 1949): 5. Cissy Patterson in 1930 had become editor of what was then the Hearst-owned *Washington Herald*. Patterson was at the time the only female editor of a prominent newspaper in the country, and the morning *Herald* became a great success under her leadership. In 1939, Patterson arranged to purchase from William Randolph Hearst both the *Herald* and the evening *Washington Times*, and she then merged the two dailies into the dominant newspaper in the nation's capital. Though sharing some of Robert McCormick's political views, Cissy Patterson grew increasingly estranged from her cousin, and before she died in July 1948 she took the unexpected step of leaving the paper in her will to seven of the company's executives. Patterson's daughter, Felicia Gizycka, initially contested the will, and when she dropped this ultimately unsuccessful effort, Robert McCormick purchased the paper from the editors who had inherited it. See Amanda Smith, *Newspaper Titan: The Infamous Life and Monumental Times of Cissy Patterson* (New York: Alfred A. Knopf, 2011), 316–18, 369–74, 515–16, 568–69.

11. "Col. McCormick Takes Over at Times-Herald," *Editor & Publisher* 84, no. 15 (Apr. 7, 1951): 10; "Post Buys Times-Herald; One Morning Daily for D.C.," *Editor & Publisher* 87, no. 13 (Mar. 20, 1954): 9; Lloyd Wendt, *Chicago Tribune: The Rise of a Great American Newspaper* (Chicago: Rand McNally, 1979), 684–85.

12. George Brandenburg, "McCormick Will Sets Plan for Tribune Management," *Editor & Publisher* 88, no. 16 (Apr. 9, 1955): 7, 72; "Control of Tribune Rests with 2 Groups," *Editor & Publisher* 88, no. 16 (Apr. 9, 1955): 10; "Tribune Co. Reports Continued Progress," *Editor & Publisher* 89, no. 22 (May 26, 1956): 14.

13. Ray Erwin, "Flynn Leads New York News with Patterson's Policies," *Editor & Publisher* 89, no. 25 (June 16, 1956): 13, 56; advertisement, *Editor & Publisher* 92, no. 28 (July 11, 1959): 27.

14. George Brandenburg, "Tribune Co. Buys Chicago American from Hearst," *Editor & Publisher* 89, no. 44 (Oct. 27, 1956): 11; "A Conversation with 'Jack' Knight," *Editor & Publisher* 99, no. 35 (Aug. 27, 1966): 9–10, 55; George Brandenburg, "Field Acquires Chicago News Control from Knight Group," *Editor & Publisher* 92, no. 2 (Jan. 10, 1959): 9, 59.

15. "Chi Trib Gets Ready to Print American," *Editor & Publisher* 92, no. 25 (June 20, 1959): 60; "Chicago's American News Office Moves," *Editor & Publisher* 94, no. 6 (Feb. 11, 1961): 77; George Brandenburg, "Wood Tells of Expansion of Tribune Co. Properties," *Editor & Publisher* 94, no. 51 (Dec. 23, 1961): 14; "2 Chicago Papers Plan Single Plant," *Editor*

& Publisher 92, no. 31 (Aug. 1, 1959): 10; "Don Maxwell Tells Chi. Tribune's Aim," *Editor & Publisher* 95, no. 27 (July 7, 1962): 12.

16. "Chicago Today to Fold; Tribune to Go 'All Day,'" *Editor & Publisher* 107, no. 35 (Aug. 31, 1974): 43; Wendt, *Chicago Tribune*, 761–64.

17. *Editor & Publisher International Yearbook, 1946* (New York: Editor & Publisher, 1946), 40; *Editor & Publisher International Yearbook, 1960* (New York: Editor & Publisher, 1960), 80, 82; Eli Noam, *Media Ownership and Concentration in America* (New York: Oxford University Press, 2009), 142.

18. "Chicago Tribune Buys 2 Papers in Florida," *Editor & Publisher* 96, no. 27 (July 6, 1963): 10; Jerome H. Walker, "Orlando Sentinel and Star Join Chicago Tribune Family," *Editor & Publisher* 98, no. 36 (Sept. 4, 1965): 9–10; Wendt, *Chicago Tribune*, 741–42.

19. Ontario Paper Company, "Report to the St. Lawrence Seaway Authority," July 25, 1958, 3, vol. 25793 (acquisition no. 2006-00392-5, box 504), folder Ontario Paper Company Historical and Background Material, pt. 1, QOPC, LAC; clipping, "Quebec Paper Mill Starts $54m Expansion Program," *Heavy Construction News*, July 15, 1968, vol. 25789 (acquisition no. 2006-00392-5, box 500), folder Publicity Baie Comeau No. 4 Paper Machine, ibid.; interview with Bernard Baril, Sept. 30, 2015, transcript in author's possession; notes for Mr. Schmon for Investment Bankers Meeting, June 4, 1974, Overview Section, 3–4, vol. 25737 (acquisition no. 2006-00392-5, box 448), folder Investment Bankers, QOPC, LAC; John Storm, "New Mill 'Saved Our Life,'" *St. Catharines Standard*, May 13, 1983, 9.

20. H. T. Fisher, memorandum, Jan. 13, 1964, 1–2, vol. 25355 (acquisition no. 2006-00392-5, box 66), folder Diversification Report, QOPC, LAC.

21. Notes for Mr. Schmon for Investment Bankers Meeting, June 4, 1974, Financial Review Section, 1, Marketing Section, 1; J. S. Glass to R. C. Knutson, Aug. 12, 1976, vol. 25737 (acquisition no. 2006-00392-5, box 448), folder A-O General, pt. 20, QOPC, LAC.

22. Claude Ryan, "La philosophie (?) sociale de M. Lesage," editorial, *Le Devoir*, Aug. 2, 1962, 4; Sean Mills, *The Empire Within: Postcolonial Thought and Political Activism in Sixties Montreal* (Montreal and Kingston: McGill-Queen's University Press, 2010), 3–4, 63–64.

23. Paul Duclos, "La municipalisation de la Côte Nord: Enterprise sociologique," *L'Aquilon*, Jan. 31, 1955, 2; "Des étrangers dans la cuisine," *Le Devoir*, Mar. 15, 1957, 14; La Vigie, "C'est l'temps qu'ça change à Franquelin," *La Côte-Nord*, Nov. 2, 1960, 3; Jacques Pigeon, "Quel sort attend Lesage sur la Côte Nord?," *La Presse*, July 27, 1962, 2.

24. Albert Deschenes, "Union Solidarity Well Demonstrated at Baie Comeau on March 31st," vol. 25365 (acquisition no. 2006-00392-5, box 76), folder Quebec North Shore Co. Local 361, pt. 2, QOPC, LAC.

25. Memorandum, "Detail of Events Covering the Period of Illegal Walk-Out by Papermakers on Tuesday," June 15, 1965, 1–2, vol. 25365 (acquisition no. 2006-00392-5, box 76), Local 101 Brief, pt. 12, QOPC, LAC; "Second Paper Mill Closed," *St. Catharines Standard*, June 16, 1965, 1, 2; "Paper Mills Silent," *St. Catharines Standard*, June 17, 1965, 1, 2; "800 Strike in Canada; Close 2 Paper Mills," *Chicago Tribune*, June 17, 1965, 16; "Ontario Paper's Supply of Pulpwood Threatened," *St. Catharines Standard*, June 21, 1965, 9; John M. Lee, "Newsprint Strike Looms in Canada," *New York Times*, July 1, 1965, 39–40; "Paper Company Employees Back after Walkout," *Chicago Tribune*, July 18, 1965, D3.

26. The Thorne Group, report, 1966, 2, 4–5, 7, vol. 25373 (acquisition no. 2006-00392-5, box 84), folder Grievances, pt. 14, QOPC, LAC.

27. Tom Nevens, "Production Halted at Ontario Paper," *St. Catharines Standard*, Sept. 29, 1973, 1; Tom Nevens, "One Paper Mill Strike Averted, the Other Is Now in Its Third

Day," *St. Catharines Standard*, Oct. 1, 1973, 30; "Thorold Paper Mill Resumes Production," *Niagara Falls Review*, Oct. 26, 1973, 21; "Strike Is Closing Ontario Paper Co.," *Chicago Tribune*, Sept. 17, 1975, E7; "Canadian Paper Unions Ratify 3-Year Contract," *Chicago Tribune*, Jan. 22, 1976, C8; "Tribune Co. Tells Loss of $304,000 in Quarter," *Chicago Tribune*, Apr. 24, 1976, G7.

28. J. A. Carpenter, memorandum, Nov. 29, 1976, 1–2, vol. 25377 (acquisition no. 2006-00392-5, box 88), folder Strike Thorold Third Mill, pt. 6, QOPC, LAC.

29. John Sawatsky, *Mulroney: The Politics of Ambition* (Toronto: Macfarlane Walter & Ross, 1991), 5, 14–15, 19–20, 102, 107–8, 342–43.

30. Dimitry Anastakis, *Auto Pact: Creating a Borderless North American Auto Industry, 1960–1971* (Toronto: University of Toronto Press, 2005), 5, 172.

31. Task Force on the Structure of Canadian Industry, *Foreign Ownership and the Structure of Canadian Industry: Report of the Task Force on the Structure of Canadian Industry* (Ottawa: Privy Council Office, 1968), 1, 360–61; Richard A. Preston, "Introduction: A Plea for Comparative Studies of Canada and the United States and of the Effects of Assimilation on Canadian Development," in *The Influence of the United States on Canadian Development: Eleven Case Studies*, ed. Richard A. Preston (Durham, NC: Duke University Press, 1972), 6–7.

32. Kari Levitt, *Silent Surrender: The Multinational Corporation in Canada* (1970; repr., Montreal and Kingston: McGill-Queen's University Press, 2002), 3, 32–33, 142.

33. Patrick Brown, Robert Chodos, and Rae Murphy, *Winners, Losers: The 1976 Tory Leadership Convention* (Toronto: James Lorimer, 1976), 95; Michael T. Kaufman, "Canada's New Tory Leader: Bilingual and Stylish," *New York Times*, June 13, 1983, A3.

34. Rae Murphy, Robert Chodos, and Nick Auf der Maur, *Brian Mulroney: The Boy from Baie-Comeau* (Toronto: James Lorimer, 1984), 213–14.

35. Christopher Waddell, "Policy and Partisanship on the Campaign Trail: Mulroney Works His Wonder, Twice," in *Transforming the Nation: Canada and Brian Mulroney*, ed. Raymond Blake (Montreal and Kingston: McGill-Queen's University Press, 2007), 20; Nicholas Kristof, "Canada-U.S. Trade Tie," *New York Times*, Dec. 11, 1984, D24.

36. Brian Mulroney, *Memoirs: 1939–1993* (Toronto: McClelland & Stewart, 2007), 16, 20. Mulroney also noted that, as prime minister, he told a version of this story at a public event and quipped that, in receiving these fifty dollars, "the Mulroney family became the first direct beneficiaries of American foreign aid." Mulroney noted that the "audience roared with laughter, but Canada's dour economic nationalists failed as usual to see the humour, and denounced the incident as another appalling example of Canadian 'subordination' to the wicked Americans" (21).

37. Robert Bothwell, *The Penguin History of Canada* (New York: Penguin, 2006), 478. This performance remained a sore point for many Canadians for years after. In a major academic treatment of the Mulroney years published in 2007, Raymond Blake noted in his introduction of the event that it was "an image that disturbed many Canadians at the time and continues to 'stick in their craw' some twenty years later." Blake, introduction to *Transforming the Nation*, 9. Mulroney completed a trifecta of sorts when he sang the song for President Donald Trump in February 2017 at a benefit held at the president's Palm Beach estate. As one report noted, the "former prime minister got a standing ovation led by the Trumps" after his performance. See Robert Fife, "Mulroney Draws Praise from Trump for Canada, Trudeau," *Globe and Mail*, Feb. 20, 2017, A1.

38. John Herd Thompson and Stephen J. Randall, *Canada and the United States: Ambivalent Allies*, 3rd ed. (Athens: University of Georgia Press, 2002), 277.

39. Gregory Inwood, *Continentalizing Canada: The Politics and Legacy of the Macdonald Royal Commission* (Toronto: University of Toronto Press, 2005), 6, 35; Michael Hart, with Bill Dymond and Colin Robertson, *Decision at Midnight: Inside the Canada-US Free-Trade Negotiations* (Vancouver: UBC Press, 1994), 72; Michael Hart, "The Road to Free Trade," in *Free Trade: Risks and Rewards,* ed. L. Ian MacDonald (Montreal and Kingston: McGill-Queen's University Press, 2000), 24–25.

40. "Voices of Reason on Trade," editorial, *Chicago Tribune,* Oct. 27, 1985, A12; "Good for U.S., Good for Canada," editorial, *Chicago Tribune,* Oct. 7, 1987, 22; *The President's Report,* Nov. 1988, 3, vol. 25786 (acquisition no. 2006-00392-5, box 507), folder President's Reports 1983–92, QOPC, LAC.

41. Andrew H. Malcolm, "Canada's Historic Decision: Mulroney Pulls No Punches," *New York Times,* Nov. 17, 1988, A1, A18.

42. John F. Burns, "Mulroney to Push Approval of U.S.-Canadian Pact," *New York Times,* Nov. 23, 1988, A12; Michael Hart, *A Trading Nation: Canadian Trade Policy from Colonialism to Globalization* (Vancouver: UBC Press, 2002), 389–95; "The Promise of Free Trade," editorial, *Chicago Tribune,* Aug. 13, 1992, D24.

43. Clyde H. Farnsworth, "Mulroney Quits Post in Canada, Urges 'Renewal,'" *New York Times,* Feb. 25, 1993, A1, A12; Maxwell Cameron and Brian Tomlin, *The Making of NAFTA: How the Deal Was Done* (Ithaca, NY: Cornell University Press, 2000), 193, 202–7; William Watson, *Globalization and the Meaning of Canadian Life* (Toronto: University of Toronto Press, 1998), 8.

44. Andrew Malcolm, "Man in the News: A Triumphant Canadian; Martin Brian Mulroney," *New York Times,* Nov. 23, 1988, A1, A12; Graham Fraser, "His Town Is Symbol of Nation, PM Says," *Globe and Mail,* July 2, 1987, A1; Mulroney, *Memoirs,* 21.

45. Interview with Christine McMillan, Sept. 30, 2015, transcript in author's possession; D. Craig to C. A. Sankey, May 17, 1955, 1–2, vol. 25759 (acquisition no. 2006-00392-5, box 470), folder Vanillin Plant General, pt. 3, QOPC, LAC.

46. Casey Bukro, "Newsprint Firm Tells Progress in Fight on Pollution," *Chicago Tribune,* Oct. 10, 1970, sec. 1, 9; Eugene Griffin, "Thorold Mill Wages Pollution Fight," *Chicago Tribune,* Nov. 12, 1970, sec. 3, 9; "The Bonus Values in a Vanillin Plant," *Canadian Chemical Processing* 44, no. 9 (Sept. 1961): 48–50, 59.

47. "Col. R. R. McCormick—a Builder of Canada," editorial, *Thorold News,* July 28, 1950, 1; "Arthur A. Schmon Dies in Chicago," *St. Catharines Standard,* Mar. 19, 1964, 1, 10.

48. The speeches are collected in Ontario Paper Company, *In Tribute to Robert McCormick,* undated, vol. 25345 (acquisition no. 2006-00392-5, box 56), folder Special Correspondence Colonel McCormick, pt. 26, QOPC, LAC. There are no page numbers in the volume.

49. R. M. Harrison, "Now," *Windsor Daily Star,* July 30, 1956, 17.

50. "Feature Report: Great New Sinews for the Canadian Economy," *Financial Post,* June 7, 1958, 49, 56–57.

51. Boris Miskew, "Spirit of Publisher Evident in Pulp Area," *Peterborough Examiner,* Dec. 3, 1969, 9; Boris Miskew, "Quebecers Envy Success of Baie Comeau," *Peterborough Examiner,* Dec. 10, 1969, 37; Robert Parisé, *Géants de la Côte-Nord* (Quebec: Éditions Garneau, 1974), 59, 61, 64; Pierre Cousineau, *Baie-Comeau, 1937–1987* (Baie-Comeau: Gestion Sportive et Culturelle de Baie-Comeau, 1987), 57.

Chapter 9 · The Problem of Paper in the Age of Electronic Media

1. Christopher Sterling, *Electronic Media: A Guide to Trends in Broadcasting and Newer Technologies, 1920–1983* (New York: Praeger, 1984), 222–23, 236–37.

2. *Historical Statistics of the United States, Earliest Times to the Present, Millennial Edition*, ed. Susan B. Carter, Scott Sigmund Gartner, Michael R. Haines, Alan L. Olmstead, Richard Sutch, and Gavin Wright (New York: Cambridge University Press, 2006), tables Dg267–74 and Dg287–92.

3. For bagasse, see "Paper Making from Cane Waste to Get First Practical Test Today," *New York Times*, Jan. 27, 1950, 35, 40; "Bagasse Paper Run Stands Up in Test," *New York Times*, Jan. 28, 1950, 20; "Bagasse Paper Test Termed Successful," *Editor & Publisher* 83, no. 5 (Feb. 4, 1950): 24. For peat, see "Paper Makers Interested," *Editor & Publisher* 6, no. 30 (Jan. 12, 1907): 3. For straw, see "Independent Mills Combine," *Editor & Publisher* 6, no. 34 (Feb. 9, 1907): 2; "Inventor Claims $30 Newsprint from Straw," *Editor & Publisher* 71, no. 3 (Jan. 15, 1938): 8. For corn stalks, see "Clarence Brown to Head Cornstalk Firm," *Editor & Publisher* 61, no. 49 (Apr. 27, 1929): 104; George Manning, "New Senate Farm Bill Gives Aid to Cornstalk Paper Industry," *Editor & Publisher* 61, no. 52 (May 18, 1929): 28. For kenaf, see "Fresno Bee Pushes Kenaf as Source of Newsprint," *Editor & Publisher* 113, no. 23 (June 7, 1980): 78; Jim Rosenberg, "Alternative Fiber Sources for Newsprint," *Editor & Publisher* 129, no. 3 (Jan. 20, 1996): 22–26.

4. Arthur Schmon to Charles B. James, Mar. 1, 1932, 1, vol. 25355 (acquisition no. 2006-00392-5, box 66), folder De-inking and Old Waste, pt. 4, QOPC, LAC; Robert U. Brown, "N.Y. News, Chi. Tribune to Run '5% Scrap' Newsprint," *Editor & Publisher* 77, no. 1 (Jan. 1, 1944): 7.

5. W. K. Voss, "The De-inking of Newsprint," Jan. 25, 1943, vol. 25355 (acquisition no. 2006-00392-5, box 66), folder De-inking and Old Waste, pt. 3, QOPC, LAC; Arthur Schmon to Robert McCormick Sept. 18, 1943, vol. 25355 (acquisition no. 2006-00392-5, box 66), folder De-inking and Old Waste, pt. 2, ibid.

6. Charles Storch, "Newsprint Price Cuts Not Always Good News," *Chicago Tribune*, June 17, 1983, D13; Edward Clifford, "Firms Take Advantage of Weaker Dollar to Gear Up to Meet Pulp, Paper Demand," *Globe and Mail*, Mar. 30, 1981, B15; *Quebec and Ontario Paper Company Ltd.: A Profile*, undated (ca. 1988), 7, vol. 25809 (acquisition no. 2006-00392-5, box 520), folder Quebec and Ontario Paper Company Ltd.: A Profile, QOPC, LAC.

7. Memorandum, 1982, box 178, folder Newsprint, 1966–96—General, NYT, NYPL.

8. Thomas Moore, "Why Tribune Co. Is Feeding the Chicago Cubs," *Fortune*, June 28, 1982, 44; Celeste Huenergard, "Production Plant Plans Unveiled by Chi Tribune," *Editor & Publisher* 112, no. 26 (June 30, 1979): 10; Bill Gloede, "In Chicago, the Tribune Has Built the Newspaper Plant," *Editor & Publisher* 115, no. 39 (Sept. 25, 1982): 42.

9. Eugene Katz and Alan Barth, "The Facsimile Newspaper," *American Mercury* 61 (Oct. 1945): 409–10.

10. Abe Chanin, "First Electronic Paper Prospers in England," *Editor & Publisher* 113, no. 36 (Sept. 6, 1980): 27.

11. Lewis Titterton, "Radio and Its Progeny," *American Scholar* 10, no. 4 (Autumn 1941): 498; Hugh Beville, "The Challenge of the New Media: Television, FM, and Facsimile," *Journalism Quarterly* 25, no. 1 (Mar. 1948): 3–11.

12. Silas Bent, "Radio Steals the Press' Thunder," *Independent* 119, no. 4023 (July 9, 1927): 49; Silas Bent, "The Future Newspaper," *Century Magazine* 117 (Jan. 1929): 346;

Philip Schuyler, "What of Newspapers in This Radio Age?," *Editor & Publisher* 59, no. 48 (Apr. 23, 1927): 26.

13. Sol Taishoff, "Facsimile Looms as Press Rushes into Radio," *Broadcasting* 9, no. 7 (Oct. 1, 1935): 7–8; "Radio Plans to Print Newspapers in Homes by Facsimile Process," *Editor & Publisher* 66, no. 35 (Jan. 13, 1934): sec. 2, 6.

14. "Groundwork Laid for Control of Facsimile, a Scientific Reality," *Editor & Publisher* 69, no. 23 (June 6, 1936): 11; Alfred Goldsmith, "Facsimile Broadcasting," in *Radio Facsimile*, ed. Alfred Goldsmith, Arthur Van Dyck, Charles Horn, Robert Morris, and Lee Galvin (New York: RCA Institutes Technical Press, 1938), 345; D. G. Park, "Says Facsimile Will Never Take Place of Newspapers," *Editor & Publisher* 73, no. 14 (Apr. 6, 1940): 10; "Crosley Will Build Finch Radio Sets," *Editor & Publisher* 72, no. 2 (Jan. 14, 1939): 6.

15. Ruth Brindze, "Next—the Radio Newspaper," *Nation* 146, no. 6 (Feb. 5, 1938): 134; Robert Coe, "Facsimile Must Be Speedier, Says Expert," *Editor & Publisher* 72, no. 23 (June 10, 1939): 12; "Facsimile 'Paper' Given Delegates," *Editor & Publisher* 72, no. 6 (Feb. 11, 1939): sec. 2, 10; "Herald Tribune Has Facsimile Paper at N.Y. Fair," *Editor & Publisher* 72, no. 18 (May 6, 1939): 14; "FCC Permits Commercial Facsimile," *Editor & Publisher* 73, no. 22 (June 1, 1940): 8; David Z. Shefrin, "The Radio Newspaper and Facsimile" (unpublished MA thesis, University of Missouri, 1949), 83–85.

16. Larry Wolters, "Facsimile Opens Up Visual Age in Broadcasting," *Chicago Tribune*, Mar. 6, 1938, SW6; Jerry Walker, "Reader Survey Starts Talk about Facsimile," *Editor & Publisher* 78, no. 31 (July 28, 1945): 70; George Brandenburg, "McCormick Sees Postwar Expansion of Newspapers," *Editor & Publisher* 78, no. 36 (Sept. 1, 1945): 9.

17. Larry Wolters, "Coming—Radio Printing Press for Your Home," *Chicago Tribune*, Apr. 18, 1946, 6; Larry Wolters, "Radio Tribune Makes Debut; It's Facsimile," *Chicago Tribune*, May 12, 1946, 3.

18. "Tribune Sends Facsimile for Public Today," *Chicago Tribune*, May 25, 1946, 3; "Chicagoans Can See a Newspaper Come In over Radio Today," *Chicago Tribune*, June 4, 1946, 1; Ken Clayton, "Facsimile Will Need More Than Engineers," *Quill* 35, no. 9 (Sept. 1947): 8–9.

19. Thomas Whiteside, "A Ticker on Every Breakfast Table," *New Republic* 115, no. 10 (Sept. 9, 1946): 294; Burton Hotaling, "Facsimile Broadcasting: Problems and Possibilities," *Journalism Quarterly* 25, no. 2 (June 1948): 139, 143.

20. Jennifer Light, "Facsimile: A Forgotten 'New Medium' from the 20th Century," *New Media & Society* 8, no. 3 (2006): 367–68; Noah Arceneaux, "Radio Facsimile Newspapers of the 1930s and 40s: Electronic Publishing in the Pre-digital Era," *Journal of Broadcasting & Electronic Media* 55, no. 3 (2011): 354; Mary Koehler, "Facsimile Newspapers: Foolishness or Foresight?," *Journalism Quarterly* 46, no. 1 (Spring 1969): 34.

21. Beville, "Challenge of the New Media," 7; Jerry Walker, "Facsimile Offers Peg for FM Application," *Editor & Publisher* 79, no. 17 (Apr. 20, 1946): 108; Jerry Walker, "History of Facsimile Experiment Recorded," *Editor & Publisher* 82, no. 45 (Oct. 29, 1949): 42; Jonathan Coopersmith, "The Failure of Fax: When a Vision Is Not Enough," *Business and Economic History* 23, no. 1 (Fall 1994): 278.

22. Jonathan Coopersmith, *Faxed: The Rise and Fall of the Fax Machine* (Baltimore: Johns Hopkins University Press, 2015); M. J. Peterson, "The Emergence of a Mass Market for Fax Machines," *Technology in Society* 17, no. 4 (1995): 469–82; Mark Fitzgerald, "Chicago Tribune Folds Fax Paper," *Editor & Publisher* 123, no. 33 (Aug. 18, 1990): 13.

23. William S. Hedges, "What Television Will Mean to Newspapers, Explained by Chicago News Broadcast Chief," *American Press* 48, no. 10 (July 1930): 5; Jerome Walker, "Television Seen as Adjunct to Dailies," *Editor & Publisher* 63, no. 3 (June 7, 1930): 11.

24. Ben Bagdikian, *The Information Machines: Their Impact on Men and the Media* (New York: Harper & Row, 1971), xxxii–xxxiv, 192–93.

25. Craig Tomkinson, "Seers Nix Home Fax Papers, See Widespread Computer Use," *Editor & Publisher* 104, no. 12 (Mar. 20, 1971): 9.

26. Sandra Duerr, "Dow Jones to Cable Wall Street Journal," *Editor & Publisher* 114, no. 39 (Sept. 26, 1981): 11.

27. "Two-Faced Technology," editorial, *Editor & Publisher* 114, no. 24 (June 13, 1981): 6; Andrew Radolf, "Knight-Ridder to Test Home Electronic Info System," *Editor & Publisher* 113, no. 15 (Apr. 12, 1980): 7–8.

28. "Viewtron Test Started by Knight-Ridder," *Editor & Publisher* 113, no. 30 (July 26, 1980): 18–19; "Newspaper Companies Set to Expand Videotex Systems," *Editor & Publisher* 115, no. 18 (May 1, 1982): 18; Andrew Radolf, "Knight-Ridder Looks to Electronic Future," *Editor & Publisher* 115, no. 17 (Apr. 24, 1982): 20; "Not Yet!," editorial, *Editor & Publisher* 119, no. 13 (Mar. 29, 1986): 6.

29. "N.Y. Times Remains Committed to Print," *Editor & Publisher* 115, no. 34 (Aug. 21, 1982): 30; "Not Yet!," 6.

30. Pablo Boczkowski, *Digitizing the News: Innovation in Online Newspapers* (Cambridge, MA: MIT Press, 2004); Anthony Smith, *Goodbye Gutenberg: The Newspaper Revolution of the 1980s* (New York: Oxford University Press, 1980).

31. Robert McCormick, "The W-G-N Story, Part I: The Station and Facilities," *Chicago Tribune*, Feb. 14, 1954, 24–25.

32. The Tribune was certainly not alone in this. Indeed, hundreds of newspapers around the country operated radio stations, and their activities were instrumental in developing the basic structures of the broadcasting industry and the federal policies regulating it. For a full account of this history, see Michael Stamm, *Sound Business: Newspapers, Radio, and the Politics of New Media* (Philadelphia: University of Pennsylvania Press, 2011).

33. Larry Wolters, "WGN-TV Starts Telecasting Tomorrow," *Chicago Tribune*, Apr. 4, 1948, 1, 6; Jerry Walker, "N.Y. News, with WPIX, Shows It to Sweeney," *Editor & Publisher* 81, no. 26 (June 19, 1948): 48; George Gent, "WPIX's Night before Christmas: Nothing Stirring but a Yule Log," *New York Times*, Dec. 9, 1966, 49.

34. Roy Rowan, "Secrets of the Tribune Tower," *Fortune*, Apr. 5, 1982, 70.

35. Smith, *Goodbye Gutenberg*, 52; Rowan, "Secrets of the Tribune Tower," 66, 72, 74.

36. Andrew Radolf, "Survival Still Top Priority for Embattled N.Y. News," *Editor & Publisher* 115, no. 12 (Mar. 20, 1982): 24, 26; "The *News* Survives," editorial, *Editor & Publisher* 115, no. 19 (May 8, 1982): 8; "Lessons from the Daily News," editorial, *Editor & Publisher* 115, no. 40 (Oct. 2, 1982): 4; "Murdoch Offers to Buy News, Cook Says No," *Editor & Publisher* 115, no. 16 (Apr. 17, 1982): 10; "Editor Foresees Strong Profits for N.Y. News," *Editor & Publisher* 115, no. 49 (Dec. 4, 1982): 18; George Garneau, "Zuckerman Takes the Helm," *Editor & Publisher* 126, no. 3 (Jan. 16, 1993): 12, 43; Rowan, "Secrets of the Tribune Tower," 66.

37. "Prospectus on Tribune Co. Paints Profitable Picture," *Editor & Publisher* 116, no. 36 (Sept. 3, 1983): 10, 12.

38. "Tribune Names New Chief of Ontario Paper," *Chicago Tribune*, June 28, 1984, sec. 2, 5; Charles Storch, "Newsprint Price Cuts Not Always Good News," *Chicago Tribune*, June 17, 1983, D13.

39. *The President's Report*, Feb. 1984, 3, vol. 25796 (acquisition no. 2006-00392-5, box 507), folder President's Reports 1983–92, QOPC, LAC; *The President's Report*, Nov. 1989, 1, vol. 25796 (acquisition no. 2006-00392-5, box 507), folder President's Reports 1983–92, ibid.; *The President's Report*, Nov. 1991, 1–2, vol. 25796 (acquisition no. 2006-00392-5, box 507), folder President's Reports 1983–92, ibid.; *The President's Report*, Aug. 1991, 2, vol. 25796 (acquisition no. 2006-00392-5, box 507), folder President's Reports 1983–92, ibid.; *The President's Report*, Year End Update, Feb. 1992, 1, vol. 25796 (acquisition no. 2006-00392-5, box 507), folder President's Reports 1983–92, ibid.; *The President's Report*, Aug. 1992, 1, vol. 25796 (acquisition no. 2006-00392-5, box 507), folder President's Reports 1983–92, ibid.

40. Casey Mahood, "Quno Files Public Offering," *Globe and Mail*, Feb. 5, 1993, B7; "Tribune Co. Will Take Some Paper Mills off Balance Sheet," *Editor & Publisher* 126, no. 15 (Apr. 10, 1993): 30; Harvey Enchin, "Tribune Co. Cuts Stake in Newsprint Maker Quno," *Globe and Mail*, Mar. 24, 1994, B9; Peter Kennedy, "Tribune Examines Ways of Unloading Quno Stake," *Financial Post*, Sept. 12, 1995, 5; Casey Mahood, "Quno Ripe for $1 Billion Buyout," *Globe and Mail*, Sept. 12, 1995, B1; Paul Bagnell, "Smart Money Will Stay with Forest Products," *Financial Post*, Nov. 21, 1995, 24; "Deal Planned to Sell Maker of Newsprint," *New York Times* (Dec. 25, 1995), 58; Susan Yellin, "Newsprint an 'Ideal Fit,'" *Financial Post*, Dec. 27, 1995, 3; "Donohue Buys Quno Operations," *Globe and Mail*, Mar. 1, 1996, B3; Oliver Bertin, "Quno Deal Donohue's Gain," *Globe and Mail*, Dec. 27, 1995, B1.

41. Mark Fitzgerald, "Madigan Named CEO of Tribune Co.," *Editor & Publisher* 128, no. 19 (May 13, 1995): 20; interview with John Houghton, Oct. 1, 2015, transcript in author's possession; interview with Adrian Barnet, Sept. 30, 2015, transcript in author's possession.

42. Mark Fitzgerald, "A Year of Expansion," *Editor & Publisher* 130, no. 1 (Jan. 4, 1997): 7–8; Lucia Moses, "A Marriage Made in Heaven," *Editor & Publisher* 133, no. 12 (Mar. 20, 2000): 2–3.

43. "Ameritech Joins Tribune Co. as Peapod Investor," *Editor & Publisher* 127, no. 43 (Oct. 22, 1994): 38; Lucia Moss, "Buying into the Net," *Editor & Publisher* 132, no. 16 (Apr. 17, 1999): 30.

44. Scott Kirsner, "Explosive Expansion at Tribune Web Site," *Editor & Publisher* 130, no. 27 (July 5, 1997): 28–29.

45. Martha Stone, "Exodus of the Web People," *Editor & Publisher* 132, no. 36 (Sept. 4, 1999): 29.

46. "Toward a Trimmer Tribune Co.," *Editor & Publisher* 134, no. 26 (June 25, 2001): 8; Christine Haughney, "A News Giant Going It Alone," *New York Times*, Aug. 4, 2014, B1; Roger Yu, "Gannett Joins Spinoff Craze," *USA Today*, Aug. 6, 2014, 1B; James O'Shea, *The Deal from Hell: How Moguls and Wall Street Plundered Great American Newspapers* (New York: Public Affairs, 2011).

47. Ravi Somaiya, "Firing at the Los Angeles Times Focuses Discontent," *New York Times*, Sept. 20, 2015, A1; Martha White, "Tribune Publishing's New Name, 'tronc,' Puzzles Marketing Experts," *New York Times*, June 20, 2016, B3; Lee Schafer, "Just What the 'tronc' Were They Thinking?," *Minneapolis Star Tribune*, June 9, 2016, 1D.

48. Milton Rockmore, "How Will Newspapers Be Affected by CATV?," *Editor & Publisher* 113, no. 16 (Apr. 19, 1980): 82, 84; "Relevancy in the '80s," editorial, *Editor & Publisher* 113, no. 17 (Apr. 26, 1980): 6.

49. Pew Research Center, "Newspapers Fact Sheet," www.journalism.org/fact-sheet/newspapers/; Pew Research Center, *State of the News Media 2015* (Washington, DC: Pew Research Center, 2015), 27; interview with John Houghton, Oct. 1, 2015, transcript in author's possession.

50. Jim Rutenberg, "Yes, the News Can Survive the Newspaper," *New York Times*, Sept. 4, 2016, B1.

Conclusion · Media Infrastructures, Old and New

1. Ontario Paper Company, *A Company Not Like Others* (Thorold: Ontario Paper Company, 1985), 5; interview with Don McMillan, September 30, 2015, transcript in author's possession.

2. Nathan White, "After 75 Years, the Mill Stops Running," *Telegraph-Journal*, Jan. 31, 2008, A1.

3. Murad Ahmed, "Cold Storage," *Times*, Mar. 1, 2012, 20; Tung-Hui Hu, *A Prehistory of the Cloud* (Cambridge, MA: MIT Press, 2015); James Glanz, "Power, Pollution and the Internet," *New York Times*, Sept. 23, 2012, A1; James Glanz, "Data Barns in a Farm Town, Gobbling Power and Flexing Muscle," *New York Times*, Sept. 24, 2012, A1; Lisa Parks and Nicole Starosielski, introduction to *Signal Traffic: Critical Studies of Media Infrastructures*, ed. Lisa Parks and Nicole Starosielski (Urbana: University of Illinois Press, 2015): 1–27.

4. Alex Roslin, "www.energy.hog.com," *Montreal Gazette*, June 4, 2011, B1; Naomi Powell, "Nordic Countries Gain Appeal as Locations for Data Centers," *International Herald Tribune*, Apr. 30, 2012, 16; Ahmed, "Cold Storage," 20; Ben Arnoldy, "Plugging the Internet into Clean Power," *Christian Science Monitor*, Nov. 29, 2007, 3; Nathan Vanderklippe, "Kelowna Touted for High-Tech Storage," *Vancouver Sun*, June 28, 2008, E5.

5. Jonathan Stoller, "It's Good to Be Cold," *Globe and Mail*, Dec. 20, 2012, B4.

6. "To Get Google, Get Energized," editorial, *Telegraph-Journal*, July 20, 2009, A6.

Page numbers in *italics* refer to graphics or tables.